D1268805

BLUEPRINT FOR A CELL:

THE NATURE AND ORIGIN OF LIFE

# BLUEPRINT FOR A CELL:
# THE NATURE AND ORIGIN OF LIFE

CHRISTIAN DE DUVE

NEIL PATTERSON PUBLISHERS
CAROLINA BIOLOGICAL SUPPLY COMPANY
BURLINGTON, NORTH CAROLINA

BLUEPRINT FOR A CELL:

THE NATURE AND ORIGIN OF LIFE



Printed in the United States of America

Library of Congress Cataloging-in-Publication Data

de Duve, Christian.
Blueprint for a cell : The nature and origin of life /
Christian de Duve.
p. cm.
Includes bibliographical references and index.
ISBN 0-89278-410-5 : $19.95
1. Life—Origin. 2. Cells. I. Title.
QH325.D4 1990 577—dc20 90-62090

Printing and binding are by Arcata Graphics. The text is set in Times Roman.

Book and dustjacket design are by Malcolm Grear Designers, Inc.

Dustjacket illustration is by Francis Leroy.

NEIL PATTERSON PUBLISHERS

1308 Rainey Street
Post Office Drawer 2827
Burlington, North Carolina 27216-2827
U.S.A.

| | |
|---|---|
| *Telephone* | 800 227-1150 |
| | 919 226-6000 |
| *Facsimile* | 919 222-1926 |
| *Telex* | 574354 |
| *Cable* | SQUID |

TO JANINE

# Table of Contents

# Preface

*The origin of life might well become a new centre of abstract quarrels, with schools and theories concerned, not with scientific predictions, but with metaphysics.*

FRANÇOIS JACOB (20, page 306)

Novelists say that their characters often acquire a life of their own, to the point of taking over their creator and rewriting the plot. The American writer Ray Bradbury has even made this observation into a rule. If you work at it sufficiently, he counsels in *Zen and the Art of Writing,* "the time will come when your characters will write your stories for you" (26, page 130). It may be revealing of the nature of this book that this is exactly what happened to my character: the living cell.

I had set out to write what I hoped would be a concise, unadorned, almost abstract description of the basic blueprint common to all living cells, drawn in as simple and nontechnical terms as the subject allowed. By the time I had completed a first draft, the cell was in command, forcing me to re-enact with it the history of its birth. I undertook this unanticipated excursion into the past with the naive optimism of the neophyte, but soon found myself engaged in an increasingly arduous expedition through many strange lands, still largely uncharted and almost totally impenetrable. What had started as a carefree jaunt turned into an authentic voyage of exploration and discovery, with all the efforts and frustrations such a voyage entails, with also, as a compensation, the rare illuminating flash that, true or illusory, conveys a joy of a kind I had not thought I would experience again. When I had reached, not the end of my quest, but as far as I could arrive with the means at my disposal, the size of the book had doubled and its tone was so completely altered that the first part had to be largely rewritten in a different language. I thereby lost many of my intended readers. As to those I have gained, they may well reproach me for not having sufficiently freed the new version from its original elementary form. To the ones and to the others, I can only express my regrets. Blame the cell.

Not the least demanding part of my journey was wading through stacks of books, reviews, and specialized articles, written by a wide gamut of scientists and thinkers, ranging from cosmologists to philosophers and theologians, by way of theoretical and experimental physicists, chemists of all kinds, geologists and other Earth scientists, as well as every possible brand of biologist, each looking at the problem with the expert knowledge, biases, and limitations of their discipline. What makes the origin of life a uniquely fascinating problem is that it is so fundamental and has so many facets that it attracts the interest of all and can be encompassed by none. I have done my best to do justice to everyone but have done so no doubt in a very unsatisfactory fashion, imposed by my own personal limitations. I hope that those whom I failed to cite, or cited incorrectly, will forgive me.

This explosion of interest in the origin of life is recent. Not that humans have not always been intrigued by their origins. But, because of lack of knowledge, the problem was perforce confined to the realm of the mythic. The situation has changed considerably in the last decades. There are several reasons for this. First, our understanding of the structural and functional organization of living cells has advanced so much that we are now able to enunciate in concrete and relatively precise terms the problem of their initial generation. We know enough of the blueprint to validly attempt to reconstruct the manner in which it first materialized. Further, we have available numerous geochemical and other clues to the physical circumstances under which life arose. We can also map the process to some extent by the traces it has left on the Earth and elsewhere in the cosmos, and, especially, by the dated clues it has scattered throughout the molecular structures of the biosphere. Here, probably, lies the main reason the origin of life has become such a popular subject. Thanks to the advances of comparative biochemistry and molecular biology, almost every discovery that clarifies some aspect of the nature of life automatically also sheds light on its origin. That is why, when I started to write on the one topic, I could not escape addressing the second.

Although arrived at by a developmental process rather than by design, the organization of the book has remained as initially planned. In the first part, I have followed my original objectives of delineating the basic blueprint that I view as common to all forms of life and of describing the manner in which it is realized in prokaryotic and in eukaryotic cells. With an eye on the second part, I have endeavored to organize this panorama of life within a broad phylogenetic context, to the extent that known facts and current surmises allow this kind of exercise. In so doing, however, I have kept strictly to the level of the cell. Topics such as population genetics, cladism, gradualism, punctuated equilibrium and other attempts at revising Darwin, the anthropic principle, the Gaia hypothesis, or the extinction of the dinosaurs, are beyond the scope of this essay, let alone the competence of its author. Within those self-imposed limits, I have tried to make the best use of the information that extant organisms have to offer about their history in order to identify the properties of their last common ancestor and to retrace the evolutionary pathway whereby they arose from it. Clearly, many aspects of this reconstruction remain speculative. Some are controversial. But, on the whole, the edifice stands on fairly safe ground, resting, as it does, on the secure foundations of modern biochemistry, cell biology, and molecular biology.

The second part of the book is very different. Almost everything in it is speculative and controversial. Groping in almost complete darkness, it tries to retrace step by step a plausible pathway from prebiotic scratch to the fully developed cell that is pictured in the first part as the last common ancestor of all living beings. Although clearly dominated by the preoccupations of a biochemist, my script also allows for the concerns of molecular biologists, to the extent that I have been able to master them. In composing it, I have given special attention to the early events that have led to what is often referred to as the "RNA world." Many theories ignore these events or, rather, take them for granted. I believe this to be wrong and offer, as an important and necessary preliminary phase on the way to the RNA world, an elaborate "thioester world" operated by an extensive set of protoenzymes that catalyzed protometabolic reactions very similar, at least qualitatively, to the metabolic reactions of today. I believe this to be the most original part of the book.

My historical account of the origin of life is scientifically plausible and logically coherent, but it remains broken in several places by obscure or ambiguous passages, which I have not tried to hide. On the contrary, I have attempted each time to act as the devil's advocate and to probe the weaknesses of my hypotheses. The price of this effort is a certain ponderousness of style and a frequently unconvincing tone, which are liable to discourage many a reader with a hankering after brevity and simplicity. But to sacrifice a certain rigor would have been a heavier price to pay. Even so, my attempt to reconstruct our earliest origins no doubt includes many flaws that I did not detect and that others will not fail to point out. I await with interest the comments, criticisms, and suggestions that my theories will generate.

In spite of the metamorphosis the book has undergone, I have tried to remain as faithful as possible to my original vow of unadorned conciseness. I have intentionally avoided all particular, illustrative, or anecdotic details that might blur the main lines and have chosen a language that I hope will reach the widest possible readership acquainted with the basic concepts of physics and chemistry and with elementary biology. Readers familiar with modern biochemistry, cell biology, and molecular biology should have no difficulty fleshing out with concrete facts and images the abstract notions offered. Those in need of further clarification can readily find them in the general treatises, textbooks, and reviews listed at the end of the book. The rest of the references mentioned by no means represent a complete bibliography on the topic. Rather, they make up a sort of personal catalogue, a list of readings that I made to inform myself about recent findings likely to throw some light on the origin of life.

True to my dual cultural allegiance, I have wished this book to reach the English-speaking and French-speaking publics simultaneously. In this connection, I remembered the Italian aphorism "*traduttore, traditore*" and decided that I was best placed not to betray the author. I therefore appointed myself my own translator. The result proved as unexpected as it was disappointing. The language used in the first draft entirely dominated the translation, to the point of allowing several enormities to slip through into the latter. I was fortunate enough to benefit from the support of vigilant controllers on both sides of the Atlantic. Nevertheless, alert readers will have no trouble detecting in each of the two versions traces of their hybrid ancestry.

A book of this kind can be completed only with the help of many. My first thanks are due to my friend and colleague at The Rockefeller University in New York, Miklós Müller. He has gone carefully through every draft, making many useful remarks and suggestions. In addition, he has given me the invaluable assistance of his encyclopedic knowledge of the vast and infinitely varied world of microorganisms, including their taxonomy, phylogeny, and metabolism. Another long-time friend, my Belgian colleague Géry Hers, has also been of great help to me by his thorough critique, both for content and for style, of the French version. I owe to him the correction of numerous errors and ambiguities, as well as that of a veritable anthology of anglicisms. Many other colleagues have, directly or indirectly, put their expertise at my disposal and, sometimes, prevented me from making dire mistakes. Among them are Steven Benner, Peter Campbell, Hubert Chantrenne, Libor Ebringer, Jean-Michel Foidart, Henri Grosjean, John Gunderson, Bernard Horecker, Gerald Joyce, the late Tom Kaiser, Leslie Orgel, Jacques Reisse, Paul Schimmel, Rudolf Thauer, Alexander Tzagoloff, Arthur Weber, Theodor Wieland, and Richard Wolfenden. The manuscript was also read by Alan Weiner. I further thank Gunter Blobel, Manfred Eigen and Ruthild Winkler-Oswatitsch, Hans Kuhn and Jürgen Waser, Mitchell Sogin, Carl Woese, and J. Tze-Fei Wong for allowing me to use their illustrations. Needless to say, those who helped me can share no responsibility in the mistakes that remain nor in the hypotheses that are proposed, especially since I have chosen not always to follow their advice.

My thanks go also to my illustrators, Francis Leroy, who wonderfully executed all the drawings and the jacket, and Freddy Goossens, who did the same for all the computer-designed schemes and, in addition, did the page set of the French version. Thanks also to Donna Young and Teresa Arrwood, who handled the page set of the English version. I am also grateful to Paul Depovere who helped me with the ungrateful and difficult task of preparing the index of the French version. I am specially indebted to my American editor Neil Patterson, who has staunchly supported every phase in the evolution of the book, in spite of the progressive dwindling of its possible readership, and to his associate Sherri Foster, who has taken care of the final edition of the English version in admirable fashion. On their side, my Belgian editors, Christian De Boeck and Michel Jezierski, have spared no effort to ensure the book's success. I owe an immense and unrepayable debt of gratitude to my secretary at The Rockefeller University, Karrie (Anna Polowetzky). In addition to entering faithfully and patiently every successive version into the memory of her word processor, she has made numerous heavily laden trips back and forth between her office and the University Library and has ended up being so conversant with literature research and filing as to be able to finalize and verify the whole reference list single-handedly. The French version of this book was prepared with the competent and dedicated help of Monique Van de Maele, who, thrown suddenly into a sea of particularly abstruse jargon, learned to swim with remarkable rapidity. I wish to recall further the memory of a very dear friend, Helene Jordan Waddell, who had undertaken enthusiastically to edit this book, as she did my *Guided Tour* (1), but was prevented from finishing the job by an unforgiving malady. Last, but not least, I thank my wife for having suffered once again with patience and understanding the repercussions of an occupation that did not always make for a pleasant and relaxed husband.

New York and Nethen, August 1st, 1990

PART ONE

# CELLS PAST AND PRESENT

CHAPTER ONE

# THE BASIC BLUEPRINT

## THE UNIT OF LIFE

The cell is the unit of life. Many organisms consist of a single cell. Self-evidently, such a cell has all that is needed for life. The pluricellular plants and animals are made of different kinds of cells, which depend on each other to stay alive. The dependence is not absolute, however. If placed in a suitable environment, most such cells can survive, often even grow, multiply, and differentiate, away from the organism to which they belong. They, likewise, possess all that is needed for life. Consequently, if we wish to understand life, we must start by understanding cells.

The study of cells was long hampered by their minuscule size: a volume of the order of $10^{-8}$ cubic centimeters for the cells of plants and animals, as small as $10^{-12}$ cubic centimeters for bacteria. Cells were not even known to exist until the first microscopes were constructed a little over three hundred years ago. It took the electron microscope, introduced into biology at the end of the Second World War, for the details of cellular structures to become discernible. Yet, in this tiny volume are compressed all the elements that determine the most intricate and wonderful phenomenon known to us—life, including our own. Remember, we all started as a single fertilized egg cell.

Such miniaturized complexity can only be described with the language of chemistry. This is the object of biochemistry and of its recent extension, molecular biology. Thanks to the remarkable development of these two disciplines in the last decades, the basic mechanisms of cellular life are now understood to a considerable extent. A lesson of capital importance has emerged from these discoveries: *life is one*. Whatever the form in which it manifests itself, whatever its habitat and type of adaptation, it uses the same structures, the same kinds of molecules, the same chemical reactions, the same language, with hardly any variation worthy of mention.

In a certain sense, humankind has always been intuitively aware of the unity of life. Already in antiquity the distinction was made between living and inanimate beings, and life was referred to as a single entity. But this unity was attributed to a common vital principle that "animated" different forms of matter in the same way. Today, living matter itself has been revealed as one, whereas the vital spirit has disappeared from the preoccupations of biologists, to be replaced by a common mode of organization, a common blueprint. But for a few details, there are no fundamental differences at the molecular level among the cells of a human being, of an earthworm, or of an orchid, or even among such cells and an amoeba, a yeast cell, or a bacterium.

There is, however, immense variety in the manner in which the basic blueprint is executed in different kinds of cells and in their associations. To sift through these innumerable variations, bringing to the fore the basic themes on which they are built, is the object of this first chapter.

## THE SEVEN PILLARS OF LIFE

What do we mean by "all that is needed for life"? As viewed by the chemist, it means *the ability of a system to maintain itself in a state far from equilibrium, grow, and multiply, with the help of a continual flux of energy and matter supplied by the environment.* Note the importance of the environment. All living systems depend on their milieu. They owe their apparent autonomy to their ability to spontaneously channel and exploit environmental factors for self-support and development. Without such factors, which vary from one living system to another, there can be no life.

To satisfy the above definition, any living system must be able to:

1. *manufacture* its own constituents from materials available in its surroundings;
2. extract *energy* from its environment and convert it into the various forms of *work* it must perform to stay alive;
3. *catalyze* the many chemical reactions required to support its activities;
4. *inform* its biosynthetic and other processes in such a way as to guarantee its own accurate reproduction;
5. *insulate* itself in such a way as to keep strict control over its exchanges with the outside;
6. *regulate* its activities so as to preserve its dynamic organization in the face of environmental variations;
7. *multiply.*

These seven properties are both necessary and sufficient for life as we know it to exist and persist. They are common to all living organisms but are manifested in a wide diversity of complex ways, which has long precluded any attempt at unifying understanding. This is no longer so. Recent advances have revealed a small number of basic concepts of universal significance behind the diversity and complexity of living processes. A few keywords sum up these concepts: *transfer* (of electrons or groups), *coupling, complementarity, conformational change, feedback.* As will be shown in this chapter, it is possible, with these keywords as leading thread and some notions of biochemistry, to account for the seven basic elements that make up the common blueprint according to which all living cells are constructed.[1]

[1] There are almost as many "capsule" descriptions of life as there are authors who have attempted the exercise. Each looks at living organisms through glasses tinted (tainted?) by his or her area of specialization and personal biases. My own view is inspired by a semiabstract form of biochemistry that attempts, with the help of energetics, to discern basic conceptual kinships beyond the immense

## SELF-CONSTRUCTION

In the final analysis, living organisms derive their main constituents from simple natural oxides: hydrogen from water, $H_2O$; carbon from carbon dioxide, $CO_2$; nitrogen from the nitrate ion, $NO_3^-$ (occasionally from atmospheric nitrogen, $N_2$); sulfur from the sulfate ion, $SO_4^{2-}$; oxygen mostly from $CO_2$, sometimes from water or other oxides. From these, the organisms known as autotrophs, which can live entirely off mineral resources, build small organic molecules that they further assemble into macromolecules: amino acids into proteins; monosaccharides into oligosaccharides, polysaccharides, and related compounds; pentoses, purine and pyrimidine bases, and phosphoric acid into nucleotides, and these into nucleic acids; alcohols and acids of various kinds into lipids and other esters; and many others. Heterotrophic organisms utilize preformed organic molecules, which are products, directly or indirectly, of autotrophic industry. These they modify and combine in various ways to make the same kinds of macromolecules as are constructed by the autotrophs. Macromolecules often combine further into multimacromolecular complexes, out of which are built the granules, filaments, tubules, membranes, lamellae, walls, and other structural components that are found inside and around cells. Giant aggregates of this sort, sometimes reinforced by mineral deposits, make up the solid scaffoldings that plant and animal cells erect cooperatively within and around themselves, up to fashioning all the visible structures of the biosphere.

The description of all these compounds, their properties, and the numerous chemical reactions involved in their synthesis and processing are major concerns of biochemistry, hardly to be considered in a brief overview. But the principles of their construction are simple and boil down largely to two basic processes that both rest on the key notion of *transfer:* reduction, based on electron transfer, and dehydrating condensation, based on group transfer. With rare exceptions, all the reactions involved in biosynthesis fall into one or the other category, or both.

### *Reduction*

Originally synonymous with hydrogenation, reduction is now defined as the addition of electrons. A simple example of reduction is the conversion of ferric to ferrous iron:

$$Fe^{3+} + e^- \longrightarrow Fe^{2+} \tag{1}$$

In many cases, especially in biological reductions, protons are added together with the electrons, as in the conversion of an aldehyde to an alcohol:

$$R{-}CHO + 2\,e^- + 2\,H^+ \longrightarrow R{-}CH_2OH \tag{2}$$

Protons, which are ubiquitous in aqueous systems, join with electrons to make up the equivalent of the hydrogen gas that would be used in a conventional hydrogenation (reduction in the original sense).

variety of organic-chemical mechanisms. Several parts of Reference 1, in particular Chapters 7 to 10 and Appendix 2, elaborate this view and may be useful for understanding the pages that follow.

**1–1** *Importance of reduction for autotrophy.* Some overall chemical equations illustrating the high electron requirements of autotrophic syntheses.

$$6\,CO_2 + 24\,e^- + 24\,H^+ \longrightarrow \underset{\text{Glucose}}{C_6H_{12}O_6} + 6\,H_2O$$

$$2\,CO_2 + NO_3^- + 14\,e^- + 15\,H^+ \longrightarrow \underset{\text{Glycine}}{CH_2NH_2{-}COOH} + 5\,H_2O$$

$$16\,CO_2 + 92\,e^- + 92\,H^+ \longrightarrow \underset{\text{Palmitic acid}}{C_{15}H_{31}{-}COOH} + 30\,H_2O$$

$$3\,CO_2 + NO_3^- + SO_4^{2-} + 26\,e^- + 29\,H^+ \longrightarrow \underset{\text{Cysteine}}{CH_2SH{-}CHNH_2{-}COOH} + 11\,H_2O$$

We need only look at the inorganic building blocks of life to realize that autotrophic biosyntheses require a considerable amount of reduction (see Figure 1–1). Clearly, a large number of electrons (and protons) are needed to make the hydrocarbon backbones of organic molecules out of $CO_2$; to convert $NO_3^-$ to the amine group ($-NH_2$) and its derivatives, which are the main forms of nitrogen in biological molecules; to incorporate the sulfur of $SO_4^{2-}$ into thiol derivatives ($-SH$), which are found in numerous proteins and other natural substances, as such or dimerized oxidatively by a disulfide bridge ($-S-S-$). Heterotrophic organisms, which subsist on prereduced nutrients, depend on reduction to a much lesser extent, using it only for a few biosynthetic processes, such as the formation of fat from carbohydrate.

There are two important points to remember about reduction. First, the required electrons (or hydrogen atoms) do not occur freely in nature. They are always provided by a donor molecule ($DH_2$), which is oxidized in the process, while the acceptor (A) is reduced:

$$DH_2 + A \rightleftharpoons D + AH_2 \qquad \textbf{(3)}$$

Such a reaction is described as an oxidation-reduction reaction, or, more commonly, as an *electron transfer*. In most instances in living organisms, one of the partners of the transaction (sometimes both) is a part of the basic cell machinery that serves catalytically as an electron carrier.

Electron transfer can take place in either direction. The spontaneous direction is the exergonic one, i.e., that which is associated with a decrease in the free energy of the system. This direction is determined by whichever of the two reduced forms, $DH_2$ or $AH_2$, has the greatest tendency to donate[2] its electrons, that is, offers its electrons at the highest level of energy under the conditions considered. The tendency to donate electrons is measured by the oxidation-reduction potential, which is a quantity such

[2] I use the expression "tendency to donate" in the thermodynamic sense, not in any kinetic sense that would refer to "ease" of displacement.

that the lower the potential, the greater the electron-donating tendency, that is, the higher the energy level of the donated electrons.

This leads to the second important point to be made about biosynthetic reductions. To be accepted, the electrons must be provided at a high enough energy level for the transfer to be exergonic. This function is carried out in nature by the reduced forms of low-potential electron carriers, mainly the iron-sulfur proteins ferredoxins and the pyridinic coenzymes—nicotinamide adenine dinucleotide (NAD) and, more often, its phosphate (NADP).

## *Dehydrating Condensation*

Almost invariably, building blocks in a biosynthetic process are joined by removal of water:

$$X—OH + Y—H \longrightarrow X—Y + H_2O \quad \textbf{(4)}$$

Proteins arise in this way from amino acids; polysaccharides from sugars; nucleic acids from their constitutive pentoses, bases, and phosphoric acid; and so on. Such reactions reverse the hydrolytic breakdown reactions involved in digestion. In an aqueous medium, such as is found in living systems, hydrolysis is the energetically favored, exergonic direction. Dehydrating condensation is endergonic; it goes against the spontaneous direction and can only take place with a supply of energy.

In virtually every instance, the energy required by natural dehydrating assembly processes is provided directly, or indirectly, by the hydrolysis of adenosine triphosphate, or ATP (see Figure 1–2), either to adenosine diphosphate (ADP) and inorganic phosphate ($P_i$):

$$ATP + H_2O \longrightarrow ADP + P_i \quad \textbf{(5)}$$

or to adenosine monophosphate (AMP) and inorganic pyrophosphate ($PP_i$):

$$ATP + H_2O \longrightarrow AMP + PP_i \quad \textbf{(6)}$$

If these reactions were allowed to take place freely, the energy they release would be unavailable for the performance of the chemical work of assembly of X—Y. It would simply be dissipated as heat. Nature's answer to the problem is *coupling* between Reaction 4 and Reaction 5 or 6. In chemical terms, the mechanism of this coupling is very complex and varies according to the building blocks concerned. In principle, however, it is simple and universal, depending in all cases on sequential *group transfer*. Stripped of all chemical details (see Footnote 3, next page), it goes like this:

We start by representing ATP as B—O—Q, which means that Reactions 5 and 6 are now written:

$$B—O—Q + H_2O \longrightarrow B—OH + Q—OH \quad \textbf{(7)}$$

In the first step of sequential group transfer, an oxygen-bearing biosynthetic building block (X—OH) substitutes for $H_2O$ in Reaction 7:

$$B—O—Q + X—OH \rightleftharpoons B—OH + X—O—Q \quad \textbf{(8)}$$

**1–2** *ATP, the central energy carrier.*
The purine base adenine combines with the pentose D-ribose to form the nucleoside adenosine (A). Addition of one, two, or three phosphoryl groups to adenosine converts it to adenosine monophosphate (AMP), diphosphate (ADP), or triphosphate (ATP), respectively. The two terminal pyrophosphate bonds ($\beta$, $\gamma$) of ATP are "high-energy" bonds. Their hydrolysis fuels virtually all biosynthetic assembly processes and many other forms of biological work. In other nucleoside triphosphates, adenine is replaced by another base.

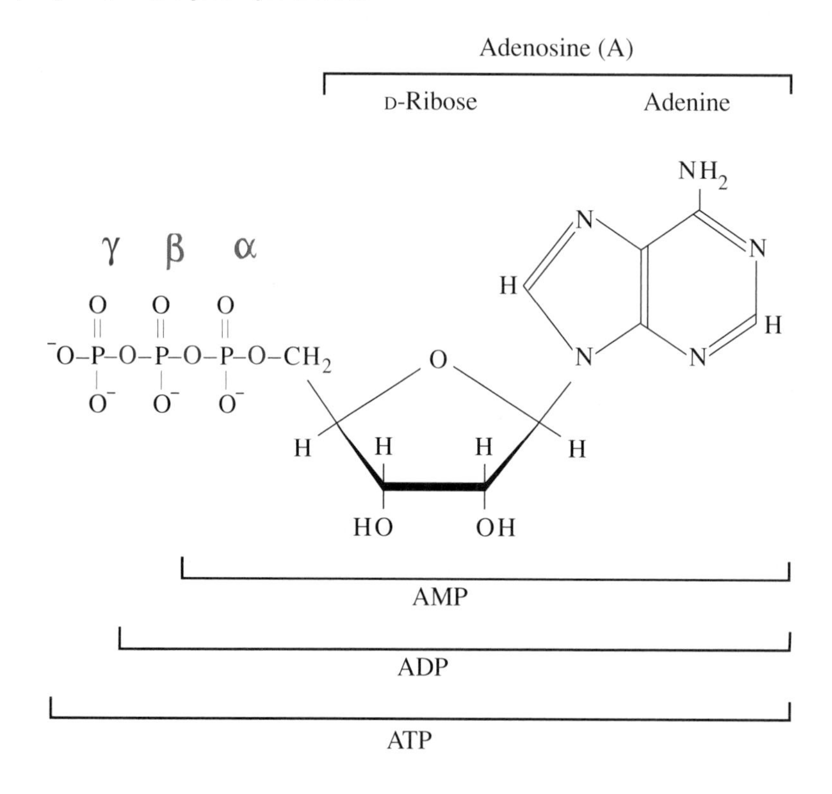


[3] I shall consider the real world a little later. However, since abstract symbolism is not to everyone's taste, the following key may be useful.

In B—O—Q, O represents the $\beta$ or $\gamma$ oxygen of ATP (see Figure 1–2), very exceptionally the $\alpha$ oxygen; B and Q are the two groups that flank this oxygen, with the transferable Q-yl group being most often the phosphoryl (left of $\gamma$) or the adenylyl (AMP-yl) (right of $\beta$) group, sometimes the pyrophosphoryl group (left of $\beta$).

X—OH may be an acid (R—COOH), an alcohol (R—OH), a sugar hemiacetal (R—CH—OH, with O bridging R and CH), a phosphorylated compound, a sulfate ion, or some other oxygen-bearing molecule. In all cases, the negatively charged or polarized oxygen is the nucleophilic attacking agent.

Y—H is also a nucleophile, but its attacking atom may be one of several, including oxygen, as in an alcohol; nitrogen, as in an amine; sulfur, as in a thiol, etc.

**1–3** *General mechanism of sequential group transfer.*
In the activation step, the Q-yl group (for example, the terminal phosphoryl group of ATP or its AMP-yl group) is transferred from the energy supplier B—O—Q (ATP) onto an oxygen-bearing biosynthetic building block X—OH. In the subsequent assembly step, the double-headed intermediate X—O—Q formed in the first step donates the activated X-yl group to the other biosynthetic building block Y—H. Note the central role of the double-headed intermediate: 1) The free energy of hydrolysis of B—O—Q supplies the free energy of assembly of X—Y by way of the free energy of hydrolysis of X—O—Q. 2) The oxygen atom of X—OH is recovered in Q—OH, again by way of X—O—Q. The great majority of biosynthetic processes can be fitted into this general scheme.

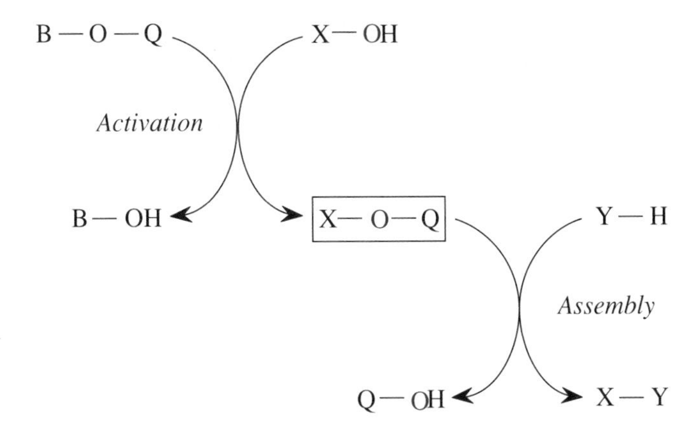


Both reactions are transfers of the Q-yl group: onto H—OH in Reaction 7, onto X—OH in Reaction 8. But note the difference. In Reaction 7, the free energy of hydrolysis of B—O—Q is lost; it is conserved in X—O—Q in Reaction 8. Reaction 7 is irreversible, Reaction 8 reversible.

In the next step, the X-yl group is transferred from X—O—Q onto the other biosynthetic building block Y—H:

$$X—O—Q + Y—H \longrightarrow X—Y + Q—OH \quad \textbf{(9)}$$

Add Reactions 8 and 9, and you obtain:

$$B—O—Q + X—OH + Y—H \longrightarrow X—Y + B—OH + Q—OH \quad \textbf{(10)}$$

This is exactly what you would get by adding Reactions 7 and 4. The difference is that water is not a participant in sequential group transfer and, as a result, the free energy of hydrolysis of B—O—Q (i.e., ATP) is not lost but is made available for the dehydrating assembly of X—Y. The key intermediate is the "double-headed" X—O—Q, which consists of two transferable groups—the Q-yl group transferred in the first step and the X-yl group transferred in the second one—joined by a central oxygen atom. X—O—Q carries the energy from one step to the other; and its central oxygen atom—which is given off by building block X—OH and ends up in Q—OH, one of the products of ATP hydrolysis—traces the cryptic transfer of water between the two steps. This is made clear by the schematic representation of Figure 1–3.

For those who prefer the real world to abstract generalizations, the following additional clarifications are in order.

1. The first step of sequential group transfer (Reaction 8) is called activation because it converts oxygen-bearing building blocks X—OH to energy-rich, transferable X-yl groups. Sugars are activated to glycosyl groups prior to glycosylation, amino acids to aminoacyl groups prior to incorporation into polypeptide chains, fatty acids to fatty acyl groups prior to esterification, nucleotides to nucleotidyl groups prior to nucleic acid synthesis, sulfate ions to sulfuryl groups prior to sulfurylation, etc. The principle is universal and is, in fact, also widely used by chemists. For example, acetylating agents, such as acetic anhydride or acetyl chloride, are none other than acetic acid activated to transferable acetyl groups.
2. The Q-yl group donated by ATP in the activation step is most often the terminal $\gamma$-phosphoryl group (activation by transphosphorylation) or the AMP-yl group (activation by transadenylylation). In a few cases, it is a pyrophosphoryl group, either transferred as such (activation by transpyrophosphorylation) or in the form of two successive phosphoryl groups (activation by double phosphorylation). Sometimes, another nucleoside triphosphate substitutes for ATP, especially in nucleotidyl transfer, which means that X—OH is activated in conjunction with uridine monophosphate (UMP), cytidine monophosphate (CMP), or guanosine monophosphate (GMP), instead of AMP.

   The nucleoside triphosphates (NTP) involved in these reactions are assembled at the expense of ATP, which therefore remains the ultimate supplier of energy:

$$\text{NMP} + \text{ATP} \rightleftharpoons \text{NDP} + \text{ADP} \qquad \textbf{(11)}$$

$$\text{NDP} + \text{ATP} \rightleftharpoons \text{NTP} + \text{ADP} \qquad \textbf{(12)}$$

3. The energetic cost of all biosynthetic dehydrating assembly processes, whatever they are, can be finally accounted for in terms of the hydrolysis of terminal $\gamma$-pyrophosphate bonds of ATP. This is obvious when activation is by transphosphorylation and Reaction 5 is the supplier of energy. When Reaction 6 supplies the energy, whether activation occurs by transadenylylation or by transpyrophosphorylation, the AMP formed is phosphorylated to ADP by adenylate kinase (actually Reaction 11, with AMP serving as NMP):

$$\text{AMP} + \text{ATP} \rightleftharpoons 2\,\text{ADP} \qquad \textbf{(13)}$$

   and pyrophosphate is most frequently hydrolyzed (exceptions are mentioned on the next page):

$$\text{PP}_i + \text{H}_2\text{O} \longrightarrow 2\,\text{P}_i \qquad \textbf{(14)}$$

   Add Reactions 6, 13, and 14, and you obtain:

$$2\,\text{ATP} + 2\,\text{H}_2\text{O} \longrightarrow 2\,\text{ADP} + 2\,\text{P}_i \qquad \textbf{(15)}$$

   The same balance is reached if activation occurs by the transfer of an NMP other than AMP, and the assembly is supported in the first place by hydrolysis of the $\beta$ bond of the NTP:

$$\text{NTP} + \text{H}_2\text{O} \longrightarrow \text{NMP} + \text{PP}_i \qquad \textbf{(16)}$$

   The sum of Reactions 16, 11, 12, and 14 gives Reaction 15.

Finally, if activation is accomplished by double phosphorylation, we also end up with Reaction 15 after the resulting pyrophosphoryl group has been released as pyrophosphate and hydrolyzed by Reaction 14.

Cases are known where pyrophosphate escapes hydrolysis and substitutes for ATP as the donor of phosphoryl groups in certain reactions (95, 145, 152, 156, 366).[4] Too infrequent to invalidate the universal role of ATP, these cases are of interest with respect to the origin of biological phosphorylations (Chapter 7).

In summary, with rare exceptions, hydrolysis of the $\gamma$ bond of ATP (Reaction 5) supports, directly or indirectly, all biosynthetic assemblies. In most instances, two such bonds, instead of one, are hydrolyzed per molecule of X — Y made when pyrophosphate is an intermediate. From the bioenergetic point of view, this means that the energy needed for the whole immense variety of biosynthetic assembly processes can be funneled through a single reaction: phosphorylation of ADP to ATP, the reverse of Reaction 5. I will come back to this point when discussing energy.

4. In my schematic outline, the activation reaction (Reaction 8) has been represented as reversible and the final assembly reaction (Reaction 9) as irreversible. This corresponds to the most frequent situation. As a rule, double-headed biosynthetic intermediates have free energies of hydrolysis equivalent to those of the pyrophosphate bonds of ATP. In the terminology introduced by Fritz Lipmann, one of the pioneers of bioenergetics, they have "high-energy bonds," which Lipmann represented by the now famous "squiggle": ~. On the other hand, many of the bonds in natural substances, such as the peptide bonds of proteins, the ester bonds of lipids, the glycosidic bonds of oligo- and polysaccharides, have distinctly lower free energies of hydrolysis. They are "low-energy bonds," represented by a straight line. The difference in free energy of hydrolysis between the double-headed intermediate and the final biosynthetic product ensures the irreversibility of Reaction 9 and, consequently, of the entire biosynthetic process.

   There are some exceptions. For example, the phosphodiester bonds of nucleic acids and of phospholipids are high-energy bonds. Nevertheless, their assembly is irreversible because it takes place by a process (NMP transfer) that involves the production of pyrophosphate. Hydrolysis of this molecule doubles the cost of the biosynthetic process (see Reactions 14 and 15) and renders it irreversible.[5]

   In some instances, the sequence of Reactions 8 and 9 is authentically reversible; it can occur in the reverse direction, sometimes even occurs predominantly in that direction, under normal biological conditions. As will be seen later in this chapter, such reactions are important for the regeneration of ATP from ADP and $P_i$ at the

[4] One such reaction is the phosphorylation of fructose 6-phosphate to fructose 1,6-*bis*phosphate, an important step of the glycolytic chain. Systems in which this reaction takes place at the expense of inorganic pyrophosphate instead of ATP (the cytosol of plant cells, for example) lack pyrophosphatase activity. Glycolysis substitutes for Reaction 14 as a means of removing inorganic pyrophosphate and thereby driving the biosynthetic processes in which this substance is produced (see Footnote 5).

[5] This result is sometimes achieved by some other exergonic pyrophosphate-consuming process, a special form of glycolysis, for example (see Footnote 4).

expense of metabolically generated high-energy bonds—in particular, thioester bonds:

$$-S \sim \overset{\overset{\displaystyle O}{\|}}{C}-$$

5. One last important point must be made. In many biosynthetic processes, the assembly reaction (Reaction 9) takes place in two steps (Reactions 17 and 18), by way of a group carrier (Cg) acting catalytically. Note that Reaction 17 is reversible: the bonds between carriers and groups are usually of the high-energy kind.

$$X-O-Q + Cg-H \rightleftharpoons Cg-X + Q-OH \quad \textbf{(17)}$$

$$Cg-X + Y-H \longrightarrow X-Y + Cg-H \quad \textbf{(18)}$$

This mechanism allows activation to be centralized and leaves the carrier to convey the activated X-yl groups to different Y—H acceptors, often in parts of the cell distinct from the site of activation. Like the electron carriers mentioned above, group carriers are important parts of the cell machinery. Together, these molecules make up the main coenzymes. They often include a vitamin as their main constituent.

Except for some additional complexities here and there, the vast majority of biosynthetic processes throughout the biosphere may be said to conform to the basic sequential group-transfer scheme sketched out above. Activation, directly or indirectly, by group transfer from ATP, participation of a double-headed intermediate, and further transfer of the activated building block to its biosynthetic acceptor, either directly or by way of a group carrier, are universals. They belong to the blueprint of life.

## ENERGY AND WORK

As just seen, the energetic cost of biosynthetic work is covered by electrons of high energy level, to the extent that reduction is required (much more in autotrophs than in heterotrophs), and by the hydrolysis of ATP to ADP and $P_i$, virtually the sole purveyor of energy, in the final analysis, for all biosynthetic assembly reactions.

Almost invariably, the hydrolysis of ATP to ADP and $P_i$ also fuels the other forms of work performed by living organisms, including mechanical work, the work of transporting ions and molecules against concentration gradients, the osmotic work of removal of water, the electric work of moving ions against electric potentials, the work of generating such potentials, the emission of light in luminescent organisms, as well as the work of energizing electrons when needed. In most such cases, the hydrolysis of ATP is obligatorily *coupled* to the performance of the work through the operation of specially harnessed hydrolyzing enzymes (ATPases) that operate in such a way that ATP is hydrolyzed only if some sort of work is carried out concomitantly, for example, if a structural element is shortened or bent, or if one or more ions are forced against an opposing electrochemical potential, or if a pair of electrons is lifted

**1–4** *The ATP cycle.*
The hydrolysis of ATP to ADP and $P_i$ (1) fuels virtually all forms of biological work (except, in most cases, biosynthetic reductions). Energy released by catabolism or provided by sunlight supports the regeneration of ATP (2). The ATP/ADP system acts catalytically in the conversion of energy to work.

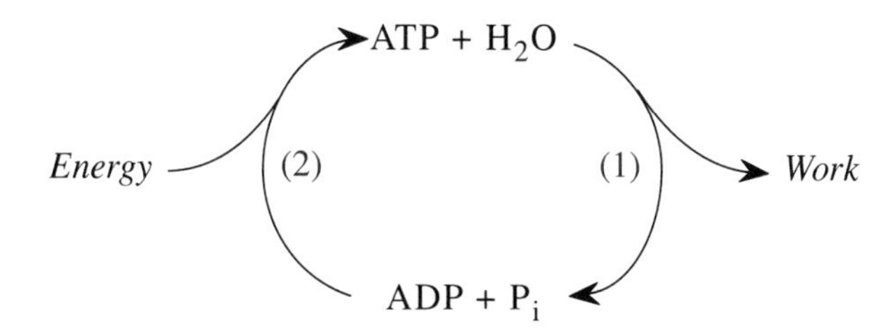


to a higher energy level. Sometimes, the work is actually supported by some high-energy intermediate form, such as a thioester bond or protonmotive force, involved in the assembly of ATP and in equilibrium with it. On rare occasions, ATP hydrolysis may be replaced or supplemented by electron transfer, for example, in bioluminescence.

Because of its universal bioenergetic role, ATP is often designated the "fuel of life." But this is a misleading term, as ATP is not provided as such by the environment. It is present in living cells in catalytic amounts, and it can play its role only if it is continually regenerated from ADP and $P_i$ (see Figure 1–4).

Almost without exception,[6] the regeneration of ATP is accomplished everywhere in the biosphere at the expense of a coupled electron-transfer reaction from a reduced donor to an oxidized acceptor of sufficiently higher oxidation-reduction potential (lower energy level) to make the required energy available. As a rule of thumb, a potential difference of about 300 millivolts is needed (for a stoichiometry of two electrons transferred per molecule of ATP assembled) to cover the cost of ATP formation from ADP and $P_i$, which is of the order of 14 kilocalories (58.6 kilojoules) per gram-molecule under the usual intracellular conditions. Such coupled reactions are frequently referred to under the name oxidative phosphorylation (*oxphos,* in

[6] Among the exceptions to this rule are a few specialized processes in which ATP assembly is coupled to hydrolysis of the substrate (see Reference 154). In reality, several of these processes may be viewed as intramolecular electron transfers. For example, there exists in certain bacteria a phosphorylation process linked to the splitting of pyruvate into acetate and formate:

$$CH_3{-}CO{-}COO^- + H_2O \longrightarrow CH_3{-}COO^- + HCOO^- + H^+ \qquad \textbf{(19)}$$

This reaction is equivalent to an oxidative decarboxylation of pyruvate in which $CO_2$ would serve as electron acceptor and, instead of being released as such, would be reduced to formate. It resembles this process by leading to the same phosphorylation mechanism involving acetyl-coenzyme A. The reaction starts with the thiolytic splitting of pyruvate by coenzyme A:

$$CH_3{-}CO{-}COO^- + CoA{-}SH \rightleftharpoons CoA{-}S{-}CO{-}CH_3 + HCOO^- \qquad \textbf{(20)}$$

Acetyl-coenzyme A then allows the assembly of ATP by way of acetyl-phosphate, as in the classical oxidative decarboxylation (see Figure 1–5, page 15).

bioenergetics jargon), alluding to the fact that the electron donor is *ox*idized (often by oxygen, but not necessarily so), while ADP is *phos*phorylated to ATP.[7] They provide the essential bridge between electron-linked and group-linked energy.

Coupled phosphorylation reactions are subdivided into two categories, depending on whether the electron donor is a metabolic substrate or the reduced form of an electron carrier.

In substrate-level phosphorylations, electrons are transferred from a substrate to an acceptor (most often $NAD^+$), and the reaction is coupled to ATP assembly by means of chemical intermediates. As a rule, these systems require no particular structural organization and are situated in a soluble compartment of the cell. A characteristic example of substrate-level phosphorylation is shown in Figure 1–5. The key feature of such reactions is that the energy-rich bond created directly by electron transfer is not a pyrophosphate bond, but a precursor of such a bond. In the example shown, it is a thioester bond between the carboxyl group of the oxidized substrate (acetate) and the thiol group of coenzyme A. Assembly of ATP at the expense of this bond then takes place by typical reversible, sequential group transfer (see Figure 1–3, page 9).

A number of important substrate-level phosphorylation reactions, like the reaction of Figure 1–5, operate via a thioester intermediate. The thiol partner of the bond is, depending on the case, an internal cysteine residue of the enzyme protein, an enzyme-bound lipoamide molecule, coenzyme A, perhaps yet other compounds. The carboxyl partner is produced by the oxidation of an aldehydic or ketonic carbonyl group belonging to the substrate. Attack on the thioester by inorganic phosphate (phosphorolysis) gives an acyl-phosphate, which reacts with ADP (or some other NDP) to give ATP (NTP), as in Figure 1–5. These facts provide one of the cornerstones of the "thioester world" to be proposed in the second part of the book (Chapter 7).

In carrier-level phosphorylations, ATP assembly is coupled to the transfer of electrons between two carriers or between a carrier and oxygen (or some other terminal electron acceptor, such as $NO_3^-$). Systems of this kind occupy strategic sites, characterized by appropriate differences in oxidation-reduction potential, in chains of up to 15 or more electron carriers. These are organized within the framework of a membrane into some sort of "bucket brigade" for electrons: each carrier sits between a neighbor of lower potential, from which it receives electrons, and another of higher potential, to which it donates them. As many as three coupled "electron falls" may be inserted in series in a single chain. Such is the case in the respiratory chain of mitochondria.

[7] Biochemists generally reserve the name "oxidative phosphorylation" for the carrier-level kind (with oxygen as final electron acceptor), whereas substrate-level phosphorylation is often presented as the main energy-retrieval mechanism (which it is) in anaerobic fermentations. This distinction tends to obscure the fact that substrate-level phosphorylation is most often coupled to an electron-transfer process, just as is carrier-level phosphorylation. The main difference between the two kinds concerns the mechanism of coupling, by chemical intermediates in one case, by protonmotive force in the other. It is also misleading to equate substrate-level phosphorylation with anaerobic fermentation, as substrate-level phosphorylation takes place in all organisms, in aerobiosis as well as in anaerobiosis.

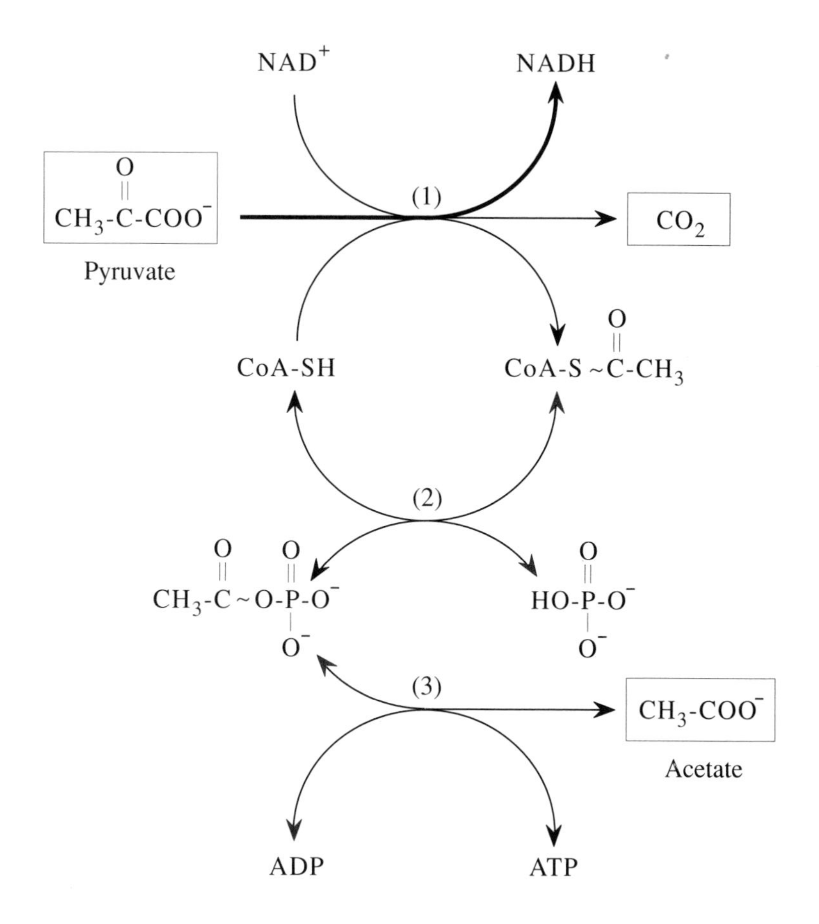


**1–5** *An example of substrate-level phosphorylation.*
The overall reaction is the oxidative decarboxylation of pyruvate to acetate and $CO_2$ (boxed reactants), with $NAD^+$ as electron acceptor (thick arrow). An important part of the free energy released by this reaction is conserved in the thioester bond of acetyl-coenzyme A (Step 1, which is really a complex reaction involving several steps and an acetyl-lipoamide intermediate). Hydrolysis of acetyl-coenzyme A allows the dehydrating condensation of ADP and inorganic phosphate to ATP by a reversible sequential group-transfer process (Steps 2 and 3), with acetyl-phosphate (the phosphorylated form of the oxidized substrate) as double-headed intermediate. Note the catalytic role of coenzyme A. In reverse (colored arrowheads), this group-transfer process would serve to condense acetate with coenzyme A at the expense of the hydrolysis of the terminal pyrophosphate bond of ATP (compare with Figure 1–3, page 9). Note that acetyl-phosphate occurs only in bacteria. In eukaryotes, the free energy of hydrolysis of the thioester bond of acetyl-coenzyme A is retrieved differently.

Electron flow through a phosphorylating respiratory chain supports ATP assembly by creating an electrochemical proton potential (see Figure 1–6). This potential generates a protonmotive force, which powers the assembly of ATP from ADP and $P_i$. The system catalyzing this reaction is a "proton pump"—i.e., a system capable of using ATP hydrolysis to force protons against an opposing potential—operating in reverse. Protonmotive force may also support other forms of work, such as active

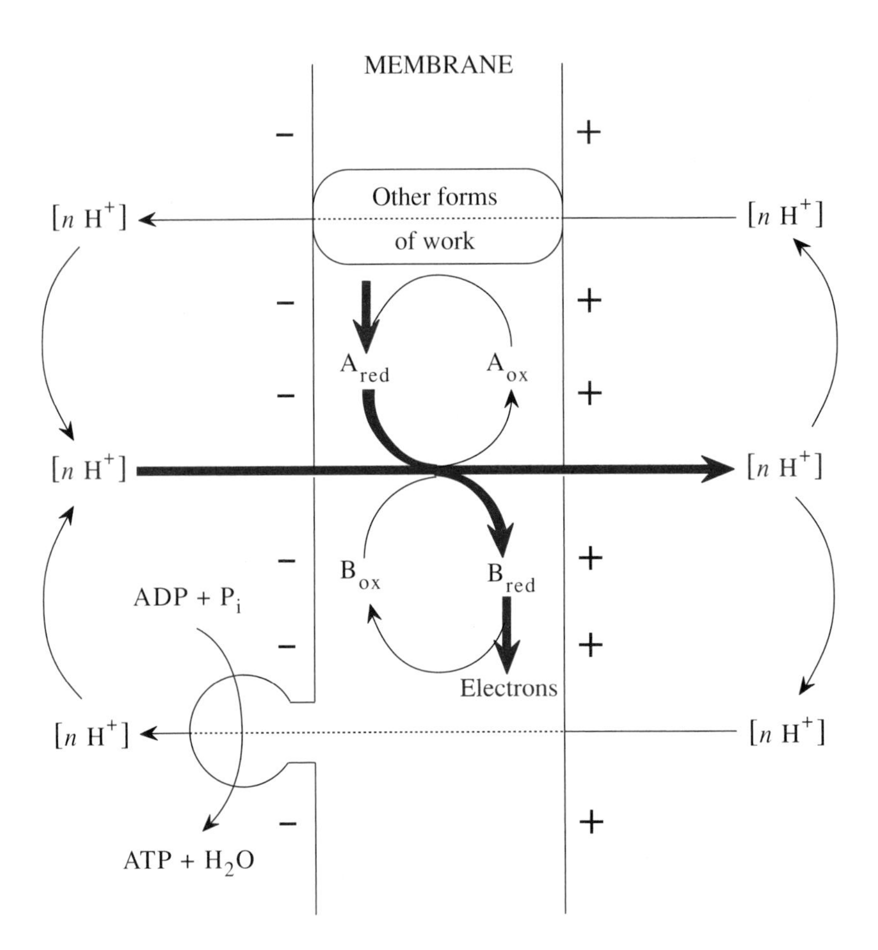


**1–6** *Carrier-level phosphorylation.*
In the middle of the diagram is shown a segment of a membrane-embedded electron-transport chain with an adequate oxidation-reduction potential span (about 300 millivolts) between the A and B carriers. Passage of a pair of electrons from A to B through this segment is obligatorily coupled with the forcible transport across the membrane of $n$ protons ($n$ = 2–4), up to the building of a steady-state proton potential of a magnitude equivalent to the oxidation-reduction potential span (600/$n$ millivolts). The protonmotive force generated in this way supports the assembly of ATP (by a system situated on knobs protruding on the negatively charged face of the membrane). It may sometimes fuel other forms of work, such as the active transport of calcium ions or the energization of electrons in another part of the chain. In each case, the potential-driven back flow of protons is obligatorily coupled with the performance of work.

transport of ions or molecules, flagellar movement in bacteria, and, as will be seen presently, energizing of electrons. For obvious reasons, such coupling depends crucially on there being no significant back-leakage of protons. Hence the absolute need for a proton-tight membrane fabric of the kind provided by a lipid bilayer (see page 31).

Many oxphos systems, whether of the substrate-level or the carrier-level kind, are reversible and may catalyze reverse electron transfer—from a donor of higher to an acceptor of lower oxidation-reduction potential—with the help of ATP hydrolysis and, in the case of carrier-level systems, also with that of protonmotive force generated elsewhere on the same membrane. Many autotrophic organisms take advantage of this reversibility to energize electrons up to the level needed for biosynthetic reductions.

In summary, electrons of high energy level are the true fuel of life. They support biosynthetic reductions and, by falling to a lower energy level through coupled phosphorylating or protonmotive systems, all other forms of biological work. With hardly an exception, living organisms are universally supported by electron fluxes passing through at least one appropriately harnessed 300-millivolt oxidation-reduction potential difference (if the electrons travel in pairs). Only the difference is critical. Absolute potential values are immaterial inasmuch as electrons can always be brought to a higher energy level, if needed, with the help of ATP or of protonmotive force (or of light, as we shall see shortly). Life-supporting electron fluxes are maintained in many different ways, depending on how input and output are ensured (see Figure 1–7).

Heterotrophic organisms, which include all animals and fungi as well as many protists and bacteria, are supplied with electrons of high energy level by their organic foodstuffs. They dispose of electrons that have dropped to a lower energy level most often by transfer to oxygen, sometimes by transfer to some other outside acceptor, such as $Fe^{3+}$, $NO_3^-$, $SO_4^{2-}$, or some other environmental oxidant. Occasionally, they manage without an outside acceptor by discharging the electrons in association with some metabolic end product (anaerobic fermentations) or with protons (hydrogen production). They pay for this advantage by strict nutritional requirements—to achieve the correct stoichiometry—and by a low yield of energy. Fermentation products, such as ethanol or lactate, leave with an abundance of electrons of high energy level, more than adequate for organisms provided with an outside acceptor. So does molecular hydrogen.

Chemoautotrophs function like heterotrophs, except that they derive their high-energy electrons from mineral donors, such as $H_2$, CO, or $H_2S$ and other sulfur compounds, and have an absolute need for an outside acceptor (usually oxygen). The electron requirements of these organisms are the heaviest found in nature. Like all autotrophs, they need a particularly large quantity of electrons for biosynthetic reductions because of the nature of their substrates (see Figure 1–1, page 6). In addition, they frequently have to spend considerable amounts of energy to upgrade these electrons by reverse transfer because their mineral nutrients rarely provide electrons at the high energy level required for biosynthetic reductions.

Phototrophs are distinguished by their ability to satisfy all their energy needs without the help of any outside electron donor or acceptor. They succeed in doing so

**1–7** *Energy-yielding electron fluxes in the biosphere.*
The diagram shows the various pathways that may be followed by electrons in living organisms. For the sake of simplicity, the span between the high and low energy levels occupied by the electrons has been represented as a single energy-yielding electron fall. In reality, up to four such falls linked in series may be traversed in succession by metabolic electrons. (1) Electron fall through one or more coupled ATP-generating transducers, which, by way of the ATP formed (or, sometimes, by way of an interposed protonmotive force), fuel virtually all forms of biological work. Every living organism obligatorily possesses at least one such system. (2) Electron inlet of heterotrophs (organic donors) and of chemoautotrophs (mineral donors). (3) Electron outlet most often used by heterotrophs and by chemoautotrophs, also by all other organisms when forced to use their endogenous electron stores (see Pathway 9). (4) Electron outlet used by organisms incapable of availing themselves of an outside electron acceptor (some obligatory anaerobes, for example), also by most other organisms when deprived of a suitable outside acceptor. Note that hydrogen production is a prerogative of only a few prokaryotic and eukaryotic microorganisms. (5) Biosynthetic reductions, incorporations, and storage. (6) Reverse electron transfer powered by ATP or by protonmotive force. (7) Electron inlet of phototrophs. (8) Photochemical energization of electrons in phototrophs. (9) Mobilization of endogenous electron reserves.

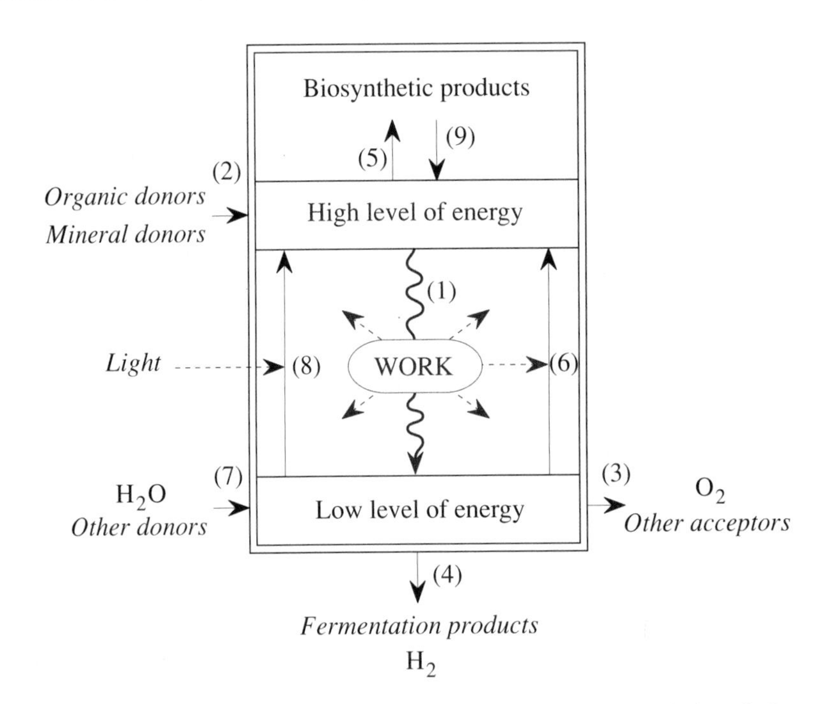


by means of an energizing photosystem that collects electrons at their exit from a phosphorylating transport chain and boosts them back to the top of the chain with the help of light energy (see Figure 1–8). In this way, electrons can cycle endlessly and convert light energy into usable protonmotive force or ATP indefinitely, without any exchange with the environment (cyclic photophosphorylation). Phototrophic organisms do, nevertheless, need outside electrons for their biosynthetic reductions. They receive these electrons mostly from water or, in the case of primitive photosynthetic bacteria, from other mineral or organic donors. They depend on light energy, directly or by way of reverse electron transfer, to render the electrons fit to support biosyn-

**1–8** *The principle of phototrophy.*
By means of a photosystem (1), light lifts electrons to a high level of energy, from which they fall through a proton-extruding electron-transport chain (2), which, by way of protonmotive force and ATP, supports all forms of work. De-energized electrons are recycled photochemically (cyclic photophosphorylation). The dotted line illustrates a rare exception (*Halobacterium halobium*), in which light generates protonmotive force without the participation of electrons.

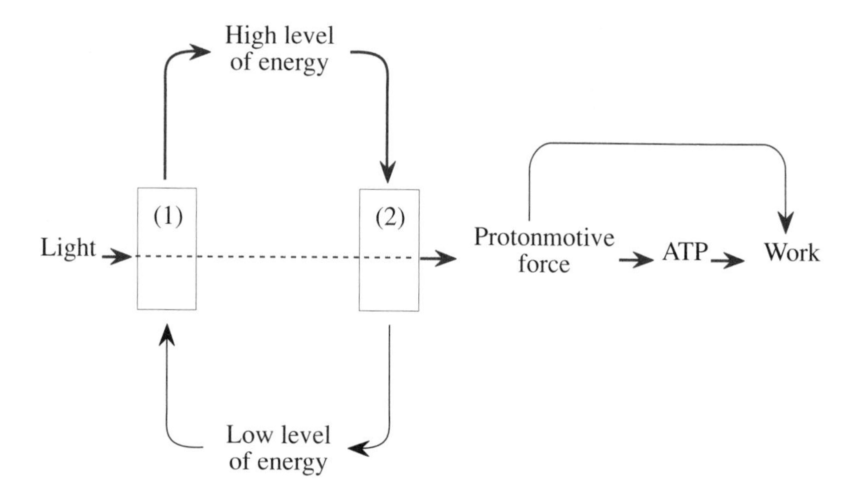


thetic reductions. These reductive fluxes usually follow a sinuous pathway that makes them pass through at least one coupled phosphorylation system. They thereby contribute to the energy supply (noncyclic photophosphorylation).

There is an extremely rare, but fascinating, exception to the above generalization. The purple halophile, phototrophic archaebacterium *Halobacterium halobium* creates a proton potential directly with the help of light energy, without any participation of electrons. More will be said about this interesting organism in Chapter 2.

All living organisms, whether autotrophic or not, consume their reserves or their own substance when their external supply of energy is cut off or inadequate (fasting animals, plants in the dark, etc.). They do this by essentially the same pathways as are followed by heterotrophs in the breakdown of exogenous nutrients. Biosynthetic products are similarly used, in a more complex fashion, within groups of organisms and, at the limit, within the whole biosphere, which may be viewed as a huge reservoir of high-energy electrons (matching the oxygen and other oxidants that were accumulated in the atmosphere and in the mineral world when the original stores were built up). Heterotrophic metabolism and endogenous consumption continually draw on this reserve and spill into the environment the extracted electrons after having used part of their energy potential. Autotrophic biosynthetic reductions compensate for this loss. Electrons thus circulate cyclically between the nonliving and the living world. Supported mostly by light energy, this circulation maintains the biospheric electron store at a relatively constant steady-state level (see Figure 1–9).

Note, however, that this steady state is crucially dependent on the green mantle of the Earth. In the Carboniferous, the golden age of forests, a large excess of electrons

**1–9** *The water/oxygen cycle.*
The diagram shows how electrons (together with protons) supplied by water are activated by light and stored in the biosphere and in fossil organic matter. Biological oxidations return the electrons to water with production of the work that supports the biosphere (and, partly, remodels the environment). The dotted line illustrates increasing man-made electron leakage. Primitive phototrophy and chemoautotrophy are not shown.

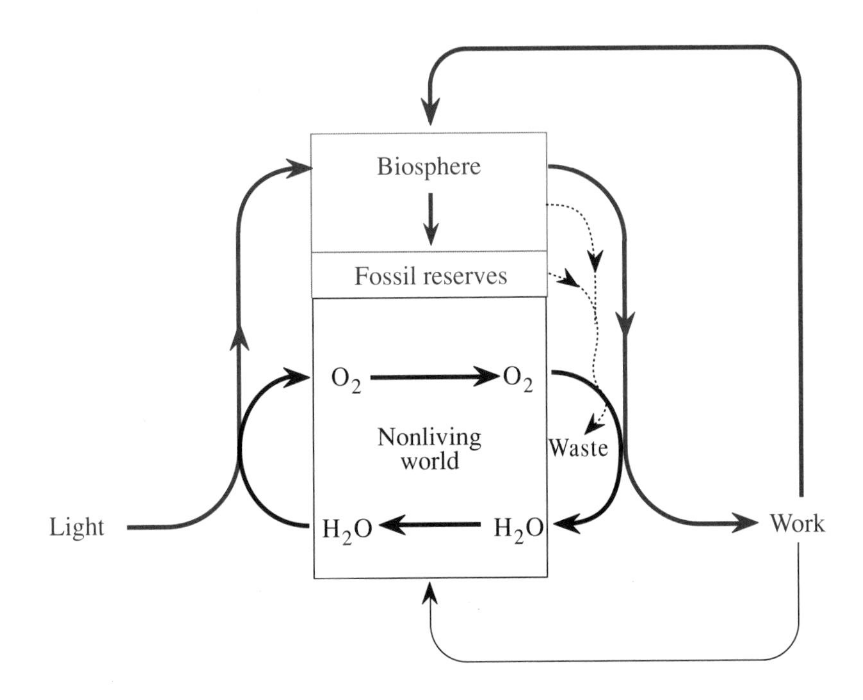


was stored and fossilized, together with huge quantities of carbon derived from $CO_2$. Today, biological electron stores, both living and fossil, are being depleted at an ever increasing pace, upsetting the balance in the other direction, with consequences, such as the release of increasing amounts of $CO_2$ into the atmosphere, that may, in the long term, be of the utmost gravity, as we are beginning to appreciate.

## CATALYSIS

Any form of independent life depends on the correct performance of thousands of unlikely chemical reactions, almost every one of which would not take place in the absence of an appropriate catalyst. The biological catalysts, or enzymes, are much more efficient and specific than are the artificial catalysts used in the laboratory and in industry. Many enzymes catalyze a single reaction or a small set of similar reactions. They do this by means of accurately molded binding sites that accommodate only the natural substrate(s) of the reaction and a few closely related analogues. These sites are strategically situated on the enzyme molecule so as to position the bound substrates appropriately in the sphere of influence of an active site that catalyzes their modification. The kind of finely adjusted molecular configurations that go into the making of an enzyme can only be realized within the framework of a complex macromolecule. Until recently, it was believed that all enzymes were proteins. This generalization has been challenged by the finding that some RNAs display catalytic activity. Considerable significance is attached to this discovery, which advocates of a protein-less "RNA world" in the origin of life have hailed as a major argument in support of their views. It will be referred to in greater detail in several of the subsequent chapters. I will come back to the specificity of enzymes when I discuss information.

Many enzymes act with the help of one or more cofactors. These may be metal ions or organic molecules of varying degrees of complexity (coenzymes) covalently attached to the enzyme protein or bound reversibly to it. Many vitamins enter into the constitution of coenzymes, most of which are either group carriers or electron carriers.

The enzymes, together with their cofactors, catalyze the multifarious chemical reactions that make up metabolism. These reactions are linked into chains of successive steps through intermediary metabolites, which are substances produced by a given step and utilized by another. In a number of cases, this organization is facilitated by the association of the enzymes into multienzyme complexes. Linear, branched, or cyclic enzymatic reaction chains are interconnected into a vast and intricate network of metabolic pathways. The nature of these pathways varies according to species. Their diversity is staggering, especially in bacteria and in plants.

Behind this diversity, however, a remarkable degree of unity can be discerned. Everywhere we look, the same basic principles are operating. The richness of metabolic pathways simply illustrates the many ways in which these principles can be given concrete expression. Furthermore, the diversity is built around a central core of reactions that are common to many, some to all, living organisms. Most metabolic reactions serve either biosynthesis or energy supply, or both. In line with the central role of transfer reactions, the majority of enzymes, probably more than 90%, are transferases, catalyzing either group transfer or electron transfer.

## INFORMATION

Any specific binding between two molecules involves information. This is why it is often said that one molecule "recognizes" another or that two molecules "recognize" each other. Thus, the specific self-assembly of macromolecules into intracellular or extracellular structural components implies information at the molecular level. So do the binding of substrates by enzymes and other catalytic systems, that of attractants and repellents by chemotactic sensors, that of hormones and other active agents by their receptors, that of newly made proteins by the translocating systems that direct them to their intracellular location, that of antigens by the corresponding antibodies, as well as many other important processes of biological regulation, communication, sorting, dispatching, or defense.

The basis of such recognition phenomena is accurate *complementarity* between two molecular structures. The lock-and-key[8] simile is often used in their description, though it may be misleading, as we are not dealing with rigid structures. Biochemical partners, when they embrace, mold themselves to each other to some extent, undergoing what Daniel Koshland has called "induced fit" (299).

### *Proteins and Nucleic Acids*

Biological locks, and often also the corresponding keys, generally consist of proteins, which are uniquely capable of assuming the multitude of three-dimensional configurations involved in biological recognition. Sometimes the information resides partly in other molecular structures, for instance, the carbohydrate side chains of glycoproteins. Even this kind of information can be traced back to proteins, since the enzymes that build the side chains are themselves proteins, which are directed by their protein substrates to the sites where the chains are to be implanted.

The information inscribed in the three-dimensional structure of proteins is entirely contained in their primary structure, that is, in the sequence of their amino acid residues. In terms of information, this sequence is often compared to a very long word, of up to several hundred letters, written with an alphabet of 20 letters—the 20 proteinogenic amino acids. Passage from the linear structures of polypeptide chains to the three-dimensional shapes of proteins occurs by a complex folding, twisting, coiling, and joining process, often interwoven with various kinds of chemical modifications, such as proteolytic trimming, closure of disulfide bonds, addition of glycosyl and other groups, etc. However complex this whole processing may be, its program may be seen as written entirely into the primary structures of the polypeptides concerned, as it depends on interactions of the constituent amino acids with each other and with surrounding molecules, including enzymes. Prominent among these interactions are, on one hand, water-excluding associations between hydrophobic residues through van der Waals forces and, on the other, electrostatic attractions, including hydrogen bonds, between hydrophilic residues and between such residues and water molecules. These interactions are influenced by a number of environmental factors, such as pH, temperature, ionic strength, presence of certain ions and molecules, presence of enzymes, and so on. The program may be written into the primary structures of the polypeptides, but its execution depends on the conditions provided by the cell. This is true of information in general; it is of value only if and where it can be understood.

[8] First proposed by Emil Fischer, who was a chemist, the lock-and-key image was subsequently adopted by Paul Ehrlich, a pioneer in the development of immunology and of chemotherapy, two fields in which the image is particularly appropriate. It has been strongly criticized by Jacques Ninio (73). In the view of this author, recognition depends not so much on highly specific complementarity as it does on a relatively unspecific sort of binding, which is subsequently corrected by rejection of the undesirable. Indeed, accuracy is often achieved in biology by the intervention of a variety of proofreading mechanisms, as will be pointed out later. Chemical locks and keys have obvious limitations. On the other hand, this is no reason for discarding the concept of chemical complementarity, which governs an immense number of important biological processes. I have identified it as one of the key concepts needed to understand life.

When it comes to the primary structures of the proteins themselves, we must look elsewhere. Proteins do not possess the information needed for their own duplication; they cannot be copied.[9] This information is encoded in specific DNA genes that dictate, by way of their RNA transcripts, the assembly of all known proteins, barring the few that are coded for directly by viral RNAs. Nucleic acid genes also command their own replication, thus completing the information circle (see Figure 1–10). A few special aspects of these well-known processes deserve to be highlighted.

First, there is the paramount importance of base pairing (see Figure 1–11), no doubt the most wide-ranging and most crucial manifestation of chemical complementarity in biology. It is mainly responsible for the three-dimensional structures of nucleic acids, from the beautifully regular DNA double helix to the various clover leaves, flowers, and other exotic configurations of RNAs. It also rules all forms of replication and of transcription, the last step of translation, and a number of more specialized processes, such as RNA splicing.

Biosynthetic mechanisms governed by base pairing all involve three components: 1) activated building blocks, 2) an informational nucleic acid template, 3) a catalytic assembly system. The activated building blocks are mononucleoside triphosphates when RNA is made, deoxymononucleoside triphosphates when DNA is made, and aminoacyl-transfer RNA (tRNA) complexes when proteins are synthesized. In terms of our general scheme (see Figure 1–3, page 9), the first two are mononucleotides

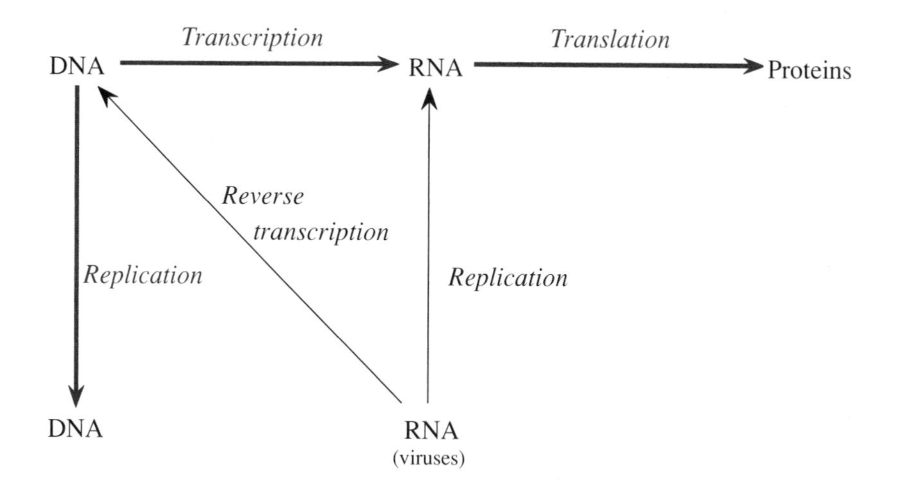


**1–10** *Biological information transfers.* DNA replication, transcription, and translation take place in all known organisms. RNA replication occurs in cells infected by some RNA viruses, under the influence of enzymes encoded by the viral genome. The RNA of retroviruses is replicated by way of DNA with the help of a virally encoded reverse transcriptase.

[9] This is a central and undisputed tenet of modern biology, firmly supported by all available evidence. According to some theories, as we shall see in Chapter 8, it may not have held true in prebiotic times.

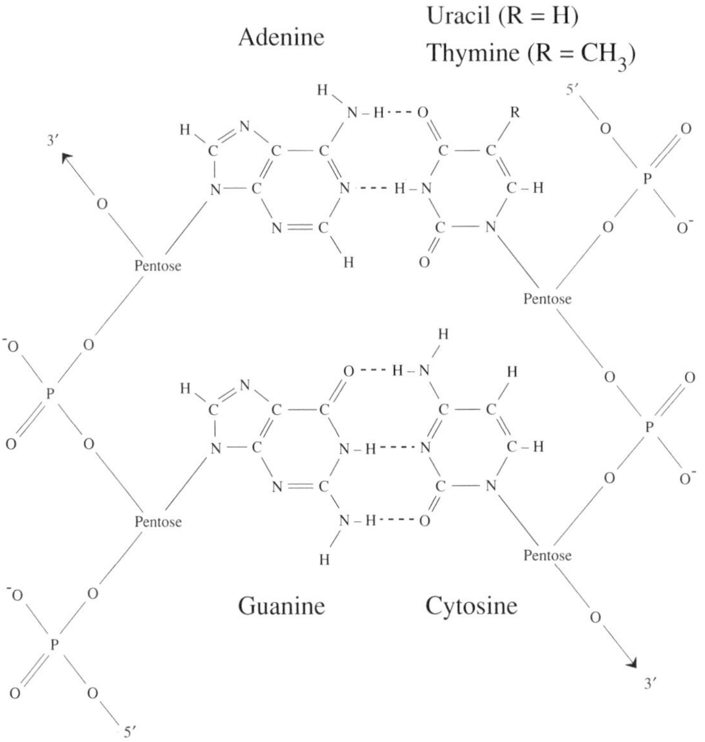

**1–11** *Base pairing, the universal cipher.* Chemical complementarity underlies pairing between adenine and uracil (in RNA) or thymine (in DNA) and between guanine and cytosine. The pairs are stabilized by hydrogen bonds. With three bridges, the GC pair is stronger than the AU(T) pair, which has only two. Pentose-phosphate backbones illustrate the antiparallel orientation of the joining strands. Many information transfers depend on this mechanism.



activated as pyrophosphates. The third are amino acids attached to tRNA carriers after activation as AMP derivatives (see Figure 1–12). The templates are DNAs in DNA replication and transcription, RNAs in RNA replication and in reverse transcription, and messenger RNAs (mRNAs) in protein synthesis. As to the catalysts, they may be viewed as incomplete enzymes that become both specific and active only when combined with an appropriately positioned nucleic acid template. Unlike the usual enzymes, they have "multiple choice" specificities, determined by the nature of the particular base or base triplet that faces the substrate-accepting site. They are locks containing a sliding part with, in nucleic acid synthesis, 4 different interchangeable positions fitting 4 different keys, and in protein synthesis, 64 positions that, altogether, fit some 40 different keys (see Figure 1–13, page 26).

All the information transfers considered above are conducted in nucleic acid language. The real act of translation—from the 4-letter language of nucleic acids to the 20-letter language of proteins—is carried out by the aminoacyl-tRNA synthetases.

**1–12** *The two steps of aminoacyl-tRNA synthesis.*
In the first step (1), the amino acid (aa) is activated by the enzyme (E) to a transferable aminoacyl group by adenylyl (AMP) transfer from ATP. The product of this reaction, the double-headed aminoacyl-AMP (aa-AMP) remains bound to the enzyme. In the second step (2), the activated aminoacyl group is transferred from this intermediate to its specific tRNA carrier. From the energetic point of view, this mechanism conforms to the general scheme of Figure 1–3, page 9. In terms of information, the first step involves the specific recognition of the amino acid, the second one that of the cognate tRNA. There are only 20 such enzymes, one for each amino acid. Each recognizes all the tRNAs specific for a given amino acid.

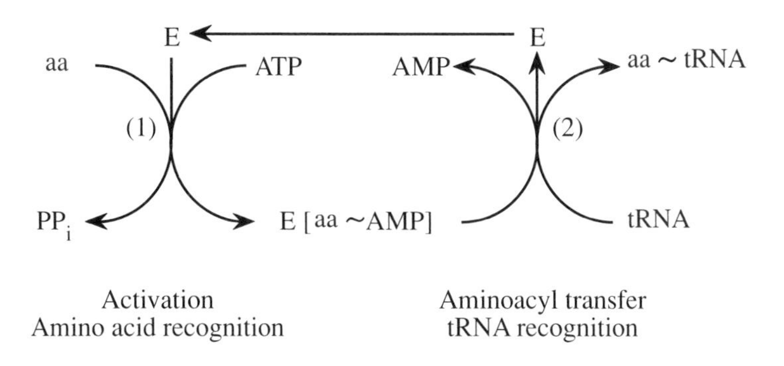


As illustrated by Figure 1–12, these enzymes catalyze two successive steps. In the first step, the amino acid is specifically recognized and activated to an aminoacyl-AMP complex (X — O — Q in our general scheme of Figure 1–3, page 9) by adenylyl transfer from ATP. In the second step, the aminoacyl group is transferred from the enzyme-bound aminoacyl-AMP to its tRNA carrier (Cg in Equation 17). This second step involves specific recognition of the tRNA. Therefore, each aminoacyl-tRNA synthetase has two specific binding sites: one for an amino acid and one for the corresponding tRNA(s). Each such enzyme bears, imprinted into its structure, one entry of the protein-nucleic acid dictionary. Remarkably, it is not the dictionary that one would have expected. This point deserves some elaboration.

The familiar dictionary is given by the genetic code, an almost universal set of correspondences between amino acids and the triplet codons displayed by mRNAs. The codons, in turn, are read by the tRNA anticodons according to the usual base-pairing rules, tempered by wobbling (see Figure 1–13). Thus, the crucial informational role of aminoacyl-tRNA synthetases is to attach amino acids to tRNAs bearing a correct anticodon. If they mismatch, the wrong amino acid will be inserted into the protein, as protein assembly itself depends exclusively on codon-anticodon recognition. In view of these facts, one would expect the enzymes to recognize the anticodons in the tRNAs. Surprisingly, this is often not so. In a number of cases, the enzymes disregard the anticodons and recognize other features in the tRNA molecules, situated closer to the aminoacyl-acceptor stem (see Figure 1–14, page 27). In other cases, the anticodon is involved to a greater or lesser extent in the recognition process. These intriguing facts have provided modern molecular biology with one of its most lively foci of research and discussion. They are better examined in the second

| Second letter | Amino acid | Codons | Anticodons | |
|---|---|---|---|---|
| | | | *E. coli* | Eukaryotes |
| U | Phe | UUU,UUC | GAA | GAA |
| | Leu | UUA,UUG | *AAA, CAA | (UAA), CAA |
| | | CUU,CUC,CUA,CUG | CAG, GAG, UAG | CAG, IAG, UAG |
| | Ile | AUU,AUC,AUA | GAU | IAU, UAU |
| | Met | AUG | *CAU | CAU |
| | Val | GUU,GUC,GUA,GUG | GAC, *UAC | IAC, UAC, CAC |
| C | Ser† | UCU,UCC,UCA,UCG | CGA, GGA, UGA | CGA, IGA, UGA |
| | Pro | CCU,CCC,CCA,CCG | *UGG, CGG | IGG, UGG, CGG |
| | Thr | ACU,ACC,ACA,ACG | GGU, UGU | IGU, (UGU), CGU |
| | Ala | GCU,GCC,GCA,GCG | GGC, *UGC | IGC, UGC, (CGC) |
| A | Tyr | UAU,UAC | *GUA | GUA |
| | His | CAU,CAC | *GUG | GUG |
| | Gln | CAA,CAG | CUG, *UUG | CUG, UUG |
| | Asn | AAU,AAC | *GUU | GUU |
| | Lys | AAA,AAG | *UUU | UUU, CUU |
| | Asp | GAU,GAC | *GUC | GUC |
| | Glu | GAA,GAG | *UUC | UUC, CUC |
| G | Cys | UGU,UGC | GCA | GCA |
| | Trp | UGG | CCA | CCA |
| | Arg | CGU,CGC,CGA,CGG | *ACG, CCG | ICG, UCG, CCG |
| | | AGA,AGG | *UCU | UCU, CCU |
| | Ser† | AGU,AGC | GCU | GCU |
| | Gly | GGU,GGC,GGA,GGG | CCC, GCC, *UCC | GCC, UCC, CCC |

† Note double listing   * Modified   (Not demonstrated, but believed to exist)

part of the book, as they have an important bearing on the origin of life and the origin of the genetic code (Chapter 8).

Another remarkable feature of nucleic acid-dependent information transfer is the breakup of many eukaryotic genes into parts that are expressed (exons), separated by parts that are not (introns). This astonishing property adds considerably to the error risk of transcription, which not only depends on the correct matching of substrates and template by base pairing, but, in addition, relies on the accurate excision of introns from pre-mRNAs and on the precise subsequent splicing of exons into mature mRNAs. Let a single base shift occur in this processing, and the message will be

**1–13** *(Opposite page.) The genetic code.* Codons and anticodons are both given in the $5' \rightarrow 3'$ direction. Reverse anticodons to verify fit. The codons shown make up the code that is obeyed almost universally by bacteria, plants, and animals. Rare deviations from this code exist, mostly in mitochondria. The anticodons from the bacterium *Escherichia coli* are from Reference 208. Most have been obtained by direct sequencing of the tRNAs; a few are derived from the corresponding DNA genes. In a number of cases, the first base of the anticodon is chemically modified (asterisk). Transfer RNAs are very rich in modified bases.

Eukaryotic anticodons are from Reference 291. Base modifications are not indicated. I stands for inosine, whose base is hypoxanthine, which, like guanine, pairs with cytosine.

The following points must be noted: 1) The code is degenerate (i.e., includes many synonyms). Of the 64 possible triplets, 61 are used to code for 20 amino acids. The three remaining triplets (UAA, UAG, UGA) are stop codons. 2) The code has a structure. Synonyms are clearly grouped together. In addition, amino acids are grouped according to the degree of hydrophobicity of their R groups, which is correlated with the nature of the middle base of the codons: U for the most hydrophobic amino acids; next C, the sister pyrimidine; and, finally, A or G for the more hydrophilic amino acids. 3) There are fewer anticodons than codons because a given anticodon often recognizes more than one codon, thanks to "wobbling" at the level of the third codon base. The first two bases of the codons are always strictly matched by the anticodons.

completely adulterated. In spite of these additional hazards, RNA splicing is carried out on a large scale in nature. Even more surprising, its practice increases with evolutionary complexity. Introns hardly exist in prokaryotes; they are rare in lower eukaryotes and numerous in higher plants and animals. The significance of these facts will be considered further in Chapter 3 and in the second part of the book.

In recent years, an increasing number of bewildering variations on the theme of split genes has come to light. Examples are the creation of genetic diversity in the immune system by the joining of DNA segments selected from sets of interchangeable pieces; the numerous translocations, insertions, deletions, transpositions, recombinations, and other DNA shufflings that can reorganize or perturb the genome;

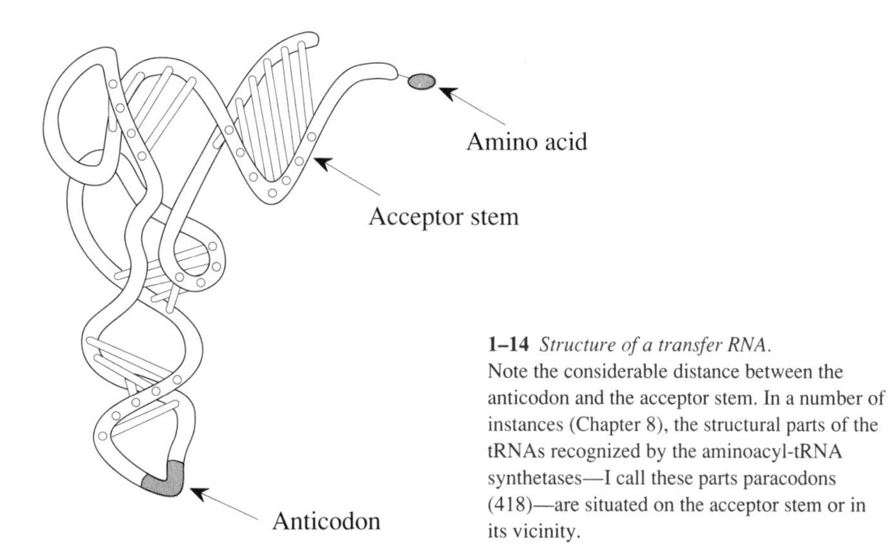


**1–14** *Structure of a transfer RNA.* Note the considerable distance between the anticodon and the acceptor stem. In a number of instances (Chapter 8), the structural parts of the tRNAs recognized by the aminoacyl-tRNA synthetases—I call these parts paracodons (418)—are situated on the acceptor stem or in its vicinity.

the alternative splicing of pre-mRNAs to give two or more different mRNAs; the splicing of an mRNA with a piece taken from another RNA (*trans* splicing); and the newly discovered phenomenon of RNA editing (Chapter 3). How these processes are directed so as to deliver correct, reproducible messages is still far from clear. Orthodoxy requires this information to be hidden somewhere in the DNA, but there are scientists who are beginning to doubt this and who suspect that some major surprise may be lurking just around the corner. Time will tell.[10]

## *The Flow of Biological Information*

DNA makes RNA makes protein makes the individual. This modern way of saying "The genotype makes the phenotype" conveniently puts in capsule form what is known of the dominant flow of biological information. However, it leaves out a number of important branches that detach from the main stream or feed back into it.

First, not all the genomic DNA feeds information into the main stream. A more or less important part of it, especially in eukaryotes, is never transcribed, even though it is faithfully replicated each time the cell prepares to divide. Some of this silent DNA is involved in the control of replication and transcription. But this is often a small fraction of the total. The remainder, according to many workers, may be little more than useless "junk," which has accumulated over the ages (Chapter 3).

Next, what is transcribed is far from being expressed totally as protein. A number of RNAs do not code for proteins but contribute directly to the phenotype by accomplishing significant functions in their own right. These noncoding RNAs include ribosomal RNAs (rRNA), tRNAs, and various other RNA molecules, usually of small size (sRNA), whose functions are beginning to be understood. Even the mRNAs are more than just protein-coding messages. Their coding part is preceded by a 5′-terminal leader sequence (often "capped"), which plays a role in initiating translation, and it is usually followed by a 3′-terminal tail, which may also be of functional importance. Furthermore, at least in eukaryotes, the primary transcription product (pre-mRNA) most often loses a considerable part of its substance in the course of the sectioning-splicing process that converts it to mature mRNA. The removed introns seem only exceptionally to play an informational role. Most often, they are simply discarded and broken down after they have been spliced off. As briefly mentioned, there are also other forms of mRNA processing that may have to be taken into account. Several of these mechanisms do not belong to the basic blueprint of life, as they are found only in eukaryotes. They will be considered in greater detail in Chapters 3 and 8.

Finally, translation products are only the raw stuff of proteins. The linear polypeptides assembled according to the instructions furnished by mRNAs most often are subjected to more or less extensive processing, in the course of which they may lose pieces of variable length, acquire carbohydrate and other groups, fold into specific configurations that are sometimes sealed by disulfide bonds, and combine

[10] As will be pointed out in Chapter 3 in reference to the particularly puzzling phenomenon of RNA editing, orthodoxy seems to be winning (see page 97).

with other polypeptides. Some polypeptides are not converted to proteins but, instead, are fragmented and trimmed to give rise to short peptides of functional significance, including hormones and neurotransmitters.

Doubling back from the expressed phenotype to the stream that commands its expression are many control lines of fundamental importance. These include all the enzyme activities and other regulatory influences that affect the chemical state (methylation, for example), topology, replication, and transcription of DNA, as well as those that condition the state and processing of RNA, the synthesis, processing, and tridimensional organization of polypeptides, the modification of proteins, and many other phenomena that exert some effect on the main flow of genetic information.

Third partner in this crisscross of instructions is the environment, including other living cells or organisms. The environment most often acts by way of the phenotype but occasionally exerts a direct effect on the genotype. Biological activity, in turn, affects the environment, sometimes in a profound manner.

Thus, as illustrated in Figure 1–15, a highly intricate dialogue continually takes place between genotype and phenotype and, mostly by way of the phenotype, between the cell or organism and the environment. The main flow of information emanating from the genome is thereby constantly modulated, qualitatively and quantitatively, resulting in such important temporal modifications of the phenotype as adaptation and, especially, development. Regardless of the complexity of signals, however, the final decision rests with the genome. Nothing can bring forth what is not in the genes.[11] This, at least, is the present creed.

## *The Accuracy of Biological Information Transfer*

A remarkable characteristic of biological information transfer mechanisms is their accuracy. The error rate of DNA replication, for example, is on the order of $10^{-9}$, or one mismatched nucleotide in one billion. This is orders of magnitude less than could be expected from the flimsy chemical complementarity relationships involved. Accounting for the extraordinary precision of biological information transfer is proofreading, a mechanism, based on a subtle combination of enzyme specificities and kinetics, that causes wrongly inserted molecules to be excised before they are sealed into place by a subsequent step. A number of mechanisms also exist that repair injuries caused to DNA by such agents as ultraviolet light, ionizing radiations, or mutagenic chemicals. Hereditary continuity is crucially dependent on the efficacy of proofreading and repair. Even with an error rate as fantastically low as $10^{-10}$, one mistake, on average, would still be made each time the genome of a human cell is duplicated. Considering that some cells, such as those that give rise to white blood cells, divide many thousands of times in the lifetime of an individual, one can easily realize the kind of havoc that would result if the error rate were much higher. Yet, perfectly accurate replication would also be disadvantageous, as it would oppose those hereditary changes (mutations) without which there can be no evolution, no diversification, no adaptation to new conditions. The precision requirements of

[11] With the proviso, of course, that genes can undergo alterations.

**1–15** *The flow of biological information.* The diagram shows the main pathways of expression of the genetic information contained in DNA (blue), together with the many retroactive effects whereby the expressed information influences DNA expression and replication (broken arrows). Interactions with the environment are also shown.

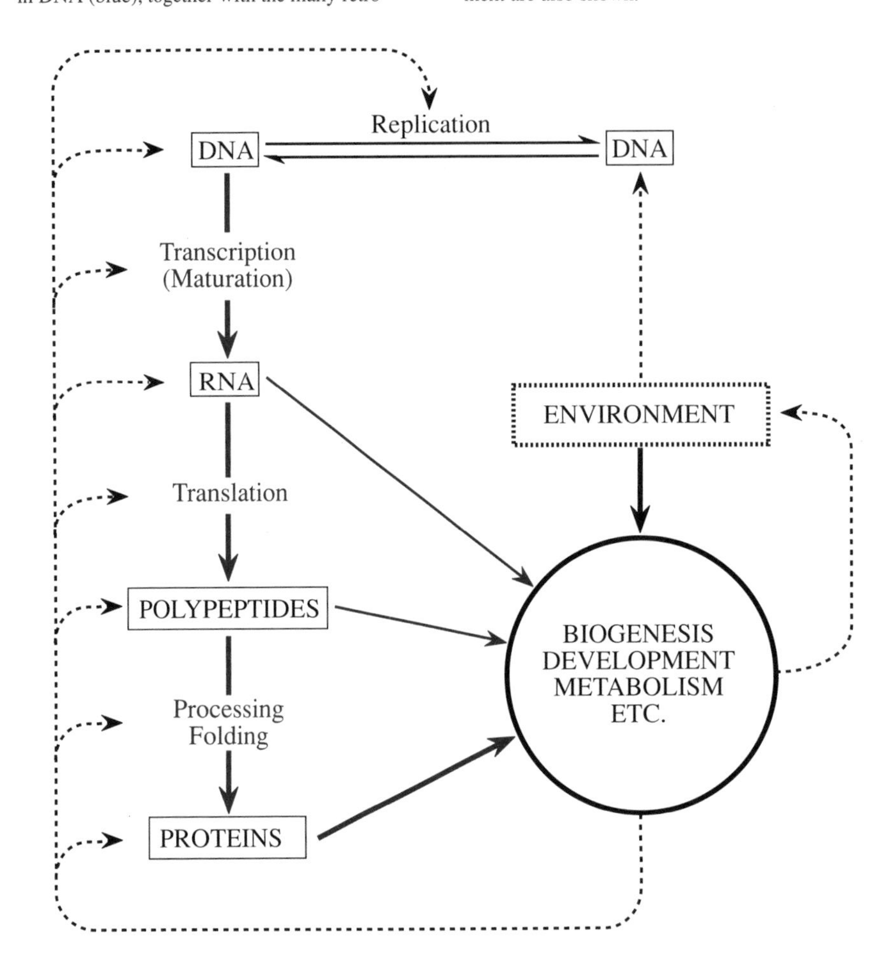


transcription and of translation are less critical than those of replication but are still stringent enough to require proofreading. Otherwise, errors are likely to spill over into essential parts of the translation machinery and to precipitate what Leslie Orgel has called an "error catastrophe."[12]

[12] This theory was originally proposed by Orgel as a possible explanation of cellular aging (290). Somatic mutations that affect components of the translation machinery and cause a decrease in the fidelity of this process are expected to spread translation errors autocatalytically to the point of endangering cell viability.

### *Homomorphic Information Transfer*

Chemical complementarity does not account for all forms of biological information transfer. Some patterns are authentically copied according to what may be called "homomorphic" principles. The distribution pattern of cilia on the surface of paramecia, studied by Tracy Sonneborn,[13] provides an early example of homomorphic growth. Of much wider significance is the growth of membranes. As far as is known, membranes never form de novo by self-assembly of their constituents; they always grow, in essentially homomorphic fashion, by accretion, that is, by the insertion of additional constituents into preexisting membranes.[14] The corresponding patterns are transmitted from generation to generation by way of the cytoplasm—mostly of maternal origin in sexually reproducing organisms—in the form of samples of the main kinds of cytomembranes found in the organism.

## CONTAINMENT AND ISOLATION

### *The Lipid Bilayer*

All living cells are surrounded—and many are further partitioned—by membranes of unique physical and functional properties. The main fabric of these membranes is the lipid bilayer, a continuous sheet, no more than 5 to 6 nanometers thick, made of two layers of amphiphilic molecules—mostly phospholipids, glycolipids, and sulfolipids—joined laterally, as well as tail to tail, by van der Waals interactions between their hydrophobic hydrocarbon chains. Protruding on the two faces of this bimolecular film are the hydrophilic heads of the lipid molecules. Sealed structures of this sort (liposomes) readily form when amphiphilic lipid mixtures are subjected to sonication in an aqueous milieu. Lipid bilayers have remarkable properties, fundamentally related to their biological role. It is fair to state that life as we know it could not possibly exist without the lipid bilayer.

Lipid bilayers are fluid structures of almost limitless flexibility. They behave as two-dimensional liquid crystals within which the lipid molecules can move about freely in the plane of the bilayer and reorganize themselves into almost any sort of shape without loss of overall structural coherence. Furthermore, lipid bilayers are self-sealing arrangements that automatically and necessarily organize themselves into closed structures, a property to which they owe their ability to join with each other by fusion and to divide by fission, while always maintaining a closed, vesicular shape

[13] The work of Sonneborn and his role in promoting cytoplasmic inheritance are discussed in an interesting historical context in Reference 24.

[14] This is a superficial way of putting things. As will be seen, the membrane constituents of eukaryotes originate in specific sites and then move to their final location, mostly by lateral diffusion, sometimes otherwise. The factors that cause them to become established at a given locus are still poorly understood; but they are undoubtedly much more complex and "heteromorphic" than the superficial description suggests. Behind homomorphism probably lies a great deal of straightforward complementarity. Even so, it is probable that a membrane pattern must exist to be reproduced. Membrane synthesis is homomorphic in this sense.

(see Figure 1–16). Also, they are essentially impermeable to water-soluble molecules and ions.

Structures of this sort are ideally suited to surround living cells by a clearly defined boundary capable of effectively and uninterruptedly containing and insulating the cells, even while these undergo highly stressful changes in shape. Lipid bilayers further serve to delimit distinct functional areas inside cells and participate in a variety of vesicular transport processes.

### *Membrane Proteins*

By itself, the lipid bilayer is an inert structure that can neither mediate nor control the multiple exchanges of matter and information that must necessarily take place between the cells and their environment and between different intracellular compartments. These functions are accomplished by integral membrane proteins (see Figure 1–17) embedded in the bilayer by one or more hydrophobic polypeptide segments or anchored to it by a hydrophobic appendage, for example, a fatty acid.

Functionally, these proteins include transporters, which facilitate the passive diffusion of their substrates; active transport systems, or pumps, which can move their substrates forcibly against a concentration gradient, with the help of ATP or of some other source of energy (protonmotive force, exergonic cotransport); and transloca-tors, of as yet poorly understood design, involved in the specific transfer of proteins

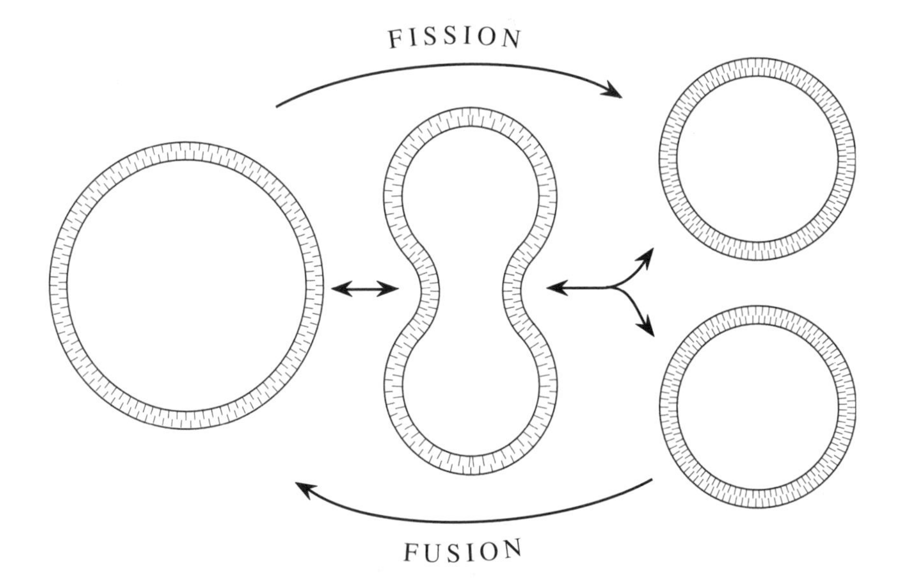


**1–16** *Fusion and fission of lipid bilayers.* Lipid bilayers form flexible, impermeable, self-sealing, vesicular structures capable of dividing into two such vesicles or, alternatively, of joining with another, without at any moment losing their structural continuity. Many biological processes depend on this property.

**1–17** *Structure of biological membranes.* All biological membranes are made of a lipid bilayer into which are inserted various proteins (blue) that mediate the exchanges of matter across the membrane as well as a number of other functions.

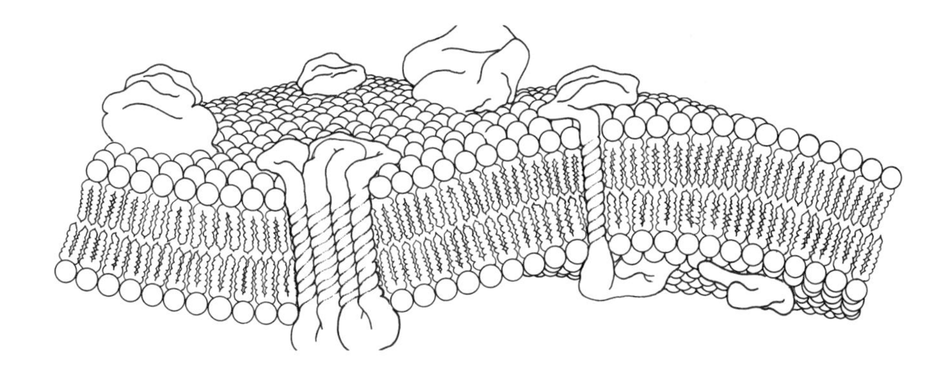

and other large molecules across membranes. When situated on the cell surface, these various transport systems allow cells to take in foodstuffs, to excrete waste products, and to maintain a specific internal ionic milieu adapted to their needs. Some pumps, through the ionic imbalances they create, generate the membrane potentials that cover the cell surface with electric charges and determine numerous biological phenomena of major importance, including the operation of the nervous system. Intracellular transport systems similarly mediate exchanges of matter between different subcellular compartments and allow the creation of a specific local milieu within each compartment. Protein translocators play an important role in the transfer of newly made proteins from their site of synthesis to their final intracellular or extracellular abode.

Receptors make up a second important group of integral membrane proteins. There are many different kinds of such receptors. They all possess one or more binding sites for specific ligands, projecting on one face of the membrane, and an effector site, situated within the membrane or on its other face. Receptors characteristically respond to the occupancy of their binding sites by undergoing a *conformational change* that triggers the effector site to set off some specific phenomenon or process. This could be the opening or closing of an ion channel, the collapse of an electric membrane potential, the stimulation of some metabolic activity, the initiation of DNA replication and of cell division, the evocation of some motile response, or a reorganization of the membrane structure.

Membranes built with a lipid-bilayer structure also serve an important function in energy metabolism. As already mentioned, they provide the proton-tight substratum that is indispensable for the activity of all the systems that function by way of protonmotive force (see Figure 1–6, page 16). These include the electron-transport chains and the associated systems that accomplish carrier-level phosphorylations, as well as the light-gathering and phototransducing complexes of phototrophic organisms. In addition, membranes shelter many enzyme systems, especially those acting on water-insoluble substrates.

## SELF-REGULATION

It was mentioned at the beginning of this chapter that living systems maintain themselves in a state far from equilibrium. This implies that they are intrinsically capable of offsetting the characteristic instability of such a state. As we know, they can do this even in the face of considerable variations in temperature, humidity, availability of food, or other environmental conditions.

Underlying this autonomy lies a closely interwoven network of structural and dynamic relationships linking together all the different parts of living systems and all the different processes that take place in them. This network rests critically on the concept of *feedback,* mostly of the negative kind, that is, the retroactive loop, also widely used in man-made regulatory systems, whereby an effect tends to correct the disturbance that causes it. Biological feedback loops, however, are multidimensional and interconnected in a way that even our most elaborate computer networks cannot emulate. They rely extensively on chemical complementarity and on conformational change. Several levels of sophistication may be distinguished in the regulatory network of living systems.

### *Regulation by Mass Action*

At the simplest level—already highly complex—there is the automatic buffering by mass action that is inherent in any multireaction system. Metabolic reactions are linked together by intermediary metabolites, that is, substances that are produced by one reaction and consumed by another. Such a link is not simply qualitative; it is quantitative, by virtue of the law of mass action. This is best shown if we first ignore, for the sake of simplicity, the manner in which reaction kinetics are affected by the participation of enzymes. Let us assume a system in dynamic equilibrium (steady state), in which the rate of production or supply of an intermediate suddenly increases for some reason or other. This will result in an increase in the intermediate's concentration, which in turn will cause its rate of consumption to speed up until it becomes equal to the rate of production. In this way, a new steady state, characterized by a new concentration of the intermediate, will automatically become established. This concept can be generalized. In a linear reaction chain, all the steps are subject to the same automatic regulation, so that their individual rates will tend to be equalized by way of the rise or fall of the concentrations of the intermediary metabolites. In systems of more complex topology, such as a branched chain, the responses to disturbances will be correspondingly more complex; but the outcome will be the same: mutual adjustments of rates through mass effects. Remembering that virtually all metabolic reactions are interconnected into a single network, we can readily see how such a system will tend to maintain its dynamic coherence in the face of perturbations that modify some reaction rates.

In such a network, particularly important regulatory roles will be played by substances that participate in several reactions and thus link together several pathways. Certain soluble cofactors—for example, those that participate in electron transfer, most prominently NAD—stand out in this respect, as they may link together up to several hundred distinct reactions. At the center of the web, we find ATP, ADP, and $P_i$, in the form of the concentration ratio, sometimes called phosphate potential:

$$\frac{[ADP][P_i]}{[ATP]} \qquad (21)$$

Any increase in this ratio, caused, for example, by an enhanced consumption of ATP for the performance of work, will, by making more ADP and $P_i$ available for ATP formation, tend to accelerate the coupled electron-transfer reactions that support this formation. In turn, the consumption of electron suppliers (foodstuffs or reserves) and acceptor (oxygen) will be increased. The reverse will happen if the ratio decreases. Called respiratory control, this fundamental regulatory process automatically adjusts the energy expenditure of living organisms to the amount of work they perform.

### *Regulation by Way of Enzymes*

In reality, the relationships just outlined are more complex than expected from a simple application of the mass-action law because of the intervention of enzymes, which may limit, to a greater or lesser extent, the effects exerted on reaction rates by changes in substrate concentrations.[15] At the limit, an enzyme working at maximum capacity because all of its substrate-binding sites are occupied cannot be made to work faster by any increase, however large, in substrate concentration. Thus, there may be bottlenecks in the metabolic network, just as there are in other dynamic networks, such as roadways.

The participation of enzymes renders possible a new set of regulatory influences considerably more elaborate than the gross mass-action effects. The catalytic activities of enzymes can be increased by a variety of substances (activators) and decreased by others (inhibitors). Some of these effectors, as they are called, act directly on the catalytic site or on a substrate-binding site. Others combine with a separate site, called allosteric, in such a way as to induce a change in the conformation of the enzyme molecule and thereby alter some kinetic characteristic of the catalyzed reaction, such as its rate constant or the affinity of the enzyme for some substrate. Here is yet another important example of the lock-and-key arrangement, but one linked, as with receptors, to the production of a specific effect by a transducing conformational change caused by occupation of the lock.

Many intermediary metabolites are activators or inhibitors of enzymes, which, themselves, may be related or unrelated to the metabolic chains to which the metabolites belong. Thus, a second network of interrelationships involving the

[15] The basic law in this connection is the Michaelis-Menten equation, which relates the velocity $v$ of an enzymatic reaction to its maximal velocity $V$ by a hyperbolic function of the substrate concentration $[S]$:

$$v = V\frac{[S]}{K_m + [S]} \qquad (22)$$

in which $K_m$, the Michaelis constant, is the dissociation constant of the enzyme-substrate complex.

modulation of enzyme activities and specificities is superimposed upon the network of mass-action effects. Enzymes catalyzing metabolic bottlenecks are particularly subject to this kind of regulation, with the result that a small change sometimes has far-reaching consequences.

Dynamic networks of the kind described serve to maintain the internal coherence of living systems and condition the responses of these systems to external perturbations. Sometimes minor disturbances, such as the appearance of a few molecules of a hormone in the surrounding medium, may elicit very important responses thanks to the operation of specific receptors on the cell surface. These are connected to some node of the network in such a way that when they are occupied by their ligand a massive metabolic change is triggered, often by a self-amplifying cascade of reactions. Communication at a distance between cells is mediated by such systems, which play a major role in interactions between unicellular organisms and in the self-regulation of multicellular organisms. Electric disturbances brought about by nervous currents may have similar effects.

### *Regulation by Way of Genes*

Occupying an even higher level in the hierarchy of biological control are the various factors that determine which parts of the genome are expressed at any moment, and in what amounts. Cells hardly ever express all the information contained in their DNA. Those of multicellular organisms usually express only a small fraction of this information. Whatever fraction they do express is what determines the cell's specificity, as a muscle cell, a nerve cell, or any other cell type. The expression of genes is controlled at the transcriptional, at the maturational (in the case of split genes), or at the translational level, sometimes also posttranslationally, by factors that are still poorly understood but are obviously of central importance. Such factors participate in metabolic regulation and adaptation and, especially, in differentiation and development. The way in which they bind to specific DNA (or RNA) sequences represents yet another important set of lock-and-key phenomena (for a survey, see Reference 424).

Crowning it all is the sort of stupendously complex script that unfolds in space and time in the sheltered environment of a germinating seed, of an incubating egg, or of the womb, faultlessly leading from the single fertilized egg cell to the multicellular miracle that is an emerging sapling, a freshly hatched chick, or a newborn baby. It all works through the opening and closing of genetic switches controlled by the products of previously expressed genes, by the products of these products, and so on (see Figure 1–15, page 30), in a fantastic network—as yet almost totally unraveled—of interactive influences.

## MULTIPLICATION

Strictly speaking, the ability to multiply is not a requirement of the living *state*. Many cells in animals and plants remain alive during most of the organism's whole lifespan without multiplying at all. The brain cells of a human adult are a good example. In contrast, such cells could not survive without any of the six other properties listed in this chapter as being essential for life. However, multiplication does represent an indispensable condition of the *continuity* of life. Any species lacking this property is bound to die out. In this sense, multiplication belongs to the basic blueprint.

There are two important aspects to multiplication: growth and division. The first aspect is covered by the properties already mentioned, in particular the ability to instruct biosynthetic systems in such a way as to ensure the accurate replication of existing molecules. Assuming a correct and efficient growth process, there remains the need for a mechanism that will allow division of the enlarged cell into two or more viable entities. This requires a coordinated process whereby separate, continuous plasma membranes end up around cellular contents sorted in such a way that each daughter cell includes a complete copy of the genetic information, together with a sufficient quota of all the other ingredients necessary for life.

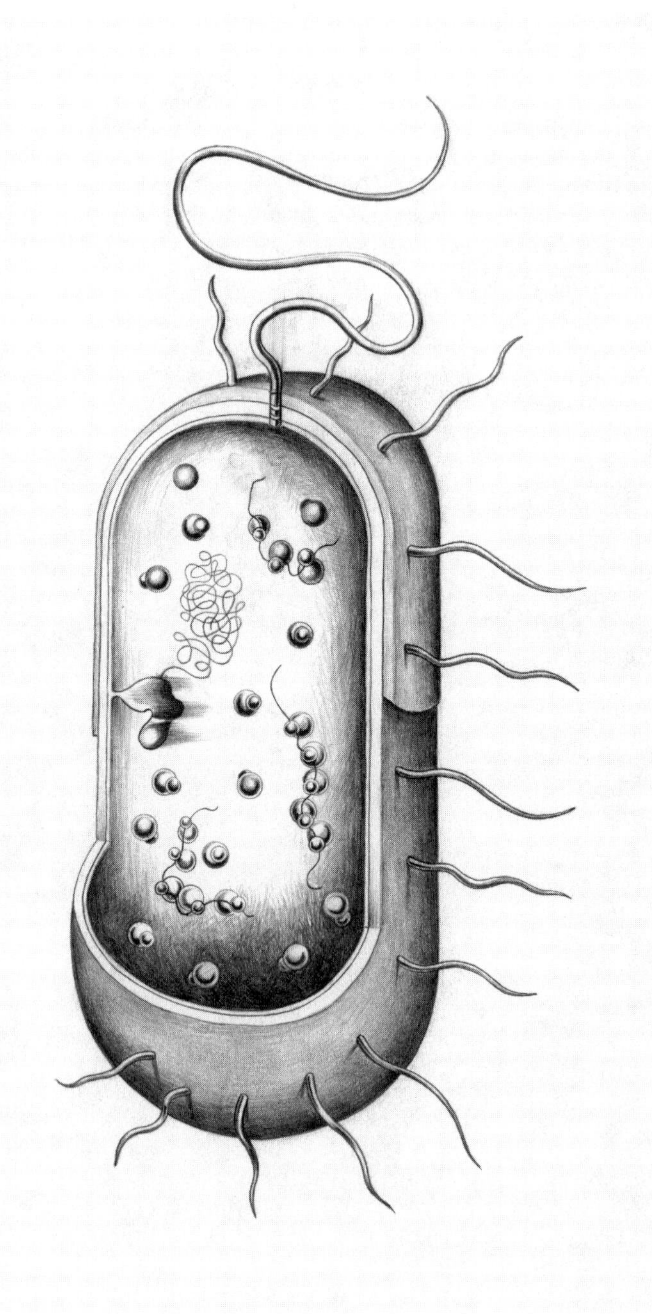

CHAPTER TWO

# THE BACTERIAL CELL

According to the best available evidence, bacteria-like organisms were already present on Earth at least 3.5 billion years ago. They remained the only forms of life for more than one billion years, until the appearance of the first eukaryotes. It is thus in a bacterium that we are most likely to find the blueprint for a cell in its most elementary form. However, we should not forget that bacteria have had a huge amount of time to evolve, innovate, and diversify. The enormously varied panorama that they offer today may hold few clues to the properties of their early ancestors or to the manner in which they arose from their ancestors. Fortunately, recent advances in biochemistry and in molecular biology are providing increasingly secure foundations for educated guesses in this respect.[1] Comparative sequencing of nucleic acids and proteins, in particular, has become a powerful tool in the phylogenetic study of bacteria.[2] As applied by Carl Woese (238, 239), this approach has upset the traditional concept of a single prokaryotic tree from which the precursors of eukaryotes allegedly branched at a relatively late date and has replaced it by a bushlike structure with three or more offshoots—including the branch leading to the eukaryotes—separating very early from a common root. Details of this structure are still hotly debated (see, for example, References 94, 120, 131, 285, 400, 438, 439, 444, 445). I will not go into this dispute, which is highly technical. I will, however, adopt as a convenient provisional classification Woese's division of the prokaryotes into two main groups, the archaebacteria and the eubacteria (see Figure 2–1). This scheme, which is accepted by a majority of workers and is receiving increasing support (236, 398, 437), is substantiated by a number of important chemical differences.

[1] For an evolutionary survey of the main bacterial properties, see Reference 94.

[2] The principle of this technique is based on the assumption—now amply verified in numerous instances—that sequence similarities between homologous genes in different organisms reflect a common origin. Comparison of the sequences then allows the phylogenetic relationships between the organisms to be reconstituted as a treelike structure in which the distance between any two genes from their closest common branching point is a function of the number of mutational events needed to account for the differences in their sequences. The same approach can be applied also to the gene products: RNAs and, more ambiguously because of coding, proteins. In practice, the technique has many pitfalls, and there is as yet no consensus on an optimal mathematical methodology for handling the data. Nevertheless, as increasing numbers of sequences of the same gene in different organisms, and of different genes of the same organisms, are compared, the derivations become increasingly secure. Today, comparative sequencing has become by far the most powerful tool for probing the biological past.

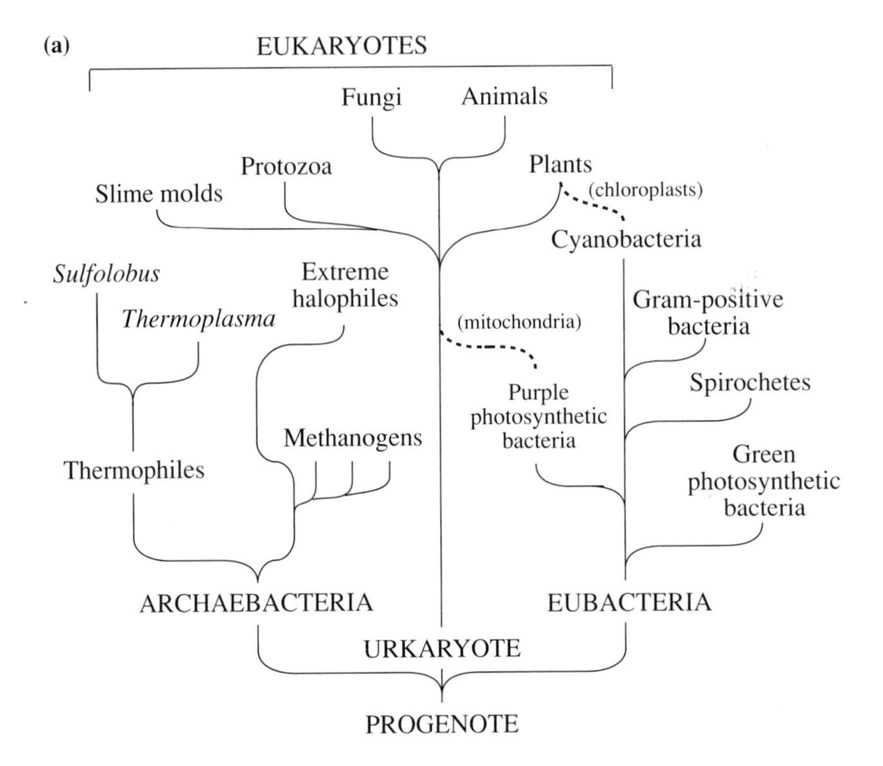
(a)
EUKARYOTES
Fungi
Animals
Protozoa
Plants
Slime molds
(chloroplasts)
Cyanobacteria
Sulfolobus
Extreme halophiles
Thermoplasma
(mitochondria)
Gram-positive bacteria
Purple photosynthetic bacteria
Spirochetes
Methanogens
Thermophiles
Green photosynthetic bacteria
ARCHAEBACTERIA
EUBACTERIA
URKARYOTE
PROGENOTE

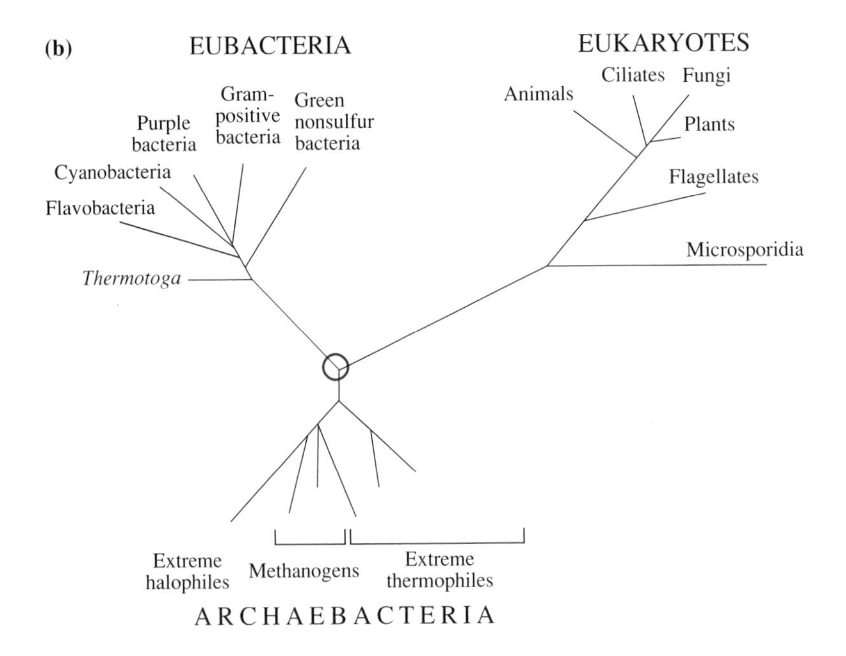
(b)
EUBACTERIA
EUKARYOTES
Ciliates
Fungi
Animals
Gram-positive bacteria
Green nonsulfur bacteria
Purple bacteria
Plants
Cyanobacteria
Flagellates
Flavobacteria
Microsporidia
Thermotoga
Extreme halophiles
Methanogens
Extreme thermophiles
ARCHAEBACTERIA

(c)

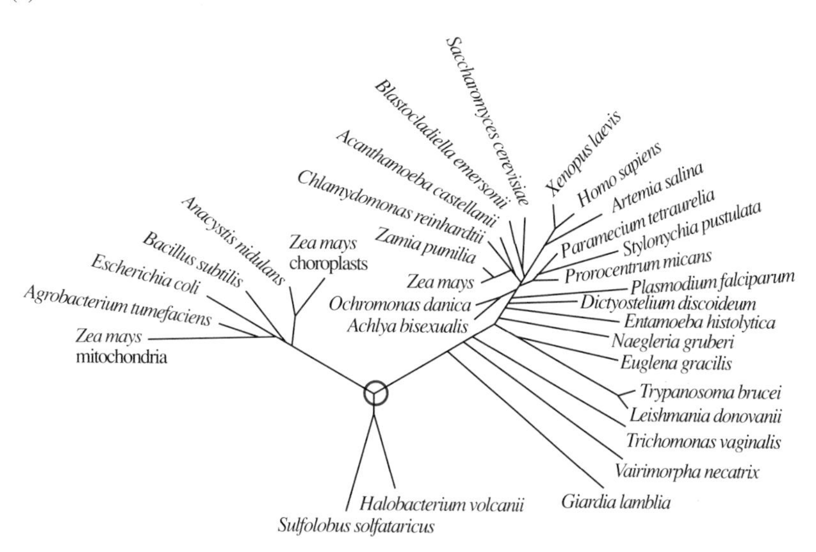


**2–1** *Phylogeny of living organisms.* Phylogenetic trees derived from comparative sequencing of 16S ribosomal RNA. **(a)** Woese's 1981 version of a rooted tree with three major prokaryotic branches separating very early from a common ancestor (according to Reference 238). Two branches, the archaebacteria and the eubacteria, lead to the two main bacterial taxonomic groups. The third branch, initiated by the urkaryote (see Chapter 3, page 58), is taken to be ancestral to the eukaryotes, formerly believed to have branched at a much later date from a single prokaryotic tree. Dotted lines represent endosymbiotic events (Chapter 3).

**(b)** Woese's 1987 version with the same three main branches but unrooted (from Reference 239). The reason for this change is that more refined theoretical analysis of the methodology has shown that the position of the common ancestor cannot as yet be determined. It could be at the central node (circled in blue), as in the earlier tree, or on any one of the three spokes radiating from it. Other authors have proposed differently structured trees (120, 131, 400). **(c)** Updated version of the unrooted, three-kingdom tree, according to Mitchell Sogin (236).

The new phylogenetic notions have not undermined the concept of the unity of life. Behind the great diversity of bacteria lie a number of fundamental traits that add up to what is essentially a single blueprint almost certainly inherited from a common ancestor. Even in its simplest form, this blueprint is already of formidable complexity, the outcome, no doubt, of a very long succession of developmental steps. Yet, it may well correspond to the minimum needed for life. Many bacteria enrich it further with specialized additions, which are, presumably, the fruits of later evolutionary innovations. The main structural features of bacteria are illustrated schematically in Figure 2–2, page 42.*

* (Added in proof.) In a recent paper, Woese and coworkers (292a) have proposed a new taxon, designated "domain," above the level of kingdom. They identify three domains, the Bacteria, the Archaea, and the Eucarya, corresponding to the Eubacteria, the Archaebacteria, and the Eukaryotes, respectively.

**2–2** *The main structural features of a gram-negative bacterial cell.*
Gram-positive bacteria lack an outer membrane and a periplasmic space.

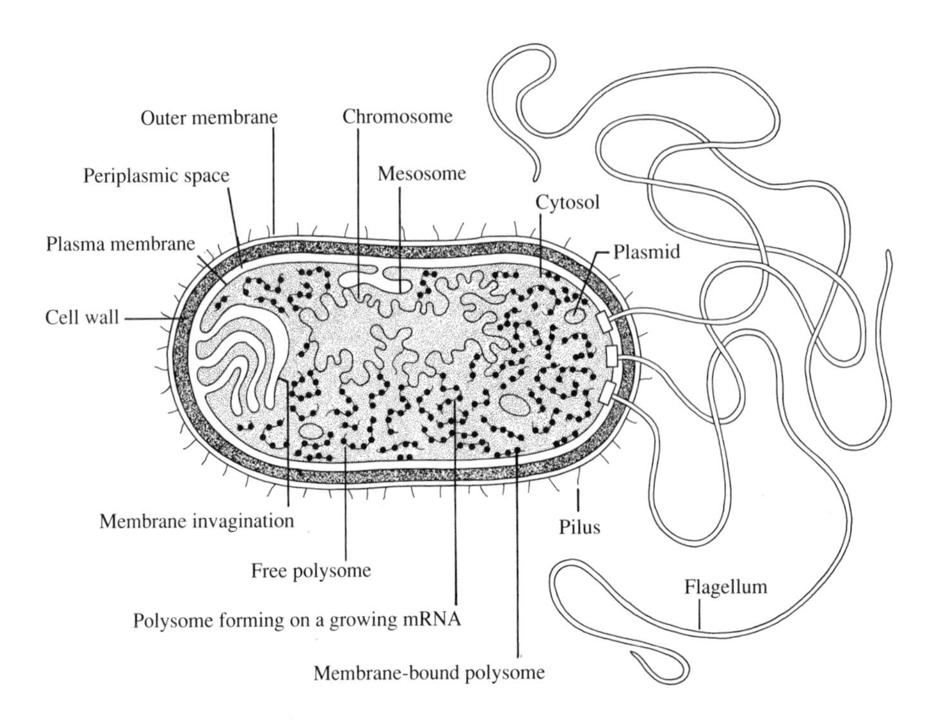


## THE CELL WALL

Most bacteria are surrounded by an inert, sturdy casing, of which the main component in eubacteria is murein, a tightly cross-linked peptidoglycan polymer, one of the rare natural components containing D-amino acids. Murein is not found in the cell wall of archaebacteria. Some archaebacteria have a pseudomurein built somewhat like murein, but with L-amino acids. The main function of the cell wall is structural containment. Bacteria maintain an intracellular osmotic pressure that often exceeds that of the surrounding medium and, but for their cell wall, would burst as a result of osmotic water intake. The cell wall also provides mechanical protection and serves as a coarse filter regulating the exchanges of matter between the cell and its environment.

[3] Based on the ability of a preparation to retain crystal-violet staining against iodine bleaching under specified conditions, the Gram test depends essentially on the abundance of murein in the cell wall. Bacteria with a thick murein capsule, such as *Staphylococcus aureus,* are positive. Those, like *Escherichia coli,* that have a thin coat of murein are negative. The terms "posibacteria" and "negibacteria," proposed by Thomas Cavalier-Smith (120) for the two classes, have not yet caught on.

The outer surface of the cell wall often bears additional components. These may play important roles, for instance, in cell-cell recognition or attachment or in determining the capacity of a pathogen to evade immune defense. The coating of the cell wall is particularly complex in gram-negative bacteria,[3] where it includes an authentic membrane made of lipids and proteins.[4] In such species, the space between the cell wall and the cell membrane proper, called periplasmic space, serves as some sort of pericellular organelle in which significant functional activities take place. It harbors a number of hydrolytic enzymes involved in digestion and also contains specific binding proteins (receptors) that act as traps for certain small molecules, often nutrients, and thereby play an important role in feeding and in adaptive behavior.

Threadlike appendages are often found protruding on the surface of the cell wall. They include pili, which play an important role in conjugation and the resulting transfer of DNA,[5] and flagella, the locomotor organelles of bacteria. More will be said about these later in this chapter.

## THE PLASMA MEMBRANE

The plasma membrane, or plasmalemma, is the true cell boundary. Bacteria can be stripped of their cell wall by treatment with lysozyme, which breaks down the murein peptidoglycan; or they can be prevented from building a cell wall by exposure to penicillin. The resulting naked cells, or protoplasts, may remain viable if protected against osmotic bursting by an appropriate milieu. Natural, wall-less forms, or L-forms, are known to exist as either permanent or transient bacterial entities.[6]

The bacterial plasma membrane may be seen as the prototype—and probably the evolutionary ancestor—of all the membranes found around and within cells. It is characteristically made of a lipid bilayer in which a variety of integral membrane proteins are inserted. Although organized largely as a smooth sheath, the bacterial plasma membrane sometimes exhibits a varying degree of infolding, which presages the extensive development of cytomembranes in eukaryotic cells. Such cytoplasmic

[4] This outer membrane differs greatly in composition and structure from the cell membrane proper. In particular, it contains a complex lipopolysaccharide (endotoxin) that, when released into the bloodstream following the breakdown of bacteria by the immune system, may be the cause of severe, sometimes fatal, intoxications. It is also relatively permeable, thanks to the presence of pore-forming proteins called porins.

[5] Bacteria can engage in a kind of sexual coupling called conjugation. The "male" form, designated $F^+$, possesses an additional small chromosome, or plasmid, the F (fertility) factor, which is absent in the "female" $F^-$ form. Union occurs with the help of a special sex pilus present on the surface of the $F^+$ form and allows this form to inject a copy of the F plasmid into the $F^-$ recipient, which then becomes $F^+$. If the F factor has become attached to the main chromosome by recombination, parts of the main chromosome may also be transferred in this manner. The discovery of bacterial conjugation has played a very important role in the development of molecular biology.

[6] L-forms are characterized by their great flexibility, a property that may have been important for the prokaryote-eukaryote transition (Chapter 3). They include the mycoplasmas, which are the smallest living cells known, perhaps the smallest possible, in view of the minimum needed for independent life (264, 265). Much information concerning these organisms may be found in Reference 92.

extensions of the plasma membrane are most frequently connected with intense electron transport or photosynthetic activity. There are cases, of special interest to the prokaryote-eukaryote transition (Chapter 3), where membrane expansion seems to be related to an increase in cell volume. An example is the colorless sulfur bacterium *Thiovulum majus* Hinze, a large microorganism of 10–20 micrometers, which develops in littoral sediments rich in hydrogen sulfide (415). Its plasma membrane is expanded considerably by evaginations, which press against its thin outer wall, and by a network of flattened invaginations studded with ribosome-like particles, which resembles the rough endoplasmic reticulum of eukaryotic cells (see Figure 2–3). A special infolding of the bacterial plasma membrane, called mesosome, serves as anchoring point for the chromosome and its replication machinery.

A major difference distinguishes the long hydrophobic chains of the membrane lipids in the two phylogenetic classes of bacteria. These chains consist of fatty acids attached to glycerol by an ester linkage in eubacteria, of isoprenoid alcohols bound to glycerol by an ether linkage in archaebacteria (see Figure 2–4). In some ether lipids, two molecules of glycerol are attached at each end of very long $\alpha,\omega$-dialcohols, making what amounts to a "stuck" bilayer, which is much more rigid than the usual bilayer. These features are believed to be adaptations to the harsh conditions of temperature, pH, or salinity characteristic of the habitat of many archaebacteria.

The functional components of the bacterial plasma membrane may be classified into several groups, of which some are universally distributed and others are restricted to certain classes of bacteria. In a first group are all the transporters, permeases, and pumps needed to permit the intake of essential nutrients, the excretion of waste products, and the maintenance of an appropriate intracellular milieu. Such equipment is common to all bacteria, as no organism bounded by a lipid bilayer could survive without it. Its composition, however, varies according to the kind of environment occupied by the bacterium.

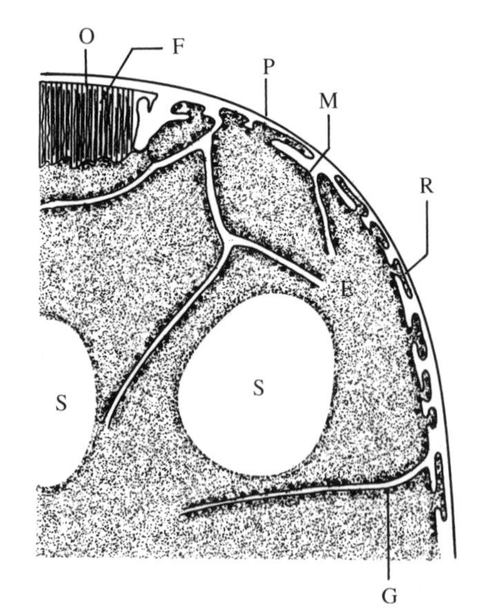


**2–3** *Expansion of a bacterial plasma membrane.*
The drawing, reproduced from Reference 415, illustrates the development of the plasma membrane of *Thiovulum majus* Hinze. Note especially the membrane evaginations, labeled R, and the endoplasmic reticulum-like invaginations (E) apparently studded with ribosomes (G). P designates the cell wall, M the plasma membrane, S the space occupied by sulfur deposits, O a mysterious "antapical organelle" containing parallel fibrils (F).

A second group of functional components of the plasma membrane includes a variety of systems involved in the manufacture and insertion of membrane constituents and in the synthesis and translocation of extracellular macromolecules. Among them are enzymes of lipid metabolism; the glycosylating and other systems that serve in the synthesis of cell-wall and outer-membrane components; and the machineries that, in conjunction with ribosomes, put into place integral membrane proteins and mediate the discharge of secretory proteins. Probably not all elements of these numerous and diverse systems belong to the basic blueprint, but one can at least define as indispensable to any membrane-bounded organism the minimum needed for the assembly of a functional membrane. Improvements to the membrane, through the acquisition of receptors or of a motile system, for example, as well as the production of extracellular structures and secretions, could have been later additions.

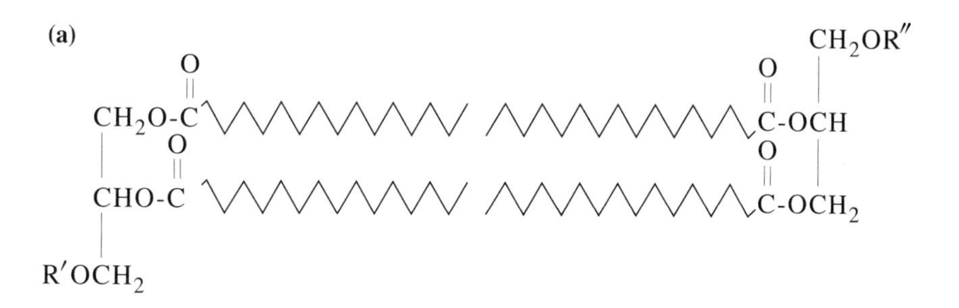


(b)

CH$_2$OR″

CH$_2$O— —OCH

CHO— —OCH$_2$

R′OCH$_2$

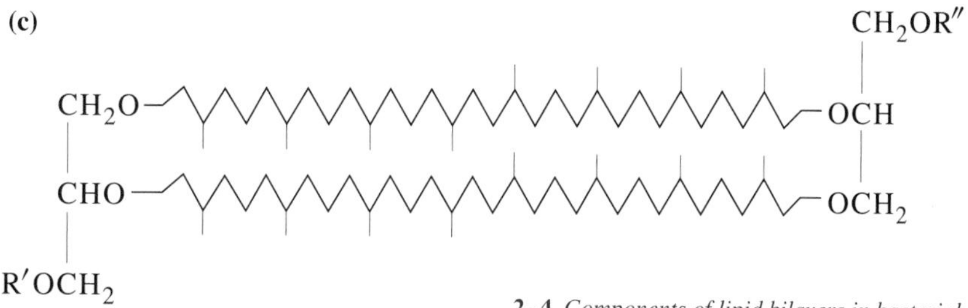


**2–4** *Components of lipid bilayers in bacterial membranes.*
(**a**) Ester-type lipids (eubacteria). (**b**) Ether-type lipids (archaebacteria). (**c**) Ether-type lipids containing $\alpha,\omega$-dialcohols and forming "frozen" bilayers (archaebacteria).

Surface receptors make up a third group of membrane components. The term "receptor" is sometimes understood to encompass the specific transport systems involved in the capture of nutrient molecules, also even the soluble binding proteins that serve to trap such molecules in the periplasmic space. Obeying more closely the definition of receptor as encountered in eukaryotes are the chemotactic receptors of motile bacteria. As will be seen in the following section, these receptors react to the binding of their ligands by a conformational change that sets in motion a chain of events ending in a modification of flagellar behavior. They also sometimes act in conjunction with soluble receptors present in the periplasmic space. One may speculate that receptors, to the extent that they are not essential to food capture, are dispensable membrane components and could, therefore, be later evolutionary acquisitions.

A very important group of membrane components includes iron-sulfur proteins, flavoproteins, cytochromes, quinones, and other electron carriers, organized into phosphorylating electron-transport chains operating by way of a protonmotive force. A great variety of such systems exists. Those of aerobic bacteria are adapted to the use of oxygen as final electron acceptor. Others deliver electrons that have dropped to a lower energy level to some other outside acceptor, such as $Fe^{3+}$, $NO_3^-$, $SO_4^{2-}$, and many others. In phototrophic bacteria, phosphorylating electron-transport chains are connected to the photosystems in such a way as to accept and exploit electrons that have been energized photochemically. Invariably in such systems, the topological organization of the carriers within the membrane is such that electron flow forces protons out of the cell and generates a membrane potential, positive outside. The resulting protonmotive force powers ATP assembly by a membrane-bound, reversible proton pump; it sometimes also supports other forms of work, such as active transport or flagellar motion.

The upper part of electron-transport chains is often reversible, so that it can also serve to energize electrons with the help of ATP or of protonmotive force generated lower down the chain. As mentioned in the preceding chapter, this mechanism of reverse electron transfer is essential in all forms of autotrophy in which electrons need an additional energy boost to sustain biosynthetic reductions. Membrane-bound, proton-extruding electron-transport chains are distributed very widely, but not universally, in the bacterial world. They are lacking in some anaerobes, which nevertheless possess a proton pump in their plasma membrane.[7]

A last important group of bacterial plasma-membrane components comprises the photosystems of phototrophic organisms. These systems are situated in the plasma membrane proper or in what appear as flat, intracellular, membranous saccules called chromatophores, which are derived from invaginations of the plasma membrane.

[7] Examples are *Clostridium pasteurianum* and *Streptococcus faecalis*, which have no respiratory chain but possess an ATP-powered proton pump, which serves to eject excess protons generated by fermentation and to support active transport processes (154, 295, 298). These facts have suggested the possibility that ATP-driven proton pumps may have preceded their electron-driven counterparts in the origin of living organisms. While this statement may be true, its premise is not. The phylogenetic position of these bacteria indicates that they are probably descended from phototrophs, not from primitive anaerobes (239).

Chromatophores are the most highly organized membranous formations of the bacterial world.

With one exception, which will be considered below, all known phototrophic bacteria are eubacteria that depend on chlorophylls or on related tetrapyrrole derivatives[8] for the capture of light, use the captured light to energize electrons, and derive their ATP from phosphorylations coupled to the downgrading of the photochemically energized electrons (see Figure 1–8, page 19). At the heart of such systems lies the reaction center, a complex of bacteriochlorophyll and protein that suffers electron destabilization when excited by light and is so intimately connected with a phosphorylating electron-transport chain as to feed the destabilized electrons immediately into it before they have a chance of dissipating their energy otherwise. The reaction center is surrounded by a patch of interconnected chlorophyll-protein complexes associated with carotenoids and other pigments; this patch serves as a "solar antenna" that catches light quanta and rapidly transfers them to the reaction center by intermolecular energy exchange (exciton transfer).

The phosphorylating arm of photosystems closely resembles the respiratory chain of aerobic bacteria. Both chains are made of the same kinds of electron carriers; both include a similarly constructed proton pump, the so-called $F_OF_1$ complex; and both support ATP assembly and other forms of work by way of a similarly oriented protonmotive force (light forces protons outwards or, alternatively, into the interior of chromatophores). The phylogenetic relationship between the two chains leaves little doubt. The most probable interpretation of these facts is that, in the history of life, phototrophy was an acquisition grafted onto a pre-existing phosphorylating electron-transport chain. As will be mentioned in Chapter 4, this event probably followed the separation of archaebacteria and eubacteria or, perhaps, signalled this separation.

Photosystems can meet all ATP requirements by cyclic phosphorylation, but they need at least one inlet accessible to outside electrons to be able to support biosynthetic reductions. These electrons are furnished most frequently by inorganic donors, for example, $H_2S$ or other sulfur compounds, as in the green and purple sulfur bacteria, which are, accordingly, classified as photolithotrophs. Sometimes the donor is an organic molecule, as in purple nonsulfur bacteria, which are photoorganotrophs. The exact "wiring" of the systems varies according to species. So does the manner in which the electrons are brought to the high energy level required by biosynthetic reductions. In green bacteria, the photosystem lifts the electrons directly up to this level, as does photosystem I in eukaryotes. In purple bacteria, the electrons receive an additional boost by reverse electron transfer. In all cases, the highest level is occupied by a ferredoxin, an electron carrier of particularly low oxidation-reduction potential belonging to the group of iron-sulfur proteins. Ferredoxin then delivers the

[8] Chlorophylls, like hemes, are built from the cyclic, tetrapyrrole, porphine ring. Their side chains are different from those of hemes, and they have a central magnesium ion in place of iron. Also important in phototrophy, especially in cyanobacteria and in some primitive eukaryotes, are the phycobilins, noncyclic tetrapyrroles similar in structure to the bile pigments that arise from the breakdown of hemes in animals.

electrons to the actual biosynthetic donor, which in bacteria is NADH. The group of iron-sulfur proteins includes many other important catalysts, among them the key enzymes of nitrogen fixation (reduction of $N_2$) and of hydrogen evolution (reduction of $H^+$), members of respiratory chains, and agents of primitive metabolic oxidations (143, 147, 228, 364, 396, 397).

A second photosystem is found in the all-important group of cyanobacteria, formerly known as blue-green algae. This photosystem II has the unique property that it can remove electrons from water, with production of molecular oxygen. It lifts these electrons to an intermediary energy level, from which they are transferred to the cyclic system built around photosystem I; they are then used for phosphorylations and for biosynthetic reductions, as in the more primitive phototrophic bacteria.

In most of these organisms, photosystem II uses open tetrapyrroles, called phycobilins, as light-sensitive pigments (see Footnote 8), but species that use chlorophyll *b,* as do green plants, are also known.[9] It is generally agreed that the emergence of photosystem II started the progressive appearance of oxygen in the terrestrial atmosphere and the beginning of aerobic life—though not, by any means, the beginning of other forms of respiration using electron acceptors different from oxygen. Such forms must have been plentiful in the primitive world. Exactly when cyanobacteria first appeared in evolution is not clear. The indications are that they started to flourish about two billion years ago and then proliferated rapidly to the point of raising the oxygen content of the atmosphere to its present level in less than half a billion years. This period could, however, have been preceded by one—possibly of great duration—in which the released oxygen was trapped by reducing minerals.

As mentioned in the preceding chapter, the purple phototroph *Halobacterium halobium* enjoys the hitherto unique distinction of being able to convert light energy directly into protonmotive force without the participation of electrons. Unlike the other phototrophic bacteria, it is not a eubacterium but an archaebacterium; and it uses bacteriorhodopsin, a carotenoid-containing protein related to the visual pigment rhodopsin, instead of a tetrapyrrole derivative, for the capture of light. A related molecule, halorhodopsin, also present in the plasma membrane of *Halobacterium,* uses light to drive out sodium ions (151).

## THE FLAGELLA

Many bacterial cells are motile. Their locomotor apparatus is unlike anything found in higher cells. In particular, although they bear the same name,[10] bacterial flagella are completely different from the flagella of flagellated protists or of spermatozoa. The bacterial flagellum is, almost literally speaking, a propeller, a rigid,

[9] As will be mentioned in Chapter 3, this discovery was first hailed as being of great importance with respect to the origin of chloroplasts. Contradictory evidence relating to this point has since been obtained (93, 402, 410, 414, 440, 442, 443).

[10] To avoid confusion between prokaryotic and eukaryotic flagella, Lynn Margulis (71) has proposed that the eukaryotic flagella and cilia be grouped under the term "undulipodia," borrowed from the old German literature (75). This proposal has not yet been widely adopted.

spiral rod that extends through special rings of the cell wall and plasma membrane to the inner face of the membrane where it is inserted into a kind of "turbine" responsible for the rotation. This machine, which can turn in either direction, is believed to be powered by protonmotive force, but its physicochemical mechanism is unknown.

Bacteria swim in a sort of random motion. They move in a given direction for a little while, and then the flagellar rotation is reversed, causing the cells to "tumble" and depart in a different direction. In positive chemotaxis, the frequency of tumbling is decreased so that the bacteria continue swimming for a longer time toward the attractant molecules that have signalled their presence by binding to the appropriate receptor. In negative chemotaxis, the frequency of tumbling is increased (76).

## THE CYTOSOL

The cytosol is a viscous, protein-rich sap. It fills all the spaces in the cell not occupied by particulate objects, which in bacterial cells are mostly ribosomes and the chromosome. It is the main seat of metabolism. It contains hundreds of distinct enzymes catalyzing all sorts of breakdown reactions (catabolism) and synthetic processes (anabolism) of extraordinary diversity. In terms of chemical versatility, bacteria are far from primitive. They perform many reactions that plants and, even more so, animals are unable to carry out.

## THE RIBOSOMES

There are about 15,000 ribosomes in an average bacterial cell, such as *Escherichia coli*. They fill a good part of the cell volume, reflecting the major importance of protein synthesis in bacteria, which may double in size and divide in as little as 20 minutes.

The bacterial ribosome is a huge molecular edifice of some 55 proteins and three rRNAs, organized into two subunits of unequal size. Most of the ribosomal constituents have been completely sequenced, and much detailed information is already available concerning their mode of assembly (104, 193, 218). The comparative sequencing of rRNAs, especially of the 5S component of the large subunit and of the 16S component of the small subunit, has become a valuable tool in the phylogenetic study of bacteria (236, 239) and of their distant eukaryotic relatives. Much is also known of the manner in which ribosomes combine with mRNAs and construct polypeptide chains from aminoacyl-tRNAs, under the general aegis of codon-anticodon recognition and with the help of a number of soluble cofactors and GTP. Surprisingly, despite such a wealth of information, the respective roles of the protein and RNA components in the functioning of ribosomes are still poorly understood (see References 100, 101).

A characteristic of bacterial ribosomes is that they often start translating a message while it is still being transcribed. This is rendered possible by the fact that translation starts at the 5′ end of the RNA molecule, whereas the growth of the molecule takes place by the addition of nucleotide units to its 3′ end. Thus, the two processes do not impede each other. Two other peculiarities explain why such a combined transcription-translation process occurs only in bacteria. First, the bacterial chromosome is unfenced. In eukaryotes, the nuclear envelope separates chromosomes physically from ribosomes, which are in the cytosol. Second, bacterial genes are not split by introns, as are many eukaryotic genes; they are transcribed and translated colinearly, without processing of the mRNA.

Considerable interest has been devoted in recent years to the mechanisms that mediate the insertion of proteins into the plasma membrane as well as the translocation of those proteins that are to be incorporated into pericellular structures, secreted into the periplasmic space, or discharged outside the cell (47, 164, 170, 186, 199, 207; see also Chapter 3). These mechanisms resemble in many respects those, better known, that perform analogous functions in eukaryotes. They include typical instances of cotranslational transfer dependent on cleavable N-terminal "signal" sequences very similar to those involved in cotranslational translocation across the endoplasmic reticulum. They also comprise cases of posttranslational transfer, with or without processing of the molecule, that resemble the targeting of proteins to mitochondria and other eukaryotic organelles. When two membranes have to be bridged, as in the insertion of protein components into the outer membrane of gram-negative bacteria, translocation seems to occur across zones of adhesion between the two membranes, as it does in mitochondria and chloroplasts (357). There are also a number of significant differences between the bacterial and the eukaryotic systems. These similarities and differences are of great interest with respect to the origin of eukaryotic cells and will be examined in the next chapter.

## THE CHROMOSOME

Bacteria have a single, circular chromosome lying in direct contact with the cytosol. Its DNA is associated with basic proteins, though without the elaborately structured organization of eukaryotic chromosomes. Its length is about 1 millimeter, which corresponds to three million base pairs, enough to code for one million amino acids, or 2,000 polypeptides of 500 amino acids (about 55,000 molecular weight), not counting the various structural and functional RNAs, which also have to be coded for. This number—2,000 polypeptides—is of the order of the estimated number of bacterial genes. Inasmuch as genes are separated by linker sequences that include various regulatory signals, the bacterial genome contains hardly any "junk" DNA, in contrast with the genomes of higher eukaryotes. It looks as if it has been "streamlined"—by natural selection, according to prevalent views—to ensure the most rapid rate of multiplication. Indeed, DNA replication hardly ever ceases in growing bacteria.

In principle, the process of replication is very simple: the two strands of the DNA duplex separate, and each serves as a template for the synthesis of a complementary strand to yield two identical duplexes, of which one strand is of parental origin and the other is newly made (semiconservative replication). In practice, the process is one of great complexity, requiring more than 20 enzymes and other specific proteins to cope with all the topological and mechanistic problems that are raised by the double-helical association of the two template strands, by their antiparallel orientation, by the fact that DNA polymerase needs priming with a short RNA strand to become effective, and by the requirement for proofreading. This whole machinery is most likely combined into a large, multienzyme complex, the replisome.

Contrary to the schematic representation of replication given in most textbooks, replisomes do not cut through the DNA template in opposite directions, opening increasingly distant replication forks at both ends of a widening "bubble." Replisomes are stationary and anchored to the plasma membrane, probably at the level of the mesosome. The DNA does the moving. Starting from the origin of replication, it is reeled symmetrically through twin replicating systems and exits from the complex in duplicated form. Upon completion of the process, the two circles are disentangled, and the newly made strand of each is circularized by ligation. By this time, the daughter chromosomes are attached to different mesosomes, which ensures separation of the two chromosomes upon cell division. Bacterial chromosomes have only one origin of replication. The whole chromosome is a single replication unit, or replicon.

In bacteria, DNA replication hardly impedes transcription and vice versa. As a process, transcription is simpler than replication. It usually concerns only one of the two DNA strands and, except for initiation and termination, is carried out essentially by a single enzyme, RNA polymerase. This enzyme is not stationary but moves along the DNA template. Its progress requires the assistance of topoenzymes, which allow rotation of the DNA as the RNA polymerase threads its way through the template's double-helical configuration. Many transcription units operate simultaneously on the same chromosome, often on the same gene, on which they follow each other in regularly spaced order, from the 3′ to the 5′ end of the coding strand.

Quantitatively, the main role of transcription is the synthesis of functional RNAs, mostly tRNAs and, even more, rRNAs, of which there is an enormous demand in growing bacteria, requiring almost continual transcription of multiple copies of the same genes. These genes code for large precursor molecules that include, inserted between a 5′-leader and a 3′-trailer sequence, several RNA species separated by spacers. In *E. coli,* for example, a single precursor includes the three rRNAs and two tRNAs; another contains seven tRNAs. These precursors are processed by nucleases that cut them in pieces and trim each piece accurately to size. Interestingly, one of these nucleases, ribonuclease P, which is responsible for the 5′ trimming of tRNAs, is a complex of a protein and an RNA molecule. This RNA can, by itself, catalyze the reaction almost as efficiently as the complex. This is one of the recent findings that have demonstrated that RNA can act as a catalyst.

A quantitatively small, but qualitatively immensely important, part of DNA transcription is devoted to the synthesis of mRNAs. Bacterial genes coding for mRNAs (therefore for proteins) differ from eukaryotic genes in two important respects: 1) they are never split into exons separated by introns; 2) they are often grouped as strings of several genes, or operons, usually coding for enzymes belonging to the same metabolic pathway. Operons are controlled by a single promoter and transcribed as polycistronic mRNAs bearing several distinct translation units. As first shown by François Jacob and Jacques Monod (180), the transcription of operons is often blocked by repressor proteins, which bind to a specific region of the DNA (operator) in such a way as to render the promoter inaccessible to RNA polymerase. Repressors lose their affinity for the operator and cease to block it when they are occupied by certain ligands. These are called inducers because, by derepressing the corresponding genes, they induce the production of certain enzymes. Most often, the induced enzymes are required for the metabolism of the inducer, which makes the derepression of their genes adaptive. In terms of our general concept, repressors may be viewed as allosteric receptors bearing two mutually exclusive "locks," one for the DNA operator, the other for the inducer. Constitutive genes are not controlled by an operator or are controlled by an operator for which there is no repressor, as a result of a mutation, for example.

Enhancers of transcription also exist. One is a protein likewise fitted with two "locks," as are repressors, but with very different properties. Binding of the protein to DNA accelerates transcription, and it binds to DNA only when occupied by its second ligand. This ligand is cyclic 3′,5′-AMP, an important regulatory molecule, made from ATP by adenylate cyclase. Cyclic AMP also fulfills many functions in eukaryotes, for example, in intercellular signalling by means of hormones and other factors. It is the prototype of so-called second messengers.

It has already been mentioned that translation of an mRNA often starts before its transcription is completed. We have also seen that transcription goes on in multiple areas of the chromosome while the latter is undergoing replication. Thus, bacteria offer a remarkably compact and interdigitated combination of the three main biosynthetic processes involved in genetic information transfer (see Figure 1–10, page 23).

A notable property of the bacterial genome is its plasticity. Bacterial genes participate in a variety of transpositions and recombinations, and they can also be exchanged between cells in several ways, including conjugation (see Footnote 5, page 43), transduction (mediated by a virus or a phage), transformation (entrance of naked DNA),[11] or plasmid acquisition. Genetic engineering copies nature much more closely than was originally believed.

## PLASMIDS

Plasmids are frequent, though not obligatory, components of bacterial cells. They are small circles of extrachromosomal DNA that can be exchanged between cells upon conjugation and can also be made to enter cells under special conditions. They are normally replicated and expressed by the machinery of the cells in which they reside.

[11] The purification of the pneumococcus transforming factor and its chemical identification as DNA by Oswald Avery, Colin MacLeod, and Maclyn McCarty in 1944 played a decisive role in establishing that DNA, not protein, is the genetic material. McCarty has given a sober historical account of this epoch-making discovery (23).

Their replication often occurs independently of chromosome replication, so that they can be present in multiple copies. Plasmids are dispensable; the genes they carry are not essential for life. But they may markedly affect the properties of their host cells, including the cells' ability to survive. For example, the F factor (see Footnote 5, page 43) is a plasmid. It carries a number of genes necessary for conjugation, including those coding for the sex pili proteins. Some plasmids provide their host cells with the ability to metabolize special compounds. Others carry genes that code for highly poisonous toxins or for proteins that mediate resistance to certain antibiotics, thus conferring redoubtable properties to the bacteria that harbor the plasmids. In addition, because of their mobility and tendency to engage in genetic recombination, plasmids facilitate the spread of such properties. In recent years, plasmids have acquired a major practical importance as vectors for genes to be introduced into bacterial cells by genetic engineering.

## CELL DIVISION

Bacterial cell division depends essentially on the formation of a plasma membrane furrow. This constriction develops between the duplicated chromosomes so that one is included in each of the two daughter cells. The fact that bacterial chromosomes are anchored to the plasma membrane makes this separation possible without the construction of an elaborate mitotic apparatus. However, the actual mechanism of furrowing, which involves a coordinated expansion and infolding affecting the plasma membrane and the cell wall simultaneously, is not known.

CHAPTER THREE

# The Eukaryotic Cell

## THE PROKARYOTE-EUKARYOTE TRANSITION

Eukaryotes comprise all pluricellular animals, plants, and fungi, as well as the unicellular protists. To speak of *the* eukaryotic cell in the face of such diversity hardly seems warranted. What is there in common among a human neuron, a spinach leaf cell, a yeast, and an amoeba that could justify their all being lumped together into a single class? Quite a lot, in fact (see Figure 3–1), which may be summed up schematically as *additions*[1] to the prokaryotic type of organization, including:

1. An elaborate *cytomembrane system,* which allows the cells to reach many times the size of prokaryotes and which includes the *nuclear envelope,* by definition the hallmark of eukaryotes.
2. *Cytoskeletal elements* and associated *motor systems,* which serve to build the internal props and machineries required by a bulky, compartmentalized cell—among them, in particular, the *mitotic apparatus,* which is common, in one form or another, to all eukaryotic cells.
3. A complex organization of the *genome,* including elaborately structured chromosomes, nucleoli, split genes, RNA splicing systems, and a strict physical separation between transcription and translation.
4. Cytoplasmic metabolic *organelles,* comprising the mitochondria, present in all eukaryotic cells with the exception of a few protists; the microbodies, also widely distributed; and the plastids, found in algae and in plant cells.

These characteristics not only connect all extant eukaryotic cells, they also can be traced back in time: first, for pluricellular organisms, to single egg cells; thence, through the germ line, to organisms of progressively decreasing complexity; finally, to unicellular ancestors. These in turn, together with all present-day protists, are believed to be derived from an ancestral protoeukaryotic line, which was long viewed as a relatively late offshoot of a prokaryotic branch.

[1] Although it is basically correct to view eukaryotes as prokaryotes *plus* something, we should not forget that there may be minuses as well, in terms of metabolism, for example. For all our superiority, we remain indebted to the humble bacteria that inhabit our gut for our supply of several vitamins. Resident bacteria also give ruminants and a number of other animals the ability to utilize cellulose.

**3–1** *Eukaryotic organization.*
**(a)** *Below.* Hypothetical biflagellated protist illustrating the main components of animal cells. **(b)** *Next page.* Green plant cell.
Possible descendants of endosymbionts are shown in blue.

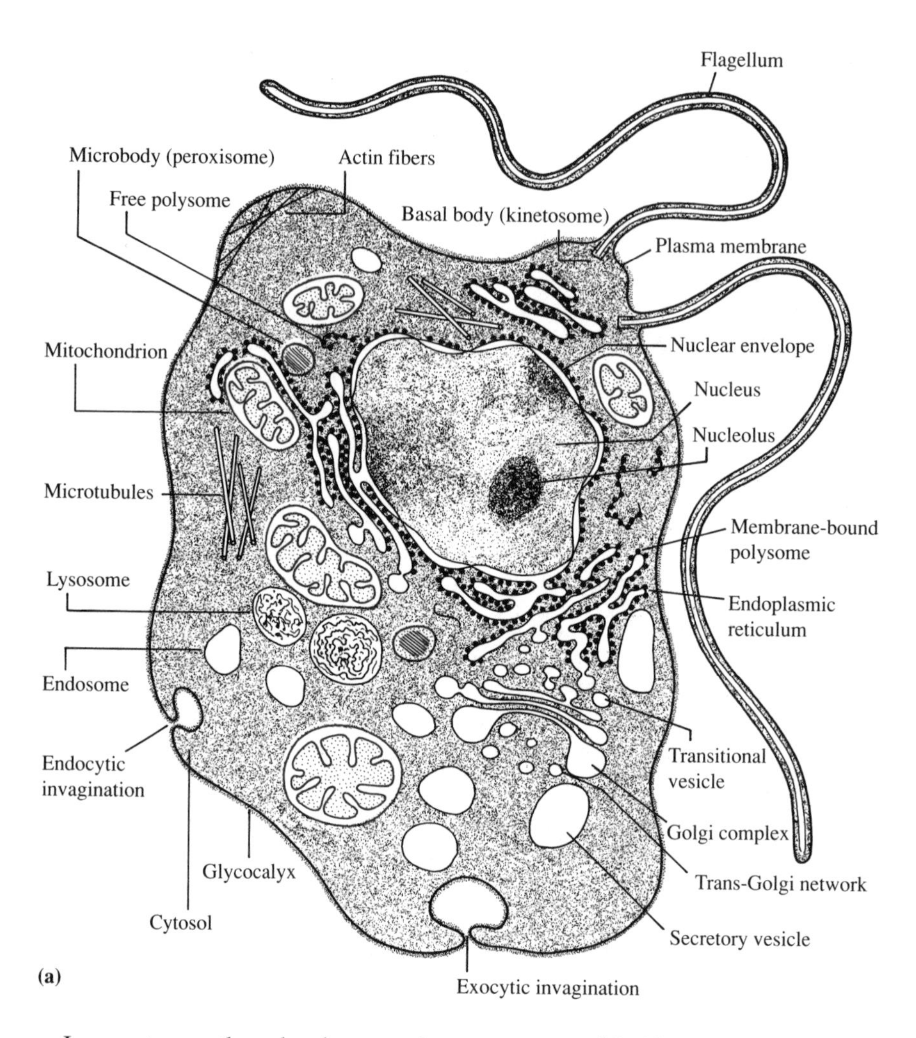


In recent years, three developments have come to modify this picture. First, thanks largely to the work of Woese, the origin of the eukaryotic line has been shifted back to a primitive *urkaryote* (see page 58), which detached very early from the two prokaryotic branches, perhaps as far back as about 3.5 billion years ago, close to the time when prokaryotes separated into archaebacteria and eubacteria (Chapter 2).

Second, comparative rRNA sequencing has shown the protists to be a much older and more diverse group than was originally believed (236, 395, 407, 411). This group may date back more than two billion years; and it may have evolved, in parallel with the prokaryotes, for as long as 1.5 billion years, before the big radiation that resulted in the multicellular plants and animals.

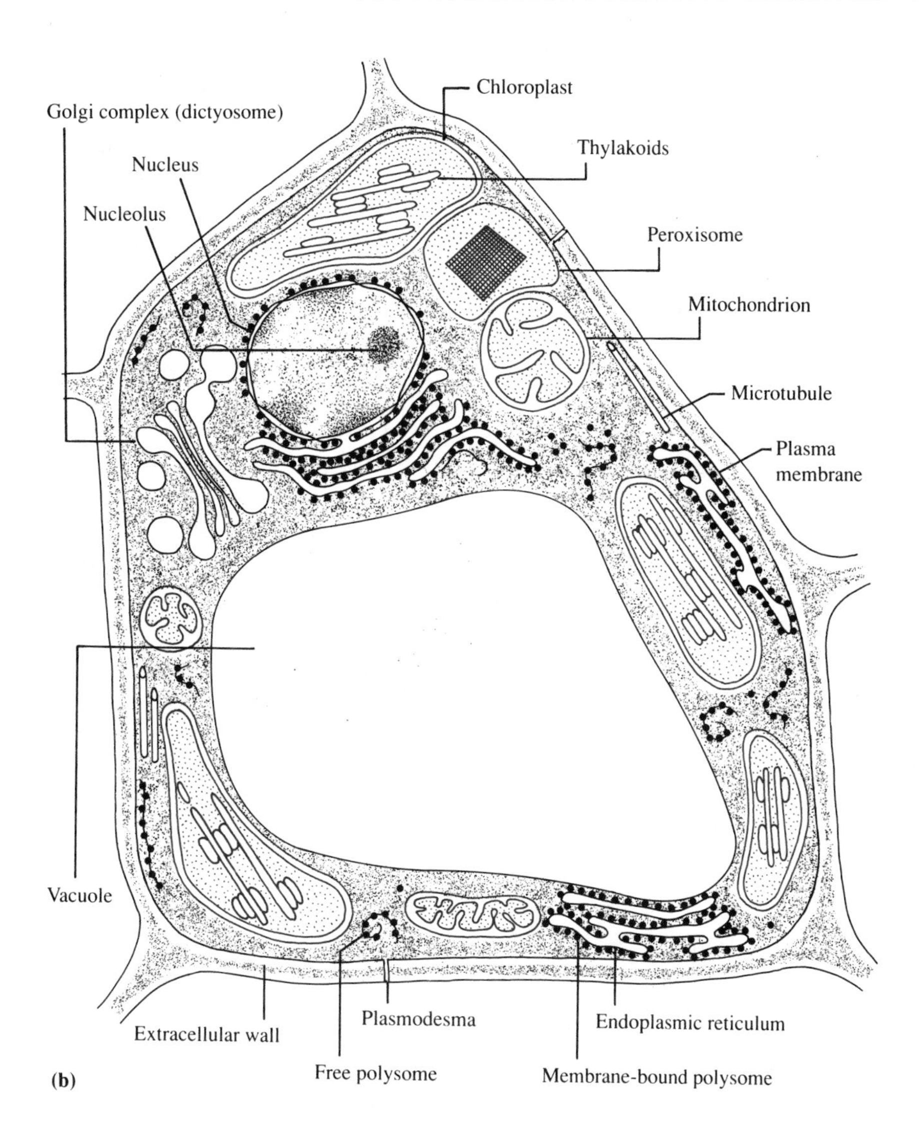


**(b)**

A third major development is the essentially conclusive demonstration that mitochondria, chloroplasts, and perhaps other cytoplasmic organelles are descended from bacteria that established themselves as *endosymbionts* in a host cell.

An important lesson gained from the new perspective is the very long time during which each of the three lines—archaebacteria, eubacteria, and eukaryotes—evolved on its own, undergoing the profound modifications that distinguish them today. We would have to know the nature of these modifications to be able to go back to origins. We are still very far from this. For all we know, the common ancestor of the three lines could have been very different from present-day prokaryotes. It could even have been closer to eukaryotes in some respects. But this argument should not

**3–2** *(Next page.) Phylogeny of the eukaryotes.* Derived from recent phylogenetic data (see Figure 2–1c, page 41), the diagram retraces, against an approximate time scale, the main steps of the transition discussed in this chapter. The blue circle identifies the central node of Figure 2–1c and can serve as a marker in recognizing the relationship between the two figures. It is seen that the root (ancestral cell) has been placed on the branch radiating from the central node to eubacteria (see text). A few representative protists are shown to illustrate the exceedingly long time during which this group evolved. Trypanosomatids and trichomonads belong to separate lines. They are shown together because of the uncertainty as to their relative ages. The older of the two are the trichomonads, according to 16S-rRNA sequencing (236), and the trypanosomatids, according to 28S-rRNA sequencing (395).

be turned around into an absurdity. It would be unreasonable to visualize the first form of life as endowed with the complex organization of today's eukaryotes and all bacteria as its degenerate descendants. The great majority of eukaryotic characters that appear as additions to the prokaryotic blueprint are undoubtedly just that: evolutionary *acquisitions,* developments that were added to a simpler and more prokaryote-like type of organization. It is legitimate to view them as such and to search for plausible mechanisms to account for their emergence.

A schematic view of the main stages of eukaryotic evolution, as presently visualized, is shown in Figure 3–2. The following intermediates are distinguished:

1. The *progenote.* This term, introduced by Woese (238), is applied here to the first primitive protocell. In Chapter 9, I shall attempt to reconstruct the properties of this mysterious entity, which must have been very different from any form of life known today.
2. The *ancestral cell.* I use this term to designate the last common ancestor of all existing living beings (see Chapter 4). Note that the ancestral cell corresponds to the as yet unlocalized *root* of the phylogenetic tree (see Figures 2–1b, page 40, or 2–1c, page 41). In Figure 3–2, the ancestral cell has been put tentatively on the branch radiating from the central node to eubacteria, because of the many reasons pointing to a closer relationship of eukaryotes with archaebacteria than with eubacteria. This choice agrees with the localization of the root in the phylogenetic tree recently published by Woese and coworkers (292a). See Footnote, page 41.
3. The *urkaryote.* Also coined by Woese (238), this term, which is derived from the Germanic root *Ur,* primitive, designates the initiator of the line that led to the eukaryotes. In agreement with phylogenetic data, it is shown as detaching very early from the prokaryotic branches. For reasons just mentioned, I have chosen to have it branch off the archaebacterial line after separation of the eubacterial line. Later in this chapter, I shall mention some of the traits that link eukaryotes more closely to archaebacteria than to eubacteria.[2] They include the involvement of

[2] Eukaryotes also have some properties in common with eubacteria (120). To explain this fact, it has been proposed that the urkaryote was a chimeric organism that arose from the fusion of a eubacterial with an archaebacterial ancestor (293). The time at which the two kinds of genes were acquired is important in this respect. An urkaryote of archaebacterial origin would have had plenty of later opportunities to gain eubacterial genes by horizontal transfer, especially after the adoption of eubacterial endosymbionts.

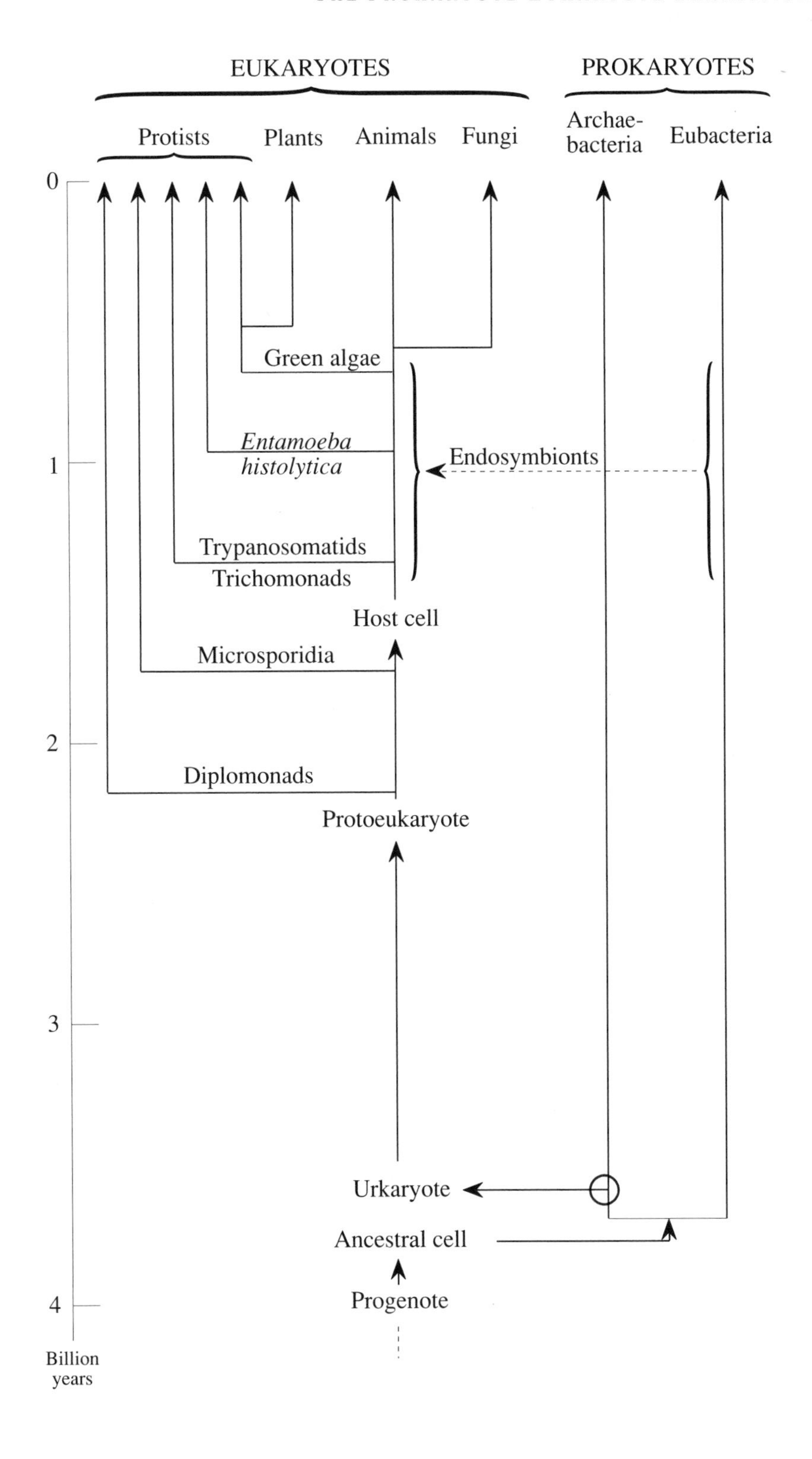
EUKARYOTES
PROKARYOTES
Protists
Plants
Animals
Fungi
Archae-
bacteria
Eubacteria
0
1
2
3
4
Billion
years
Green algae
*Entamoeba*
*histolytica*
Endosymbionts
Trypanosomatids
Trichomonads
Host cell
Microsporidia
Diplomonads
Protoeukaryote
Urkaryote
Ancestral cell
Progenote

opsins as light-sensitive pigments, the use of dolichol phosphate as glycosyl carrier, certain molecular characteristics of cytomembrane proton pumps, the presence of actinlike and histonelike proteins, a number of properties of the protein-synthesizing machinery, perhaps even the occurrence of some split genes (90, 120, 135, 139, 146).

4. The *protoeukaryote*. This stage is identified from a consideration of the most ancient protists, the diplomonads (407) and the microsporidia (411), as a cell that still retains some prokaryotic characters, for example, in its protein-synthesizing machinery, but already possesses primitive cytomembrane and cytoskeletal systems, an enveloped nucleus, and a mitotic apparatus. It may lack more differentiated membrane systems, such as a Golgi complex, and is devoid of endosymbionts.
5. The *host cell*. This, by definition, is the stage at which endosymbionts with an extant progeny were first adopted. In addition to more advanced eukaryotic features, an important property of this cell may have been adaptation to oxygen.

Having thus sketched out the framework of our discussions, we can now consider the main properties of eukaryotic cells, including the mechanisms whereby these properties could have developed in the course of the long prokaryote-eukaryote transition.[3]

## EXTRACELLULAR STRUCTURES

Like bacteria, most eukaryotic cells build coverings around themselves. There are, however, no obvious connections between the two types. Eukaryotic cells do not make the peptidoglycans that are the main materials of bacterial cell walls.

Most characteristically walled in are the plant cells. They occupy individual, closed chambers made largely of cellulose, often bolstered by other polymers such as lignin. They communicate with each other by means of cytoplasmic bridges (plasmodesmata) traversing the partitions that separate the cells. In fungi, the cells are also enclosed by sturdy walls.

Animal cells sometimes also occupy individual enclosures, for example, in bone or cartilage. Quite often, however, they surround themselves only by a flimsy, fuzzy, carbohydrate-rich surface coat, or glycocalyx; and they associate together into sheets, clusters, sacs, columns, or other kinds of multicellular groupings by means of intercellular junctions of various types that link contiguous plasma membranes. The solid frameworks needed for the structural support and functional organization of these assemblages are made largely of interlocked fibers of collagen, embedded in a

[3] It is almost impossible to do justice to the innumerable data, comparisons, hypotheses, assertions, and controversies that fill the recent literature on the prokaryote-eukaryote transition. A profitable introduction to the topic is provided by the proceedings of a symposium that brought together many of the protagonists in the field in 1986, under the auspices of the New York Academy of Sciences (90). Many of the contributions include guides to the earlier literature. More recent information is to be found in the proceedings of a sequel to this meeting, held in Lyon in 1989 (91).

fine meshwork of proteoglycans and sometimes reinforced by deposited minerals. A variety of carbohydrate polymers may also participate in the construction of extracellular coverings. An example is chitin, the main component of the teguments of arthropods and other invertebrates, as well as of many fungi.

However, not all solid structures in animals are extracellular. The voluminous eukaryotic cells are also buttressed on their insides by a variety of cytoskeletal elements, and these may become so dense—at the cost of the cells themselves shriveling away and dying—as to make up very tough structures. Such a cytoskeletal constituent is keratin, the main component of the horny layer of the skin and of the many thickenings and appendages that arise from it.

Like those of bacteria, the components of the extracellular structures of eukaryotic cells are made and secreted by membrane-associated biosynthetic and translocating systems. Although similar to those found in the plasma membrane of bacteria, these systems are not associated with the plasma membrane but, rather, with intracellular membranes. As will be seen, this is a major difference between eukaryotes and prokaryotes.

## THE CYTOMEMBRANE SYSTEM

### *Origin*

Eukaryotic cells usually have very irregular boundaries, contorted by all sorts of protrusions, clefts, and folds. As a result, they have a greatly expanded plasma membrane. In addition, their cytoplasm is filled with a variety of membrane-bounded structures. These include the metabolic organelles that will be considered later in this chapter and an extensive system of membranous sacs of various shapes and sizes comprising the rough-surfaced and smooth-surfaced endoplasmic reticulum, the Golgi complex, the trans-Golgi network, secretion granules, endosomes, lysosomes, various storage and transport vesicles, and elements of the nuclear envelope. These structures are all united into a single system by their ability to directly or indirectly establish connections with each other and with the plasma membrane by means of fusion-fission processes (see Figure 1–16, page 32). The whole of this cytomembrane system forms with the plasma membrane a complex organelle, concerned mainly with exchanges between the cell and its environment and with related chemical syntheses and processings.

The existence of such an elaborate membranous network is easily rationalized as a necessary condition of the large size of eukaryotic cells as compared with bacteria. With an average surface-to-volume ratio of about 1/20 that of bacteria, eukaryotic cells would be unable to sustain exchanges with the outside at a rate compatible with their metabolic needs, were it not for the great expansion of surface area—up to fiftyfold or more—offered by plasma membrane folds and by intracellular membranes.

It is tempting to convert this physical necessity into a historical event and to visualize the cytomembrane system of eukaryotic cells as having developed from an ancestral plasma membrane of prokaryotic type by progressive expansion, infolding, vesiculation, and differentiation, in parallel with a commensurate enlargement of the cell size.

Known facts support such a hypothesis. As shown in Table 3–1, a number of functional properties of the prokaryotic plasma membrane—those not bequeathed by way of endosymbionts—are distributed over different parts of the eukaryotic cytomembrane system. These are not superficial similarities. There is, as we shall see, clear molecular evidence that the prokaryotic plasma membrane and the eukaryotic cytomembrane system are related phylogenetically and derived from a common ancestral membrane. To view this membrane as of prokaryotic type, and not of eukaryotic type, is consistent with our general model of the emergence of eukaryotes.

The postulated membrane expansion and development must have been an early event in the prokaryote-eukaryote transition, as most other eukaryotic properties could not have been acquired without it. Most likely, it had already started in the original urkaryote. Perhaps, it started even earlier and played a decisive role in setting some ancient prokaryote on its way to becoming the urkaryote.

## *The Urkaryote*

Should we try to reconstruct what properties could have made a prokaryote into a possible eukaryotic ancestor, the first characteristic that comes to mind is flexibility. Although membrane expansion can occur in a cell surrounded by a wall (see Figure 2–3, page 44), the functional adaptations that may have favored the development of the eukaryotic cytomembrane system would most likely have required the kind of surface freedom and accessibility enjoyed only by wall-less forms, such as indeed exist in nature (see Chapter 2, Footnote 6, page 43). In compensation for its lack of external support, especially if it were to grow bigger, we would expect the urkaryote to possess some sort of internal prop, that is, the seeds of a cytoskeleton.

Another possible clue is provided by the absence of any phosphorylating electron-transport chains in the eukaryotic plasma membrane and cytomembrane system, although parts of the system (e.g., the endosome-lysosome complex) have a proton pump directed intraluminally (i.e., toward what corresponds to the outside of the cell). We are reminded of some anaerobes, such as *Clostridium pasteurianum* or *Streptococcus faecalis* (see Chapter 2, Footnote 7, page 46). However, the analogy is misleading since these eubacteria are phylogenetically far removed from the eukaryotic line (238). Little doubt exists that the urkaryote was an anaerobe, since oxygen was almost certainly absent from the atmosphere in those early days. But the urkaryote could very well have been some other kind of respirer, or even a phototroph. As we know, its long-term destiny was to develop into the adoptive host of the endosymbiotic forebears of mitochondria and chloroplasts. Once equipped with efficient endosymbionts, the cells could have discarded any redundant membrane-bound energy-retrieval systems they possessed and thereby gained scope for alternative specializations.

Indeed, eukaryotic cytomembranes are not entirely devoid of electron-transport systems. Membranes of the endoplasmic reticulum contain several cytochromes (e.g., $b_5$, $P_{450}$) related to components of respiratory chains. Another interesting

**Table 3–1** Distribution of prokaryotic membrane functions among eukaryotic membranes.

| Functions of prokaryotic plasma membrane | Location in eukaryotic membranes |
|---|---|
| Molecular and ionic transport | Plasma membrane and other membranes |
| Biosynthetic translocations | |
| Proteins—cotranslational | Rough endoplasmic reticulum |
| posttranslational | Mitochondria, microbodies, chloroplasts |
| Glycosides | Endoplasmic reticulum, Golgi |
| Other | Endoplasmic reticulum, Golgi |
| Assembly of lipids | Endoplasmic reticulum |
| Communication (receptors) | Plasma membrane, other membranes (sorting) |
| Electron transport and coupled phosphorylation | Mitochondrial inner membrane |
| Photoreduction and photophosphorylation | Chloroplast inner membrane and thylakoids |
| Anchoring of chromosome (replication) | Nuclear envelope |

eukaryotic-membrane component is rhodopsin, which serves as a light-sensitive pigment in all types of eyes and is situated in infoldings of the plasma membrane of retinal cells and in membranous discs derived from such infoldings. This molecule is related to bacteriorhodopsin and halorhodopsin, which, as we saw in Chapter 2, are present in the plasma membrane of the archaebacterial *Halobacterium halobium* and have the unique property of using light energy directly for active ion transport.

This very sketchy theoretical reconstruction of the urkaryote is not too far removed from the equally blurred picture that can be deduced from experimental data. According to sequencing and other results, the closest extant prokaryotic relatives of the eukaryotic ancestor seem to belong to the group of thermoacidophilic archaebacteria (135, 139, 239). The best-studied representative of this group, *Thermoplasma acidophilum,* is indeed wall-less, seems to possess an actinlike protein, and has a primitive respiratory system (135). It is not known to have an opsin, but it is sufficiently close to halobacteria to allow for the possibility that molecules of this type might have been present in the urkaryote. However, *Thermoplasma,* like other archaebacteria, has ether lipids in its membranes, instead of the ester lipids found in eubacteria and in eukaryotes. Thomas Cavalier-Smith has proposed that archaebacteria originally had ester lipids and that they acquired ether lipids later, as an adaptation to harsh environmental conditions, after one of their relatives had already started its long journey toward becoming a eukaryote. In support of this thesis, he writes: "The fact that these [ether lipids] have not been replaced in derived mesophilic archaebacteria by fatty acid esters makes the assumption of such replacement independently in the ancestral eukaryote and eubacterium highly implausible" (120, page 23).

His proposal may well be true, but his argument in support of it is not very convincing. A priori, in view of the numerous metabolic roles of acyl-coenzyme A derivatives, it does not seem very plausible that the ability to make ester phospholipids should represent such an unlikely trait that it could be discovered only once. It seems just as likely that mesophilic archaebacteria kept their original ether lipids because they derived other advantages from them besides resistance to harsh conditions. For example, their membrane proteins, which were adapted to ether lipids, could have had exigencies that were not readily fulfilled by ester lipids. I shall have more to say about this question in Chapter 4.

Ether lipids, especially those made from $\alpha,\omega$-dialcohols (see Figure 2–4, page 45), form more rigid bilayers than do ester lipids. For this reason, ether lipids may not have lent themselves readily to the development of a cytomembrane system. Therefore, acquisition of ester lipids could have been a significant event in the conversion of some archaebacterial precursor into the urkaryote.

### *Internalization of Digestion*

Assuming some genetic modification that favored membrane expansion and internalization, what sort of evolutionary advantage did it confer that furthered its selection? A likely answer to this question is intracellular digestion (122, 137, 167, 284).

The equivalent of endocytosis does not exist in prokaryotes. Such digestive processes as bacteria accomplish take place extracellularly with the help of secreted exoenzymes, which is possible only if the cells occupy a confined and stagnant environment. The periplasmic space, which separates the plasma membrane from the cell wall in gram-negative bacteria, provides a naturally confined area for this purpose; but this is of limited advantage, as this space is not accessible to exogenous large molecules. Now imagine some genetic change of the kind we have been considering that would cause the plasma membrane of some naked cell type to expand into infoldings and outfoldings. The resulting crypts would turn into preferential sites for the digestion of trapped extracellular materials by enzymes discharged locally by membrane-bound systems. Should some of the crypts close into vesicles—a very likely possibility in view of what is known of the properties of membranes—authentic primitive forms of endocytosis and of intracellular digestion would be accomplished.

Such an acquisition, even if initially haphazard, would provide a considerable evolutionary advantage likely to favor any chance mutation that would result in its further development. As Robert Wattiaux and I wrote more than 20 years ago in considering the origin of lysosomes: "The next evolutionary step may be pictured as the production of infoldings of the cell membrane, allowing the formation of internalized extracellular pockets into which captured food and secreted enzymes were trapped together. The advantages of this development are obvious. It relieved heterotrophy from the ecological requirement for a confined and relatively stagnant environment; at the same time, it made larger membrane areas available for nutritive exchanges to take place efficiently with deep-seated portions of the cell, possibly providing in this manner an opportunity for further cell growth and organization." (167, pages 471–472). I conveyed the same message later with different words: "Until then, in order to benefit of their exoenzymes, the cells had to rely on extracellular digestion. Unless they had other means of subsistence, they were practically condemned to reside inside their food supply, like maggots in a chunk of cheese. Henceforth, they would be free to roam the world and to pursue their prey actively, living on phagocytized bacteria or on other engulfed materials. This development could well have heralded the beginning of cellular emancipation" (1, page 99). In the words of Roger Stanier: "The capacity for endocytosis would have conferred on its early possessors a new biological means for obtaining nutrients: predation on other cells" (137, page 27). In those of Cavalier-Smith (who then assumed the ancestral protoeukaryote to have been a cyanobacterium): "A phagocytic pre-alga could photosynthesize by day or during the Arctic summer, and phagocytose by night or during the Arctic winter (or in any dark environment)" (284, page 463). The images differ, but the main idea is the same.

## *Differentiation of Cytomembranes*

According to the proposed model, the first intracellular vesicles would have combined all in one the properties of endosomes, lysosomes, and cisternae of the rough endoplasmic reticulum. We can think of such a system as connecting intermittently with the plasma membrane, thereby engaging alternatively in the uptake of extracellular materials (primitive endocytosis) and in the discharge of digestive residues and secretory products (primitive exocytosis), with the digestion of exogenous materials and the processing of endogenous products taking place in the interval within closed vesicles. Starting from there, one can visualize a progressive dissociation of these various functions, leading to, among other things, a spatial separation of import and export processes (see Figure 3–3).

In the course of this differentiation process, the plasma membrane and its endocytic invaginations would progressively lose all the systems involved in the synthe-

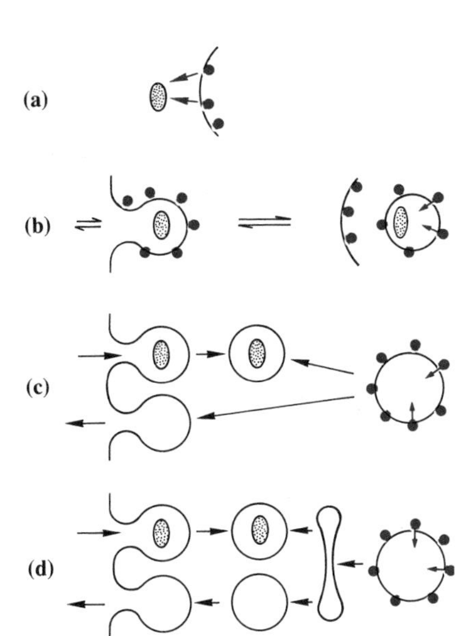


**3–3** *Some hypothetical steps in the development of the eukaryotic cytomembrane system.* **(a)** Extracellular digestion by exoenzymes discharged by plasma membrane-bound ribosomes (prokaryotes). **(b)** Reversible infolding and vesiculation of the plasma membrane allows intracellular digestion of internalized materials and subsequent excretion of residues. **(c)** Ribosome-bearing membranes move from the surface to the interior of the cell, forming a proto-endoplasmic reticulum, which secretes its products into endocytic vesicles, as well as outside by exocytosis. **(d)** Proto-Golgi sorts digestive enzymes from secretory products.

sis, processing, and translocation of secretory products, together with those that similarly mediate the formation and insertion of the main membrane constituents. All these machineries would be segregated in the membranes of special intracellular vesicles forming some sort of proto-endoplasmic reticulum. These vesicles would discharge their products both outside the cell and into endocytic vacuoles, thereby supplying enzymes for both extracellular and intracellular digestion, as well as materials for the construction of pericellular structures. At the same time, the vesicles would provide proteins and other components needed for the assembly of membranes, taking advantage of the exchanges by lateral diffusion that could take place on the occasion of membrane fusion phenomena. This separation of a proto-endoplasmic reticulum from the plasma membrane could have been very gradual; it would have progressed in almost imperceptible steps, until sufficiently refined mechanisms of vesicular transport and of membrane anchoring and recycling had emerged both to protect the proto-endoplasmic reticulum against invasion by extracellular materials and by their breakdown products and to help distinct membrane domains maintain their identity in spite of incessant mergings.

After this basic separation of import from export (or perhaps concomitantly with it), the two systems would differentiate further into subdomains connected by vesicular transport. From the proto-endoplasmic reticulum would arise the Golgi complex and the other parts of the secretory pathway. The various forms of endocytic vesicles, endosomes, and lysosomes would emerge as distinct entities on the import side. A significant element in these developments would have been the acquisition of receptors for the selective uptake, sorting, and transport of defined exogenous or endogenous materials. The receptors that serve to separate enzymes destined for lysosomes from true secretory products at the exit of the Golgi complex would have been particularly important in this respect (182, 201, 213, 356).

Our knowledge of contemporary prokaryotic and eukaryotic systems is consistent with the proposed model. As I shall mention when discussing ribosomes, the mechanisms whereby nascent secretory and membrane proteins are targeted and translocated cotranslationally in the eukaryotic rough endoplasmic reticulum are so similar to the cotranslational mechanisms that perform the same function in the prokaryotic plasma membrane as to leave little doubt concerning the existence of a phylogenetic relationship between the two systems. Similarities also exist between the translocating glycosyl transferases involved in the synthesis of extracellular glycans by bacteria and in the *N*-glycosylation of proteins by the rough endoplasmic reticulum. Both depend on a lipid-soluble polyisoprenoid glycosyl carrier for translocation of their highly hydrophilic substrates across the membranes: undecaprenol, with 11 isoprene units, in prokaryotes; dolichol, with between 16 and 20 such units, in eukaryotes. Some archaebacteria, which make authentic glycoproteins, even use both dolichol and undecaprenol (149). There is also evidence that the cytomembrane proton pumps, which play an important role in the acidification of lysosomes, endosomes, and, perhaps, part of the Golgi complex (153, 188, 195), are related to the analogous enzymes present in bacterial plasma membranes. Interestingly, the kinship is closest with archaebacterial systems (235). Also supporting the model is the existence of primitive eukaryotes that seem to lack a morphologically recognizable

Golgi complex, sometimes even a rough endoplasmic reticulum, and presumably discharge their secretory products into food vacuoles by a more direct route.[4]

Assuming the model to be correct in its main features, what benefits, in addition to those of intracellular digestion, did the cells derive from an increasingly differentiated cytomembrane system? An obvious advantage is that the plasma membrane would have been relieved of a number of its functions, thereby gaining the opportunity to develop and refine others. Most characteristic of the eukaryotic plasma membrane is that it is almost exclusively adapted to communication and exchange with the outside world. It is fitted with a variety of transporters, pumps, channels, and gates; and, especially in animals, it veritably bristles with receptors of various kinds hooked onto sophisticated transducing machineries. All in all, the eukaryotic plasma membrane appears as a boundary organelle of enormous complexity, in keeping with the almost irreconcilable requirements of cells obliged to be continually or intermittently open to a variety of substances and influxes and closed to many others. Such a development would probably have been impossible with a plasma membrane cluttered by biosynthetic and translocating systems, let alone respiratory chains and photosystems. This is why, as pointed out above, the absence of energy-retrieval systems in the plasma membrane of eukaryotic cells tells us relatively little about the energy metabolism of their prokaryotic ancestor.

This type of organization as a boundary organelle is not restricted to the plasma membrane. Intracellular cytomembranes also possess specific transport systems and receptors related to their functional specializations. This is understandable. The spaces delimited by cytomembranes represent some sort of no man's land between the cytoplasm proper and the outside medium. Their contents are mostly in transit, on their way out of the cells or into them. Consequently, the exchanges between these spaces and the cytoplasm need to be controlled just as tightly as those between the cytoplasm and the environment. This is perfectly in keeping with a plasma-membrane origin.

Clear separation of export from import would have been another advantage of cytomembrane differentiation. Digestive vacuoles in which everything is cut to pieces indiscriminately are obviously not the best transit station for complex biosynthetic products. On the other hand, certain specific hydrolytic clippings are often required for the maturation of secretory products. In the present organization of the cytomembrane system, these two functions take place at the exit from the Golgi, in sites that are now distinct but retain a number of telltale signs of a common ancestry.

[4] An example of a protozoon apparently lacking a Golgi complex is *Entamoeba histolytica,* which is said even to lack a typical endoplasmic reticulum, in spite of carrying out very active intracellular digestion (77). Diplomonads, which include *Giardia lamblia*—which has been identified by 16S rRNA sequencing as a member of what may be the most ancient eukaryotic branch extant (407)—have a conspicuous rough endoplasmic reticulum, but no identifiable Golgi complex (413). The significance of such claims obviously depends on the thoroughness of the search and on the reliability of the morphological identification criteria used. The possibility of evolutionary losses also needs to be considered. This could be the case for *Entamoeba,* which presumably arose from a cell with a fully developed cytomembrane system (see Figure 3–2, page 59).

Other obvious advantages of differentiation lie in the many new functions that it makes possible. Insertion of sorting endosomes between endocytic uptake and lysosomal digestion allows certain engulfed materials to bypass the lysosomes and to be stored in the cytoplasm or conveyed to another extracellular compartment by diacytosis (transcytosis). Such insertion also permits the large-scale retrieval and recycling of receptors and membrane material. On the other hand, division of the secretory pathway into a series of distinct elements interposed between the rough endoplasmic reticulum and the plasma membrane now spreads the processing of export materials over something close to an assembly line. The necessary enzyme systems are distributed along this system in the order in which their intervention is required (47, 171, 173, 177, 192, 197, 200). Thanks to this topology, cells cut newly made export polypeptides to their final size; fold them into specific configurations; seal these configurations by means of disulfide bonds; trim and reorganize the *N*-linked oligosaccharide chains assembled in the rough endoplasmic reticulum; build new *O*-linked chains, including the giant polysaccharide chains of proteoglycans; add sulfuryl, phosphoryl, acyl, and other groups; put together large lipoprotein complexes; and administer any other finishing touches to their secretory products and membrane constituents—all in a remarkably economical and efficient manner. Furthermore, insertion of sorting stations at the end of the assembly line allows transport to lysosomes to branch from the main export pathway and allows the latter to divide further into an intermittent, regulated route and a continual, unregulated one.

As already mentioned, what makes the whole model particularly attractive is that it does not require any quantum jump. The evolutionary process can be entirely gradual. Every small step in the direction of differentiation entails a selective advantage. In fact, the model is so compelling that one may well ask why development of a cytomembrane system did not happen more frequently and why it took such an apparently long time (see Figure 3–2, page 59). Of course, we do not know how many times it happened. Much more information will be needed before we can state with certainty that all existing eukaryotes originated from a single protoeukaryotic ancestor. In addition, we do not know how many lines may have become extinct. More directly to the point, the model is not as simple as it may seem. It needs much more than mere membrane differentiation to work.

## *Supporting Mechanisms*

Developing a eukaryotic type of cytomembrane system by the envisaged mechanism (or by any other mechanism) implies the solution of a number of logistic problems. As soon as discontinuities are created by vesiculation of a previously continuous membrane, mechanisms are needed to allow the transfer of contents: from the vesicle where they first appear, by uptake or synthesis, to one where they are processed or serve in processing; from one processing vesicle to another, if sequential processing occurs; from a terminal vesicle to the outside, for discard or secretion; conversely, from the outside into a vesicle, for uptake. The general solution to these problems is fusion-fission (see Figure 1–16, page 32), an apparently simple catchword that actually covers a great deal of complexity.

To fuse or divide, membranes must be helped by some outside agency that brings them in sufficiently intimate contact for their lipid bilayers to reorganize. Furthermore, simple fusion-fission can only serve to homogenize the contents of the whole system. If the process is to be directional, it must be asymmetric, so that fusion takes place with a full container and fission removes the container (or its equivalent) either empty or loaded with other materials. If, in addition, the process is to be selective, the container involved must bear receptors on its luminal face, and the local conditions must be such that certain defined passenger molecules bind to the receptors at the loading site and detach from them at the unloading site. A typical example is the transport of lysosomal enzymes by the mannose 6-phosphate receptor from the Golgi complex, where loading is favored, to the endosome-lysosome system, where the local acidity causes unloading of the receptor (201, 213, 356). Most instances of receptor-mediated endocytosis rely on a similar pH-dependent mechanism (188). Finally, for such traffic to take place without disruption of the overall organization of the cytomembrane system, the membrane material involved in transport must be returned directly (shuttling) or indirectly (cycling) to its site of origin or replaced in some other fashion.

The transport of newly made membrane constituents poses analogous problems. Most integral membrane proteins belonging to the cytomembrane system are synthesized in the endoplasmic reticulum by membrane-bound ribosomes and are glycosylated and otherwise processed en route to their final destination, much as are secretory proteins. They travel similarly with the help of fusion-fission processes; but they do so by lateral diffusion in the plane of the fused membranes, not by transfer of luminal contents (171, 173, 177, 197). The same is probably true of many lipid components, such as phospholipids, which are assembled in membranes of the endoplasmic reticulum, cholesterol, which is likewise completed in membranes of the endoplasmic reticulum, and complex lipids, which, after assembly in membranes of the endoplasmic reticulum, are further glycosylated or sulfated in Golgi membranes. Some of these molecules may be transported to other parts of the cytomembrane system by cytosolic carriers, but for many, the main pathway is probably intramembranous (158, 165, 211).

Perhaps the most difficult problem raised by the development of a differentiated cytomembrane system is that of maintaining the differentiated state against the incessant homogenizing threats of the fusion-fission processes on which its functional activity depends. In spite of lateral diffusion, which clearly does affect many membrane constituents, each part of the cytomembrane system retains its characteristic structural proteins, its enzymes, its receptors, even its lipids. (For example, there is almost no cholesterol in the endoplasmic reticulum where it is made; it is mostly in the plasma membrane, in the endosome-lysosome system, and, to a lesser extent, in Golgi membranes.)

We know too little about the manner in which these various problems are solved in present-day organisms to even try to guess how they were solved serially and progressively in the course of the phylogenetic development of the eukaryotic cytomembrane system. But one word, at least, comes to mind as mandatory: cytoskeleton. Membranes interact with a variety of cytoskeletal elements on their cytosolic face, and it is likely that these interactions play an important role, in addition to intramembrane associations, in anchoring certain constituents in place and in preventing them from slipping away by diffusion. Furthermore, all forms of vesicular transport between different parts of the cytomembrane system, as well as to and from the extracellular environment, depend on the participation of cytoskeletal elements and associated motor systems. Whatever the details, which still largely remain to be elucidated, the development of an increasingly differentiated cytomembrane system from infoldings of the plasma membrane in the way I postulate could not have taken place without the parallel construction of increasingly elaborate cytoskeletal and motor systems. The two processes are indissociable.

### *The Nuclear Envelope*

One important aspect of the proposed differentiation model has not been considered yet, namely its possible relationship to the development of the eukaryotic nucleus. In bacteria, the chromosome is attached to the plasma membrane at the level of the mesosome. In eukaryotes, as will be seen later in this chapter, chromosomes are attached, by way of lamins and other structural elements, to special vesicles of the rough endoplasmic reticulum that are fused into a sealed, double-membranous envelope pierced by pores. It is tempting to assume that the mesosome link of the prokaryotic, ancestral plasma membrane moved inward, dragging with it the attached chromosome, as part of the postulated expansion and differentiation process. As with other segregation phenomena, an immediate advantage of this displacement would have been to relieve the plasma membrane of a burdensome function, in this case, the anchoring of the bulky DNA-replicating system. The subsequent steps that led to the elaborate construction of the eukaryotic nucleus as it is known today will be discussed later.

## THE CYTOSKELETON AND CYTOMOTILITY

Among the wealth and diversity of eukaryotic cytoskeletal elements (82a), two structures stand out as being of both general and primordial importance: actin fibers and microtubules. Seemingly present in all eukaryotic cells, from diplomonads to plants, yeast, and man, these two structures share a number of characteristic properties, even though their architectures and functional adaptations are very different and their constituent proteins are apparently unrelated genetically.

First, actin fibers and microtubules are both made of globular protein building blocks. This is a remarkable property, as virtually all other cytoskeletal structures are built from fibrous protein molecules, associated laterally and longitudinally into filaments that further combine to form characteristic three-dimensional constructions.

The principles of assembly of actin fibers and microtubules are similar. Both depend on the presence of two pairs of complementary lock-and-key arrangements in their building blocks. One, polar, allows the molecules to associate end to end; the other, lateral, joins them side to side. In actin, the building blocks are all the same—

globular (G) actin—and their lateral binding causes two single filaments to combine into an elongated, helically twisted, double-stranded string of fibrous (F) actin. Microtubules are made of two different, but closely related, proteins, $\alpha$-tubulin and $\beta$-tubulin, evolutionary siblings encoded by two descendants of the same ancestral gene. Their polar and lateral lock-and-key arrangements are reciprocal: $\alpha$ binds only to $\beta$ and vice versa. One of the two polar linkages is stable, so that the actual building blocks are $\alpha$-$\beta$ heterodimers. The lateral attachments are such that exactly 13 linear protofilaments join side by side to form a hollow, cylindrical microtubule. The primary structures made in this way usually bind additional proteins, such as tropomyosin in the case of actin and microtubule-associated proteins (MAPs) in that of microtubules. They often associate further, with each other and with other proteins, into a variety of complex assemblages.

Another important property common to actin fibers and microtubules is that their assembly takes place reversibly and by analogous mechanisms. G actin has a binding site for ATP. When polymerization is triggered, this ATP is hydrolyzed, and the resulting ADP remains bound to the F actin. ADP must be displaced by ATP for depolymerization to take place. Events are similar in the reversible assembly of microtubules, except that GTP is the source of energy instead of ATP. In both cases, assembly is a directional process, starting from stable nucleation points or organizing centers. This reversibility contrasts with the largely durable character of other cytoskeletal structures. It has far-reaching consequences in that it allows cells to dismantle certain of their supporting structures and to rebuild them again according to the same or to a different pattern. The plasticity of eukaryotic cells, some of which may undergo truly dramatic metamorphoses, depends largely on such phenomena. The periodic construction and disassembly of the mitotic spindle at each turn of the cell cycle offer a particularly revealing and significant illustration of the importance of reversible assembly in the eukaryotic way of life. Note that reversibility is not an invariable property of the structures built with actin fibers or microtubules. Some of these structures are stable, but they can regenerate when removed or destroyed. We shall see examples of them below.

Actin fibers and microtubules also have in common that they both are often associated with an ATP-driven mechanochemical transducer: actin with myosin or myosin-like molecules, microtubules with dynein or related molecules, such as kinesin. The molecular structure and mode of arrangement of the two systems are completely different, but they function similarly. The transducing molecule is anchored at one end to some object or support and is bound at the other end to either an actin fiber or a microtubule. This second binding is such that, in the presence of some triggering agent ($Ca^{2+}$ ions, for example), the transducing molecule becomes an active ATP-hydrolyzing enzyme but effectively carries out this activity only if it can bend while remaining bound at both ends. The outcome is a relative displacement (sliding) of the anchoring point and of the participating actin fiber or microtubule. This displacement can occur against a resistance and thereby accomplish work, thanks to its coupling to the hydrolysis of ATP. Many instances of intracellular movements, such as cytoplasmic streaming, saltatory movements of granules, axonal transport, chromosome displacement in mitosis, and many others, rely on repeated processes of this sort; so does cellular locomotion, from the crawling of amoebae to the swimming of paramecia and spermatozoa.

A last similarity between actin fibers and microtubules is that both can, jointly with their respective mechanochemical transducers and with a number of additional structural proteins, serve in the construction of elaborate motor units. The striated muscle fibrils of higher animals are the most complex such assemblages made from actin and myosin. Cilia and flagella are the most complex ones built from microtubules and dynein.

A characteristic and mysterious entity directly related to cilia and flagella is the centriole, which is structurally identical to the root (the basal body or kinetosome) of cilia and flagella and shares with the kinetosome the ability to initiate ciliary regeneration (102). In certain flagellated protists, flagella disappear at mitosis, leaving the basal bodies to act as centrioles. Centrioles are the main constituents of centrosomes, in which they are present in pairs surrounded by an amorphous, pericentriolar matrix. At the beginning of each mitotic division, the centrioles duplicate, and the two resulting centrosomes move apart and sprout a crown of microtubules (aster) out of which the polar microtubules of the mitotic spindle grow. These facts have sparked many speculations concerning the possible semiautonomous nature of centrioles and their evolutionary origin. Unlike microtubules, however, centrioles are not a eukaryotic fixture. They are absent in higher plants, yeasts, and other organisms unable to make cilia or flagella and are obviously not indispensable for mitotic division. An attractive explanation for the association of centriolar duplication with mitosis is that it ensures that each daughter cell will inherit the ability to generate cilia or flagella in due time or at some appropriate developmental stage.

There are virtually no clues as to the evolutionary origin of actin and microtubules. Actin has not been identified in any prokaryote, except possibly the archaebacterium *Thermoplasma,* which is the closest prokaryotic relative of eukaryotes identified so far. But the evidence is still only circumstantial (135). Lynn Margulis and her coworkers have reported the presence of molecules with tubulin-like properties, including immunological cross-reactivity, in the gram-negative eubacterium *Spirochaeta bajacaliforniensis* (117, 118, 132). They have adduced this finding in support of the hypothesis, proposed by Margulis a number of years ago, that cilia and eukaryotic flagella, as well as the seemingly self-duplicating centrioles, are the evolutionary descendants of endosymbiotic flagellated microorganisms.[5] The postulated filiation is obscure, since there is no apparent relationship, apart from their name, between prokaryotic and eukaryotic flagella. The two are built differently and function according to different principles. Margulis is, of course, aware of this and has even resurrected the term "undulipodium" to avoid confusion between the two types of flagella

[5] In *Micro-Cosmos,* written in collaboration with Dorion Sagan (72), Margulis credits the alleged ancestral spirochetes with a major evolutionary role, which, by way of microtubules, extends to the functioning of the nervous system and of the human brain. Commenting on the "spirochetal nature of intellect" (page 195), the authors write: "Could the true language of the nervous system then be spirochetal remnants, a combination of autocatalyzing RNA and tubulin proteins symbiotically integrated in the network of hormones, neurohormones, cells, and their wastes we call the human body?" (page 151).

(Chapter 2, Footnote 10, page 48). There are also indications that centrioles (or kinetosomes) may arise de novo in cells devoid of such structures, which would seem to belie the hypothesis of genetic continuity of the organelles (102). However, recent findings by David Luck and colleagues could infuse new life into Margulis's hypothesis. These workers have mapped a number of genetic flagellar anomalies in the biflagellated alga *Chlamydomonas reinhardtii* to a special chromosome associated, in two apparently identical copies, with the basal bodies (which double as centrioles during mitosis) (325, 330, 344). Unlike the genomes of mitochondria and chloroplasts, however, this chromosome seems to be inactive in its storage place and to require transfer to the nucleus for replication, transcription, and recombination.

Perhaps some day, the advances of molecular biology will provide probes into the remote past of the two main cytoskeletal hallmarks of eukaryotic cells and of their associated motor systems. In the meantime, a few remarks strike me as worthy of mention. First, actin and tubulin are both finely attuned molecular structures that could not possibly have arisen by a single lucky shot. Their development must have been gradual and favored at each step by natural selection, up to their present degree of refinement, which was probably attained in the early days of eukaryotic life. Both actin and tubulin are highly conserved molecules throughout the eukaryotic world, which suggests that they are derived from ancestral molecules that were not very different from what they are today. Thus, most of the developmental process that we would like to reconstruct must have taken place during the prokaryote-eukaryote transition (or possibly earlier, but then it left little trace, if any, in present-day prokaryotes).

My second comment is inspired by the similarities between actin and tubulin. If the two molecules are indeed unrelated, as their sequences indicate, then one is tempted to attribute their development to some sort of common deterministic sequence of events. The similarities seem too close and numerous to be purely coincidental. Nevertheless, a similar sequence of events apparently did not occur in the enormously varied prokaryotic world. This fact suggests that such a process could not have taken place in bacteria or, more likely, that it would not have provided bacteria with a sufficient advantage at each step to be favored by natural selection. Which is another way of stating that only a prokaryote on the way to becoming a eukaryote could have benefited significantly from the envisioned developmental process.

Such a view fits well with the model of a cell that is progressively increasing in size and expanding its plasma membrane into an internal cytomembrane system. As I pointed out when discussing this process, such a cell would need props to support its enlarging bulk, anchoring points to maintain the differentiated state of its membranes, and motor systems to drive its membrane-mediated forms of transport. We may thus visualize the prokaryote-eukaryote transition as the coevolutionary development of a cytomembrane system and of cytoskeletal elements in which each step forward in one supports the other and vice versa.[6] A key early step in this process

[6] The crucial importance for the prokaryote-eukaryote transition of cytomembranes and a cytoskeleton developing simultaneously in a wall-less ancestral cell has been stressed by Cavalier-Smith (120, 121). See also Reference 116.

could have been the chance appearance of proteins with self-associative properties. If appropriately oriented and sufficiently reversible, the resulting associations could have provided the initial core of all further developments.

In this discussion, the main emphasis has been put, for understandable reasons, on the actin-myosin and tubulin-dynein pairs. But other cytoskeletal structures should not be forgotten. Mention must at least be made of the large intermediate-filament family, so named because they are intermediate in thickness between the thin actin and the thick myosin filaments. Generally cell-type specific, they include the keratin of epithelial cells, the desmin of muscle cells, the vimentin of fibroblasts, the neurofilaments of neurons, as well as the lamins of the nuclear lamina. These structures are assembled from fibrillar proteins of apparently common ancestry. Another remarkable structural protein is clathrin, which possesses reversible associative properties of great functional importance for receptor-mediated endocytosis and for some forms of vesicular transport. Thanks to its threefold symmetry, clathrin can form the characteristic trellises of more or less pronounced curvature that are believed to help in pulling vesicles from membranes (160).

## THE CYTOSOL

The cytosol of eukaryotic cells is almost certainly derived from the cytosol of their prokaryotic (urkaryotic) ancestor. It shares with the bacterial cytosol several important enzyme systems that most likely go back to the earliest days of life. These systems include the glycolytic chain, the main supplier of energy under anaerobic conditions, and the pentose-phosphate pathway, a purveyor of NADPH for biosynthetic reductions and of pentoses for the synthesis of nucleotides and nucleic acids. Also found in the cytosol are most biosynthetic systems or, at least, their activation part when assembly itself takes place on a structured support, such as ribosomes for protein synthesis.

The cytosol also acts as a unifying link between the various parts of the cell, being in intimate contact with all, and it participates in numerous metabolic and regulatory exchanges with them. Several important reaction chains meander in and out of organelles by way of the cytosol, with certain steps taking place in each site. As we shall see, the cytosol is also a central crossroad for the specific allocation of newly made proteins to different parts of the cell.

A point still being debated concerns the degree of molecular organization of the cytosol. Some workers believe the cytosol to be highly organized around an elaborate "microtrabecular" network. Others see it as little more than a thick, viscous sap. Whatever the final answer to this question may be, the cytosol is not likely to be a fully disordered mixture. In such a concentrated system (proteins constitute some 20% of the cytosol in weight, and water, about 70%), containing hundreds of different proteins together with a large number of coenzymes and intermediary metabolites, a great many interactions and connections must necessarily occur. Many of these may be loose and transient, without functional significance. But others may well affect reaction kinetics in a relevant way or even lead to the formation of authentic multienzyme systems with special properties. Several such systems are known to exist, both in prokaryotes and in eukaryotes. Others are believed to exist but are difficult to demonstrate because of the ease with which they dissociate. More work is needed before the delicate and controversial problem of the structural organization of the cytosol is solved.

## THE RIBOSOMES

In essence, protein synthesis in eukaryotes follows the same steps as in bacteria, but it differs from the prokaryotic process in a number of details. The ribosomes of most eukaryotes (except those of chloroplasts and mitochondria, which are of bacterial type) are bigger than prokaryotic ribosomes. The eukaryotic rRNAs are larger, and they have an additional small 5.8S rRNA that is absent in prokaryotic ribosomes. The binding sites of the two types of ribosomes are different, to the point that animal ribosomes are unable to translate bacterial mRNAs, and vice versa, even though the genetic code itself is the same in the two groups. The two types of ribosomes also respond differently to inhibitors.

It is not clear when these differences were acquired in the course of evolution. Some may have appeared already at the prokaryote stage. The protein-synthesizing machinery of archaebacteria is closer in several respects to that of eukaryotes than to that of eubacteria (120, 146). This is one of the numerous reasons supporting an archaebacterial origin of the ancestral urkaryotic cell. Other differences could have been acquired during the further evolution of eukaryotes. Indeed, the ribosomes of primitive eukaryotes, such as diplomonads and microsporidia, have a number of prokaryotic characters (407, 411). Even yeasts are sufficiently similar to prokaryotes for many of their genes to be successfully expressed in the eubacterium *Escherichia coli* (345, 358).

Another major difference between eukaryotes and prokaryotes is the strict physical separation of translation from transcription in eukaryotes.

All nonorganellar ribosomes belong to a single pool situated in the cytosol. All start their protein-assembly activity in that part of the cell, after forming an initiation complex with the 5′ end of an mRNA and a special tRNA bearing a methionyl group which, as in archaebacteria, is not formylated. (This methionyl group is formylated in eubacteria.) Subsequent events depend, first, on the amino acid sequence of the emerging N-terminal end of the growing peptide.[7] If this sequence is of the signal variety, the ribosome soon becomes blocked by a signal recognition particle (SRP), an interesting complex of six polypeptides and a special 7S RNA molecule. Peptide elongation stops until the SRP is itself recognized by an SRP receptor, or docking

[7] The targeting of newly made proteins to their final intracellular or extracellular locations has become one of the most prominent topics of investigation in contemporary cell biology. This rapidly expanding field can be covered here only in a very summary fashion. For additional information on the subject in general, References 47, 79a, 169, 192, 212, 216, 217, 310 may be profitably consulted. Specific systems are discussed more particularly in References 164, 170, 186, 199, 207 (bacteria); 178, 214, 215 (endoplasmic reticulum); 157, 175, 210 (mitochondria); 205 (chloroplasts); 159, 184, 322a (peroxisomes and other microbodies); and 168 (nucleus).

protein, situated on a membrane of the rough endoplasmic reticulum. A reorganization of the membrane takes place, probably with the building of a tunnel (355); elongation resumes, and the product is delivered into the lumen of the rough endoplasmic reticulum or inserted into the membrane by cotranslational transfer. The signal sequence is cut off by a signal peptidase. As mentioned above, this mechanism is closely related to the bacterial mechanism of cotranslational transfer across the plasma membrane, its probable phylogenetic precursor. The two mechanisms depend on similar cleavable N-terminal sequences in their polypeptide substrates; and they even recognize each other's signal pieces.

If no signal sequence occupies the N-terminal end of the translation product, synthesis takes place entirely in the cytosol, and the finished product is discharged into this compartment. What happens next depends on what kind of topogenic sequence (310) or configuration, if any, exists in the structure of the synthesized polypeptide or protein. If there is none, the product remains in the cytosol. (There is no evidence, so far, of a cytosol-directed targeting structure.) If there is a topogenic configuration, the product is directed posttranslationally to mitochondria, to plastids (in plant cells), to microbodies, to the nucleus, or sometimes even to some cytomembranes, depending on which of these targets recognizes the configuration. The topogenic sequence is removed or not, depending on the case. I will come back to these mechanisms later.

These phenomena entail an important conclusion, already valid for bacteria but particularly relevant to eukaryotes. The genetically specified, linear sequences of polypeptide chains not only control, as we saw in Chapter 1, the three-dimensional molecular arrangements that, in turn, condition the structural and functional properties of proteins; they also dictate the intracellular or extracellular locations the proteins are to occupy. This is not all. Many proteins undergo various kinds of processing that are important in determining their final properties, sometimes their final location. The enzymes that accomplish all these modifications do not act at random. They are guided by specific structural features that label certain proteins as appropriate substrates. Even the final breakdown of proteins, in the cytosol or elsewhere, is set off by structural signals recognized by systems, often of great complexity, that brand proteins for proteolytic destruction (83). The whole life of proteins, from birth to death, is preordained—within the limits set by environmental variations—by their primary structures, which are themselves encoded by the nucleotide sequences of the corresponding genes. This last relationship, however, is not as simple and unequivocal in eukaryotes as it is in bacteria. In eukaryotic cells, the passage from DNA to mRNA takes place in several steps, which provide opportunities that do not exist in bacteria for possible modifications of the primary genetic messages.

## THE NUCLEUS

The eukaryotic nucleus is so different from the prokaryotic chromosome that it is difficult to imagine how the one could possibly have arisen from the other. In fact, even looking for such a phylogenetic pathway could be wrong. All we know is that the eukaryotic and prokaryotic genomes are each derived from a common ancestral form along lines that started diverging more than three billion years ago (see Figure 3–2, page 59). There are thus two pathways, not just one, in need of reconstruction. This remark applies to all properties of the two cell types; it is particularly relevant to the nucleus.

### *Structural Organization and Morphogenesis*

In spite of its remarkable structural complexity, the nucleus, unlike many other intracellular organelles, is not a stable assemblage. Its structure undergoes dramatic changes every time the cell enters into mitotic division. These changes include the well-known phases orchestrated by the chromosomes and, in most cells, the dismantling of the nuclear envelope. At the end of mitosis, two new nuclei appear where there was only one. The almost mystical aspects of this phoenixlike metamorphosis have long obscured what should have been an obvious fact: the eukaryotic nucleus owes its structure to an essentially spontaneous self-assembly process, entirely determined, given some source of energy, by the properties of its constituents. Experimental work has clearly demonstrated that this is indeed so (172, 190, 191, 340).

The experiment is simple. Take some naked DNA—any will do, even DNA from a phage that has never been anywhere near a eukaryotic cell—and inject it with a microneedle into the cytoplasm of an amphibian oocyte. More simply, add the DNA to a cell-free oocyte extract supplemented with ATP. Armed with appropriate tools, you can watch the stepwise formation, in some two to three hours, of a structure having all the main features of a diminutive nucleus. First, the DNA recruits histones from a preexisting cytosolic pool and combines with them to form a typical nucleosome string, which almost immediately coils into a chromatin thread. This thread then directs the assembly of a protein scaffolding around which it condenses. The resulting minichromosome finally surrounds itself by a closed wrapping looking exactly like a miniature nuclear envelope, complete with lamina, double membrane, and pores.

If ontogeny tells us anything about phylogeny,[8] we may infer from these observations that the key to the development of the eukaryotic nucleus probably lies in the stepwise appearance of proteins endowed with a property, such as DNA binding, that conferred some advantage and could be further improved by natural selection. Furthermore, to account for the lack of any similar succession of events in prokaryotes, we must assume that these proteins were of significant benefit only to an enlarging cell undergoing the kind of concerted development of a cytomembrane system and of a cytoskeleton that probably traced the passage from prokaryote to

[8] The principle "ontogeny recapitulates phylogeny" was enunciated by Ernst Haeckel, an enthusiastic follower of Charles Darwin's theory of evolution. Haeckel was struck by the fact that embryos go through developmental stages that recall features of their evolutionary ancestors (branchial clefts versus gills, for example). Guided by this principle, he constructed a grandiose "tree of life" based partly on facts, partly on the products of his rich imagination. The principle has since been branded as grossly oversimplified. It does, nevertheless, retain an element of truth and can serve as a helpful guide to speculation.

eukaryote. This advantage could have been related to the inward migration of the anchoring point of the genome on the plasma membrane as part of the general membrane expansion and differentiation process.

An early significant event in the phylogenetic history of the eukaryotic nucleus could thus have been the emergence of DNA-binding histones. Even in prokaryotes, the DNA is associated with proteins. In *Thermosplasma*, already mentioned as a possible offshoot of the archaebacterial line that led to the urkaryote, proteins with a pronounced histone character have been detected (135). However histones developed, the process must have been gradual. Like actin and tubulin, histones are highly conserved throughout the eukaryotic world. They must be finely optimized (thanks to natural selection). This is understandable. The nucleosome "cushions" are octameric assemblages made of pairs of four different kinds of histones (H2A, H2B, H3, and H4), so disposed as to form spools around which the DNA makes about 1.75 turns. The further coiling of nucleosome strings into chromatin fibers depends on another histone (H1). The proteins play the dominant role in these arrangements. The attachment of the DNA owes little, if anything, to specific nucleotide sequences and is mediated mainly by electrostatic interactions between the negatively charged DNA and the positively charged histones. These proteins, on the other hand, must fit together neatly to offer the appropriate spooling surface; yet, they must remain capable of dissociating readily when replication or transcription of the DNA makes loosening of their association necessary.

Obvious advantages of organized chromatin versus extended DNA are compactness and better protection of the genome. Topological problems in replication and transcription are likely disadvantages, reflected in the fact that these processes are much slower in eukaryotes than in prokaryotes. The balance could be negative in cells selected for rapid multiplication and positive in cells deriving their selective assets from a greater genomic flexibility and a more elaborate cytoplasmic organization. Many authors believe that the first kinds of selective forces have shaped the evolution of prokaryotes, and the second, that of eukaryotes.

Subsequent steps in eukaryotic nuclear morphogenesis also depend on proteins, but with a more specific participation of the DNA. These steps are initiated by the binding of certain specific proteins at characteristic sites on the chromatin thread, about 60,000 nucleotide pairs apart on average. One of these proteins has been identified as topoisomerase II, an important DNA-uncoiling enzyme involved in DNA replication and in other mechanisms that require topological modifications of the DNA. There are three important aspects to this binding phenomenon. First, it sets in motion the self-assembly of the structural scaffolding of the chromosome. The protein complexes distributed all along the DNA condense into a compact core, around which the chromatin thread ends up looped in the form of a tightly bundled garland, which constitutes the basic structure of the eukaryotic chromosome. Second, there is a functional implication. The arrangement most probably corresponds to the organization of replication, with each anchoring point of the chromatin to the scaffolding serving as a center for the assembly of a replicating complex, or replisome. Finally, the scaffolding further initiates the construction of an envelope around itself by binding the structural proteins (lamins) that make up the meshwork of the lamina. These proteins, in turn, bind vesicles of the endoplasmic reticulum, which fuse to form a continuous double-membranous envelope. This process is linked in an unknown manner with the self-assembly and insertion of pore complexes.

There is a suggestive, possibly significant, analogy between the anchoring of eukaryotic chromatin threads and that of prokaryotic chromosomes. In both cases, the attachment is to a membrane by way of proteins and coincides with a replicating center. An important difference lies in the size, shape, and number of replicons (a replicon is the DNA stretch that is replicated by one replisome). Bacteria have a single, circular replicon, some three-million nucleotide pairs long; it is the entire chromosome. In eukaryotes, the same length of DNA may contain more than 50 replicons linked end to end; more than 50 such lengths may be joined into the single, linear stretch of DNA that makes up each chromosome. If, as seems likely, the two arrangements are related, what kind of genomic disposition existed in the remote ancestor from which they presumably arose by divergent evolution? Before addressing this question, we must look more closely at the functional organization of eukaryotic genomes.

### *Functional Organization*

A major difference between eukaryotic and prokaryotic genomes, which accounts for a substantial part of the difference in their lengths, is the presence in eukaryotes of a much larger proportion of DNA that is not expressed as either functional RNAs or proteins. This unexpressed DNA is of two kinds. One part is transcribed and subsequently deleted. It includes the intervening sequences, or introns, that are excised from RNA precursors and the pieces that are trimmed off at both ends in the course of RNA processing. The rest of the unexpressed DNA, which may account for more than 90% of the total nuclear DNA, is interposed between individual transcription units and is not transcribed. It contains regulatory sequences involved in the control of DNA replication and transcription, as well as sequences that play a role in nuclear morphogenesis. Such functionally significant sequences may constitute only a fraction of the untranscribed DNA, however. The rest is viewed by many as useless "junk," or "ballast," including dead genes (pseudogenes) and other vestiges of erstwhile duplications, transpositions, recombinations, and mutations. Accumulated in the course of evolution and not burdensome enough to be selectively deleted, these useless parts would simply have drifted from generation to generation together with the "good" parts, carried along by the blind and indiscriminate process of replication.[9]

The prokaryotic genome contains little, if any, junk DNA and, with very rare exceptions, no introns. Exceptions concern a few tRNA and rRNA genes in ar-

[9] This view is not unanimous. The extent to which it is subscribed to is part of a wider debate among evolutionists. At one end of the spectrum are the "strict selectionists" who do not believe in the perpetuation of anything that is not useful; at the other are the "genetic drifters" who see no reason why anything that is not harmful should not be retained. In between are those who wonder how wide the neutral gap separating the harmful from the useful can be.

chaebacteria and, surprisingly, some protein-encoding genes in eubacterial bacteriophages. No split genes have yet been uncovered in eubacteria themselves. Introns are rare in lower eukaryotes, and they are particularly abundant in higher plants and animals (see Table 3–2). The mitochondria and the chloroplasts of some lower eukaryotes, though not those of higher eukaryotes, may contain split genes. This observation is intriguing in view of the probable origin of these organelles from eubacterial symbionts.

Two explanations, not necessarily mutually exclusive, have been proposed to account for these facts. One is that introns were progressively acquired by eukaryotes in the course of evolution (and, at some unknown time, by a few rare archaebacteria and phages). The other explanation is that introns were present in the common ancestor of prokaryotes and eukaryotes and that they were lost by virtually all prokaryotes and, to a varying extent, by eukaryotes, surviving best in those eukaryotic lines that gave rise to the higher plants and animals.

At first sight, the acquisition theory looks more attractive. It is consistent with the general view of eukaryotes evolving from prokaryotes by gaining new attributes. Its plausibility is enhanced by the wide-ranging occurrence of transposition and other mechanisms capable of mediating DNA insertion. One difficulty of the theory is that it requires the prior acquisition of an accurate enough splicing mechanism, since inserts almost invariably kill genes. Yet, visualizing the development of a mechanism before there was any need for it smacks of predestination, anathema to most biologists.[10]

To address this question, we must look at the mechanism of splicing.[11] In the eukaryotic nucleus, this process is catalyzed by a highly complex association of several proteins and small RNAs, which has received the incongruous name of "spliceosome."[12] This assemblage is simply too complex to have served in primeval splicing. If it had done so, we would indeed be entitled to invoke predestination. Fortunately, a much simpler precursor exists. As first shown by Thomas Cech (161, 162, 314), a number of RNA molecules, mostly rRNAs, have the ability to catalyze their own splicing, without the help of any protein. Furthermore, there are clear indications that the complex nuclear mRNA-splicing mechanisms are derived from the much simpler self-catalyzed process (206). These facts mean that the development of splicing can now be visualized without calling for some kind of foresight. One can envisage a prolonged period of evolutionary "tinkering"[13] with ribosomal and

[10] The word "predestination" means different things to different people. To some, especially its adversaries, it is akin to creationism. To others, it refers to inherent properties of the universe (possibly, but not necessarily, willed by a Creator) that make the emergence of life an obligatory phenomenon under given geochemical conditions (see, for example, Reference 59). I shall discuss this question further in Chapter 10.

[11] Good elementary descriptions of splicing mechanisms can be found in standard textbooks of molecular biology. The topic has been summarily reviewed by Phillip Sharp (206), who is generally credited with the discovery of split genes.

[12] The term "spliceosome," which combines an English and a Greek root in a regrettably untranslatable form, was conceived simultaneously on the West Coast (312) and on the East Coast (323) of the United States. The properties of spliceosomes are reviewed in Reference 334.

[13] Jacob coined the expression "le bricolage de l'évolution," the "tinkering of evolution" (29, 288).

**Table 3–2** Distribution of introns in various actin genes. Note increasing number of introns with evolutionary advance.

| Organism | Number of introns |
|---|---|
| *Dictyostelium discoideum* | 0 |
| *Saccharomyces cerevisiae* | 1 |
| *Acanthamoeba* | 1 |
| *Caenorhabditis elegans* | 2 |
| *Drosophila melanogaster* | 1 |
| Sea urchin, gene C | 2 |
| Sea urchin, gene J | 4 |
| Chicken skeletal muscle | 5 |
| Rat skeletal muscle | 5 |
| Soybean | 3 |
| Maize | 3 |

Simplified data from Zakut and coworkers (370).

other functional RNAs, without incidence on viability inasmuch as genes coding for such RNAs usually are present in several copies. Introns could have entered mRNAs later, after sufficiently precise splicing mechanisms had evolved. Possible transition forms between the two processes are found in the splicing of mRNAs by fungal mitochondria, algal chloroplasts, and phages, which lack spliceosomes and rely to a varying extent on self-splicing for mRNA processing. In all these instances, the introns include sequences that code for proteins.[14] In contrast, nuclear introns seem to carry little information, except for the parts recognized by the splicing systems.

In view of these facts, the idea that introns, like other distinctive traits absent from prokaryotes, are a eukaryotic acquisition seems eminently plausible. Nevertheless, the alternative theory that introns have a very ancient history, antedating the separation of eukaryotes from prokaryotes, has gained considerable credence.[15] It attributes the loss of introns by prokaryotes and, to a lesser extent, by lower

[14] A number of introns in fungal mitochondria (332, 337), algal chloroplasts (347), and bacteriophages (317, 354) contain long coding regions (open reading frames). The corresponding proteins include factors needed for splicing (maturases), relatives of retroviral reverse transcriptase, and an endonuclease apparently involved in the transfer of the intron from one gene to another. An intron-encoded phage protein also seems to play a role in intron transfer. For an update on this rapidly moving field, see References 196 and 433.

[15] Advocated shortly after the discovery of split genes, the theory that introns have a very ancient history has become an important, though not essential, feature of the currently popular "RNA world" (244, 245, 270, 429–432a). The most radical opponent of the theory is Cavalier-Smith, who states simply that it "holds no phylogenetic water" (121, page 61). I shall say more about this important topic in Chapter 8.

eukaryotes to evolutionary pressure dominated by the advantages of rapid replication. Under this pressure, "streamlining" of the genome by the excision of any useless part was favored by natural selection. On the other hand, eukaryotes, especially the higher ones, allegedly enjoyed the benefits of rapid genetic innovation by "exon shuffling."

In agreement with this theory, exons do not, in a number of cases, look like the sort of random cuts one would expect from the haphazard insertion of transposable elements into preexisting genes. They seem to correspond to specific domains within the protein structure or to smaller segments of such domains, called modules by Mitiko Gō and defined by him as "compact regions consisting of about 20–40 contiguous amino acid residues" (99, page 915; see also References 322, 427). Hence the view that exons are derived from "minigenes" that coded for "miniproteins" at a very early stage in the origin of life. According to this theory, proteins are the products of a modular assembly of such miniproteins, itself determined by the modular assembly of the corresponding minigenes. Evidence of such modular construction has been detected in several proteins and traced back, in some cases, to before the separation of eukaryotes from prokaryotes. That exon shuffling can occur has also been clearly demonstrated. The low-density lipoprotein (LDL) receptor is a mosaic of exons that are also present in other genes (359, 360). In the immune system, gene assembly from pieces of DNA has been turned into a unique process of gene diversification (185, 209, 361). It has, however, been argued against the theory of early split genes that splicing could not have occurred in the primitive prokaryotes because it required translation to be separated from transcription (120). But this would not be so if the ancestral splicing mechanism had been of the type found in some lower-eukaryote organelles and in phages.

Several key aspects of the splicing problem go back to much earlier stages in the development of living forms than those considered here. I shall come back to them in Chapter 8. In the meantime, let me simply mention that no consensus has yet been reached on the age of introns. There are some indications that the pendulum, which for a while had favored eukaryotic conservation and prokaryotic loss, is beginning to swing back in favor of eukaryotic gain.[16]

The two theories illuminate in very different ways the question of the prokaryote-eukaryote transition and the further evolution of eukaryotes. If introns are ancestral and if their conservation was instrumental in the progressive "complexification" of the eukaryotes, then the eukaryotic phylogenetic tree may be viewed as shaped by intron loss. As introns were deleted, and in the measure that they were, offshoots branched from the main trunk into what may be viewed, in spite of enormous diversification, as "complexification dead ends": first bacteria, then protists of various kinds, fungi, and so on. On the other hand, if introns were gained in the course of eukaryotic evolution, then the main trunk of the tree grows progressively thicker,

[16] Recently observed phenomena, such as reversible self-splicing and intron transfer between genes, have established the feasibility of specifically targeted intron acquisition (196, 307, 338, 433). Evidence of evolutionary intron insertion, as opposed to conservation, has also been uncovered, for example, in serine-protease genes (348) and in chloroplast tRNA genes (400a).

and arrest of intron acquisition is the phenomenon that causes branches to separate. In the latter event, introns could have played little role in the actual prokaryote-eukaryote transition; intron acquisition presumably was slow and progressive, and the protoeukaryote itself could still have been essentially devoid of split genes, as indicated by its most ancient offshoots. However, the apparent parallelism between abundance of introns and complexity[17] holds true in both alternatives. If this parallelism is confirmed by further studies, and if it is significant and not just coincidental, the argument that introns have furthered evolution, at least in its later stages, by allowing greater genetic flexibility, experimentation, and innovation remains equally valid, whichever view one espouses.

Several other important traits, besides split genes, distinguish the genomic organization of eukaryotes from that of prokaryotes. Prokaryotic genes are often associated as operons controlled by a single promoter and transcribed into polycistronic messages. Eukaryotic messages are almost invariably monocistronic. The majority of eukaryotic mRNAs are fitted at their 5′ end with a 7-methyl-GTP cap and at their 3′ end with a poly-A tail, additions never observed in prokaryotes. The eukaryotic nucleus also includes one or more nucleoli, elaborate structures where synthesis and processing of the principal rRNAs take place. It also harbors the battery of enzymes that administer the many chemical changes undergone by tRNAs, as well as the whole machinery needed for mRNA splicing. Only mature RNAs are delivered by the nucleus to the cytoplasm. Finally, for evident reasons, the nucleus contains all the enzyme systems that take part in the replication, topological organization, transcription, repair, methylation, and other forms of processing of DNA, as well as numerous systems involved in the regulation of these reactions. Many of these developments must have occurred during the prokaryote-eukaryote transition, as they were needed by the time the nuclear envelope was completed. There are virtually no clues as to the evolutionary history and selective driving forces of these events.

## *Nucleocytoplasmic Exchanges*

Before segregation of the eukaryotic genome could be completed, some sort of partitioning of functions between nucleus and cytoplasm had to be established, and systems ensuring the exchanges required by this partitioning had to be installed. From this division, the nucleus has retained mostly functions related to DNA and RNA synthesis and processing. It has no energy-retrieval system, no protein-synthesizing machinery, and virtually no metabolic systems other than those mentioned. A rare exception is an enzyme involved in the assembly of NAD. The nucleus relies entirely

[17] The parallelism rests on a very limited number of data so far. It is, of course, much too early to speak of an actual correlation between abundance of introns and degree of complexity. Even the term "complexity" is likely to raise the hackles on many an evolutionist's back. For my part, I share the common, if arrogant, belief that humans are more complex than sea urchins and that sea urchins are more complex than yeasts. But I understand that the point is considered debatable by some, as is the meaning of the term "complexity." I will readily adopt Richard Dawkins's definition that "complicated things [I prefer the word "complex"] have some quality, specifiable in advance, that is highly unlikely to have been acquired by random chance alone" (68, page 9).

on the cytoplasm for its supply of enzymes and structural proteins, of ATP, of coenzymes, of the nucleotide and deoxynucleotide building blocks used for DNA and RNA synthesis, and of the various other substrates and activated metabolites needed for processing. In return, the nucleus delivers to the cytoplasm the RNA products of its industry, together with all the waste and other side products left from nucleic acid synthesis and processing.

Except during the mitotic interval, when, in most cells, cytoplasm and nucleoplasm mingle freely, all the exchanges between nucleus and cytoplasm take place through the nuclear pores (172). The pore complexes are relatively permeable and may allow molecules of moderate size, even small proteins, to pass through by diffusion. On the other hand, larger proteins, as well as nucleic acids, have to be assisted across the pores. Incoming proteins are recognized through specific N-terminal targeting sequences, which may include as many as five adjacent basic residues (168). Outgoing RNAs are helped across by special proteins that enter unencumbered and exit with them.

Nuclear pore complexes are intramembrane fenestrations of intricate eightfold symmetry made of proteins, including glycoproteins. The assembly of these complexes is an entirely spontaneous process, requiring only the appropriate proteins, competent membranes of the endoplasmic reticulum, and probably ATP. In line with our general model, we may therefore view the phylogenetic development of these complexes as conditioned by the appearance of the necessary proteins. The making of a simple hole would have sufficed to start with. Elements regulating the exchanges through the holes could have been added later.

### *Mitosis, Meiosis, and Sex*

If the zone of anchoring of the ancestral prokaryotic chromosome to the plasma membrane was driven inward with the developing eukaryotic cytomembrane system, whatever mechanism ensured the separation of duplicated chromosomes by a deepening membrane furrow in the dividing ancestral prokaryote was presumably internalized at the same time. The earliest form of mitosis could thus have taken place much like its prokaryotic precursor, except for involving an intracellular envelope derived from the plasma membrane, rather than the plasma membrane itself.

Then, if extant primitive eukaryotes in any way reveal the remote past (176, 183), the next step could have involved the participation of cytoplasmic microtubules in this fission process, as occurs in some dinoflagellates and in fungi such as yeasts. In these organisms, bundles of microtubules forming a rudimentary spindle press against the nuclear envelope during mitosis or traverse this envelope through tunnels, but the envelope itself remains intact. Hypermastigotes, which are protozoa present in the gut of termites and other wood-consuming insects, offer a similar disposition, but with an additional feature that may represent an important evolutionary developmental stage: sets of short microtubules connect the anchoring points of the chromatids on the nuclear envelope to the spindle poles, possibly playing a role in separating the twin chromatids. This could be seen as an evolutionary step toward the classical form of eukaryotic mitosis, in which the nuclear membrane actually breaks down and microtubules grow directly from the chromatids. However the development took place, a key event must have been the formation of kinetochores, the microtubule-organizing centers that are attached to the chromosomal framework opposite the centromeric region. Perhaps such a development was rendered necessary by the increasing complexity of chromosomal frameworks, which, upon duplication, came to be joined by centromeric bridges that had to be pulled apart. In contrast, centrioles, in spite of being conspicuous parts of the mitotic apparatus in many cells, probably do not play an essential role in the construction or functioning of this apparatus. The amorphous matrix that surrounds the centrioles in centrosomes, not the centrioles themselves, contain the organizing centers from which polar microtubules grow.

Meiosis most likely arose as a modification of mitosis in cells that had become diploid by cell fusion or by chromosome duplication (followed by independent evolution of the two chromosomes). Prophase in such cells would have provided a ready opportunity for homologous chromatids to join along their whole length and engage in extensive recombinations. However, such a powerful means of genetic diversification could not have been put to advantage without the simultaneous appearance of a mechanism for breaking the resulting chiasmata. The mitotic apparatus was there to do the job, requiring only some adaptive modifications. In the ensuing daughter cells, centromeres still remained to be severed, which happened by an essentially normal mitotic process, generating haploid cells. Any acquisition that would have favored the fusion of haploid into diploid cells would have provided a considerable selective advantage as a means of further evolutionary experimentation and innovation. In such a way, sexual reproduction could have arisen as a long-term offshoot of diploidy, setting off what may well have been the most important mechanism responsible for the extraordinary evolutionary success of eukaryotes.[18]

Highly speculative as they may be, all these conjectures have the merit of drawing attention to the importance of structural proteins in the development of the eukaryotic nucleus. This runs as a leitmotif throughout this chapter. Whether we consider the cytoplasm or the nucleus, the long transition from prokaryote to eukaryote must have been punctuated by the appearance and evolutionary elaboration of cytoskeletal elements. The astonishing number, diversity, and associative properties of the constitutive proteins of these elements are some of the revelations of modern cell biochemistry and molecular biology.

[18] Bacteria can exchange genetic material by conjugation (see Chapter 2, Footnote 5, page 43), as well as by other, more random processes. Authentic sexual reproduction, however, is a eukaryotic invention. Thanks to the independent segregation of chromosomes at meiosis and to crossing-over, sex guarantees what Peter Medawar has called the uniqueness of the individual (30). Except for identical twins, no two individuals can have exactly the same genes. Sex is also one of the main factors responsible for the evolutionary success of eukaryotes. Speaking of evolution, Jacob has written: "Les deux inventions les plus importantes sont le sexe et la mort," "The two most important inventions are sex and death" (20, page 309). More than anything else, it is the continual shuffling of genes in myriad different combinations through sexual reproduction that has allowed the "tinkering of evolution" (see Footnote 13, page 80).

## ORGANELLES OF ENERGY METABOLISM

### *The Endosymbiont Hypothesis*

Eukaryotic cells most often contain other membrane-bounded structures besides the components of the cytomembrane system. Prominent among these structures are the mitochondria and the chloroplasts, both of which contain in their membranes important functional systems that are associated with the plasma membrane in bacteria (see Table 3–1, page 63): oxidative phosphorylation in the case of mitochondria, photoreduction and photophosphorylation in that of chloroplasts. The possibility that these organelles, too, arose by internalization and differentiation of the bacterial plasma membrane obviously comes to mind. But it raises some difficulties. Mitochondria and chloroplasts are surrounded by two membranes, and their relevant energy conversion systems are situated on the inner membrane or on internal vesicles derived from it. Furthermore, the matrix of these organelles bears no conceivable relation to the periplasmic space; it is filled with important metabolic enzyme systems of the kind that are found in the cytosol of bacteria.

Although models have been proposed that explain how such structures could have originated from invaginations of an ancestral plasma membrane, clear preference must be given to the alternative hypothesis that mitochondria and chloroplasts are both derived from bacteria that established themselves as endosymbionts within some host cell. Such bacterial endosymbionts are of common occurrence in many present-day organisms.

The endosymbiont theory of the origin of mitochondria and chloroplasts goes back at least one century. Excellent brief histories of the topic are to be found in References 138 and 225. The theory was taken up in the late 1960s by Margulis (292, published as Lynn Sagan), who added flagella to the list of possible descendants of endosymbionts (see Footnote 5, page 72). She has since become one of the most ardent proponents of the endosymbiont theory and has written several books on the subject (70–72). In a landmark paper, published in 1978, Robert Schwartz and Margaret Dayhoff (406) brought forward a number of protein and RNA sequencing data in support of the endosymbiont theory. Another major advocate of the theory is Cavalier-Smith who, after first combating it (284), now constructs highly detailed and eruditely documented endosymbiont models (119–121). A comprehensive survey of data that support the prokaryotic origin of mitochondria and chloroplasts has been compiled by Libor Ebringer and Juraj Krajčovič (226, 227). The most vigorous opponent of the endosymbiont theory was the late Henry Mahler, who, in collaboration with Rudolf Raff, wrote several detailed and strongly structured papers on the subject (234). In his latest review (233), he conceded that the theory is probably correct for chloroplasts but maintained his position in favor of an autogenous origin of mitochondria. Although Mahler probably has few followers today, his papers still deserve to be read. His critical assessments of experimental data are exemplary and may still be pertinent in a number of instances.

An endosymbiont origin of mitochondria and chloroplasts readily explains the similarities of these organelles to bacteria on the basis of straightforward linear descent. The organelles' inner membrane would be derived from the plasma membrane of the ancestral endosymbionts, and their matrix from the endosymbionts' cytoplasm. As to their outer membrane, it could have originated from the host cell or from an endosymbiont's outer membrane. I shall address this problem later.

Also consistent with an endosymbiont origin is the fact that mitochondria and chloroplasts both show unmistakable signs of limited, but distinct, autonomy. They grow and divide in the cytoplasm and even engage in some form of competition,[19] as their putative ancestors must have done. They have their own genetic system, complete with DNA genes and all that is needed for replication, transcription, and translation of the genes. The system codes for only a few proteins, but it has some typical prokaryotic properties, which clearly differentiate it from the main eukaryotic genetic system. Unlike nuclear DNA, the organellar DNA is most often circular, it is not organized into nucleosomes, and it frequently bears highly compacted information. The organellar ribosomes are smaller than the cytoplasmic ribosomes of higher eukaryotes and similar in size to bacterial ribosomes; they are sensitive to antibiotics that specifically inhibit bacterial protein synthesis, such as chloramphenicol, and are not sensitive to selective inhibitors of the main eukaryotic process, such as cycloheximide. All these facts add up to a strong case in favor of an endosymbiont origin of mitochondria and chloroplasts. This hypothesis is further supported by comparative sequencing results, which, in all cases investigated so far, have revealed a closer kinship of organellar proteins or RNAs to their bacterial homologues than to their nonorganellar eukaryotic counterparts (219, 229–232, 236). At present, the endosymbiont hypothesis is almost unanimously accepted, although phylogenetic details are still hotly debated.

In addition to the mitochondria and chloroplasts, other possible descendants of endosymbionts include the cytoplasmic organelles grouped under the name of microbodies (81, 166). This morphological term applies to nondescript cytoplasmic bodies of moderate density, surrounded by a single membrane. It encompasses: 1) the peroxisomes, versatile metabolic entities found throughout the eukaryotic world and built around a core of oxidases that reduce oxygen to hydrogen peroxide, and of catalase, which decomposes this product; 2) the glyoxysomes, a subset of peroxisomes endowed with glyoxylate-cycle enzymes; and 3) the glycosomes of trypanosomatids, organelles singularly equipped with a major segment of the glycolytic chain and believed by some authors to be distant relatives of peroxisomes. Often included among the microbodies are the hydrogenosomes, which are present in trichomonads and other facultatively anaerobic protists (even some fungi). Hydrogenosomes are characterized by the ability to oxidize pyruvate to acetate and $CO_2$, with the coupled formation of ATP, and, unique among eukaryotes, the generation of molecular hydrogen. However, hydrogenosomes have two tightly apposed membranes and could be more closely related to mitochondria than to microbodies (121).

[19] When mutant mitochondria carrying a gene conferring resistance to chloramphenicol (an inhibitor of bacterial protein synthesis, which also blocks eukaryotic organellar ribosomes) are injected into cells possessing mitochondria normally sensitive to the inhibitor, the resistant mutants (at least their DNA) rapidly become established as the sole mitochondrial population if selective pressure is exerted by addition of chloramphenicol to the culture medium (331).

Or, as is believed by Miklós Müller (150), they could belong to an entirely distinct line. None of these entities have a genetic system. However, their limiting membranes are distinct from the cytomembranes, and their mode of growth resembles that of the other metabolic organelles in several respects. The possibility of their endosymbiont origin rests mainly on these similitudes (97, 121, 159, 166). It may become testable by comparative sequencing.

The possible endosymbiont origin of eukaryotic cilia and flagella has also been suggested and has recently received some support from the finding that a large chromosome is associated with the basal body in *Chlamydomonas* (325). In this case, however, there is no limiting membrane around the structure, and it seems entirely devoid of the means for replicating and transcribing its DNA.

### *The Host Cell*

If mitochondria, chloroplasts, and perhaps other organelles are derived from endosymbiotic bacteria, what sort of cells served as their original hosts? Clearly, these cells had to be sufficiently large to accommodate their guests, and they must have had some means to capture the guests or, at least, to let them in. The primitive phagocyte assumed to have arisen through the progressive development of the cytomembrane system answers this definition perfectly. Such a cell would probably be identified today as a eukaryote rather than as a prokaryote (120, 121). In addition to an intracellular membrane network, it would have possessed associated cytoskeletal and motor systems and, probably, an enveloped nucleus and an authentic mitotic apparatus.[20] Such cells exist today. A number of known protists contain no mitochondria or any other cytoplasmic organelles, such as microbodies (120, 121, 150, 436). Although the possibility of evolutionary losses cannot be excluded, it is tempting to assume that some of these organisms issued from lines that branched from the main eukaryotic line before endosymbionts were adopted.

Most ancient in this respect, according to rRNA sequencing, are the diplomonads and the microsporidia (see Figure 3–2, page 59). These are indeed primitive protozoa devoid of particulate cytoplasmic organelles. The free-living ancestor of diplomonads (which are parasites), as reconstructed by Guy Brugerolle (413), is remarkably similar to the hypothetical ancestral host cell imagined by Cavalier-Smith and described by him as a "real and fully eukaryotic nonameboid, biciliated anaxostylean metamonad protozoan" (121, page 55).

The ancestral host cell probably still possessed a number of prokaryotic characters at the time when endosymbionts were first adopted. For example, it could, like the most ancient protists, have had bacterial-type ribosomes. This is an important

[20] This view came under challenge with the discovery of a green alga, *Nanochlorum eucaryotum*, possessing a very archaic-looking nucleus (no histones or nucleosomes, simple binary fission without spindle or microtubules), yet equipped with both a mitochondrion and a chloroplast, as well as with a typical Golgi complex (416). It has now been established by 16S-rRNA sequencing that this organism is a close relative of other green algae with much more advanced properties, such as *Chlamydomonas*, and that evolutionary attrition must have been the cause of its minimal features (405).

point. In discussions of the endosymbiont hypothesis, it is sometimes forgotten that not only the endosymbionts, but also their host cells must have been derived from bacteria. The example of early protists tells us that certain differences between the organelles and their host cells—between their ribosomes, for example—may not have existed at the time of adoption and may be the result of subsequent divergent evolution.

Margulis does not see the ancestral host cell, at least that of mitochondria, as a phagocyte. She attributes the initial penetration of the mitochondrial progenitors to aggressive invasion of a prokaryotic victim by some "fierce predator," rather than to passive uptake. She implicates engulfment by a phagocyte only in the adoption of the precursors of chloroplasts and considers the development of phagocytic properties to be conditioned by the prior acquisition of mitochondria. She gives no compelling argument in support of this claim.[21]

At one time, Margulis also made the assumption, shared by a number of workers, that the original host cell was an anaerobe, which owed the ability to use oxygen to its aerobic captives (70). This seems unlikely. An obvious implication of the endosymbiont hypothesis is that host and prey must have shared the same habitat. It seems unlikely that the ancestors of mitochondria, and impossible that those of chloroplasts, could have occupied an oxygen-free niche. An anaerobic phagocyte could not have met its prey. Keep in mind also that few oxygen-free niches can have remained on the Earth's surface at the time endosymbiont adoption presumably first took place. Therefore, we need a host cell capable of utilizing oxygen metabolically or, at least, of defending itself against the toxicity of oxygen (287). The host cell could even have been a full-blown aerobe, complete with carrier-level phosphorylation. As was pointed out in the discussion of the cytomembrane system, the absence of electron-transport chains from the plasma membrane of present-day eukaryotes is no reason to assume that such chains were not present in the ancestral host. They could have been lost after adoption of the mitochondria had made them redundant or, in some species without mitochondria, as a result of parasitism.

Judging from the age of the most ancient extant protists that contain organelles presumably derived from endosymbionts (see Figure 3–2, page 59), as much as half a billion years may have elapsed between the first protoeukaryote and the first host cell. This enormous gap could be only apparent and reflect the scarcity of available data or the occurrence of evolutionary losses or both. Or it could indicate that a primitive, diplomonad-like protoeukaryote was unable to adopt endosymbionts and required a long period of evolutionary modification to become a competent host cell. Or, more likely, this gap may simply denote that endosymbiont adoption was an

[21] In her 1981 book (71), Margulis defended the aggressive invasion theory based on the argument that "pinocytosis and phagocytosis have never been seen in prokaryotes" (page 196), and she proposed as a model the behavior of bacterial predators such as bdellovibrios. She further suggested that it was the acquisition of mitochondria that allowed the development of phagotrophy and that this development was followed later by the uptake of the ancestors of chloroplasts, this time by phagocytosis. She still defends the same theory in her more recent 1986 book with Sagan (72). However, her argument neglects the fact that the primitive phagocyte that I and others postulate would not have looked like a prokaryote but, rather, like a fairly typical eukaryote.

essentially aerobic phenomenon, which did not take place until oxygen appeared in the atmosphere. All endosymbionts known today by their progeny must have been adapted to oxygen.

## *The Guests*

Much attention has been devoted to the identification of the closest extant relatives of the bacterial ancestors of cytoplasmic organelles. According to sequencing results, the mitochondrial ancestor was related to the forebears of purple, nonsulfur bacteria (123, 128, 406, 412). This does not necessarily imply, as is sometimes assumed, that it was a phototroph and that it lost its photosystem and became aerobic after adoption. As Woese has pointed out (239), the group of purple bacteria comprises a rich variety of phototrophic and nonphototrophic species, presumably all derived from a phototrophic ancestor. Therefore, it is perfectly possible that the mitochondrial ancestor was an authentic aerobe, similar, for example, to *Paracoccus denitrificans,* which has been termed a "free-living mitochondrion" (128).

It is generally accepted that the ancestor of chloroplasts was an oxygen-producing cyanobacterium. For a while, a favorite in this vast group was the genus *Prochloron* because it has chlorophyll *b* instead of phycobilins as the main photosensitive pigment of its photosystem II. The initial enthusiasm for this theory is no longer shared by all authors (see Chapter 2, Footnotes 8, page 47, and 9, page 48).

The identity of the possible precursors of microbodies is entirely conjectural, as is their bacterial nature itself. Because of the enzyme systems found in microbodies, one thinks of very ancient bacteria lacking carrier-level phosphorylation systems dependent on protonmotive force. Obligatory anaerobes of clostridium type could have inaugurated both the glycosomes and the hydrogenosomes. As to the peroxisomes (97), their mode of respiration recalls that of those primitive, cytochromeless, aerobic bacteria that derive little benefit from using oxygen as an electron acceptor but that, at least, are not destroyed by oxygen, as are strict anaerobes, because they can detoxify the peroxides that arise from it (144).

Were endosymbiotic guests adopted only once (monophyletic origin) or more than once (polyphyletic origin) for each line of organelles? This question has stirred considerable debate, which, however, seems to be drawing to a close. Endosymbiosis is a common phenomenon, judging from the present-day world. The kind of integration that has converted endosymbionts into organelles is probably not a unique, or even a highly unlikely, development; it has happened to mitochondria, chloroplasts, and possibly some microbodies, and may even be in the process of happening on this very day. Most remarkable in this respect are the "cyanelles," photosynthetic inclusions that are found in a variety of lower organisms and appear as intermediates between endosymbiotic cyanobacteria and chloroplasts. Cyanelles are genetically integrated into their host cell, very much as are chloroplasts, but retain several typical bacterial characters, including a murein wall and unstacked chromatophores in place of thylakoids. Opinions differ, however, as to the significance of cyanelles. Hainfried Schenk and his coworkers (134) see them as established relatively recently and on their way to becoming full-fledged organelles, whereas Cavalier-Smith (286) considers them as descendants of the original (and single) ancestor of all chloroplasts. A case in which endosymbiotic adoption is actually being witnessed started in 1966, when an unidentified gram-negative bacterium infected a strain of *Amoeba proteus*. After behaving as a pathogen for some time, the bacterium has started to establish a symbiotic relationship with its host, which it now supplies with a protein necessary for its survival (127). In view of these facts, it seems indeed very plausible that endosymbiotic adoption occurred more than once in each line, in which case the monophyletism-polyphyletism issue concerns, not the event itself, but the survival of the resulting progeny. As will be pointed out below, there are such bewildering variations in the organization of the genomes of organelles, especially of mitochondria, as to make a common origin at least questionable.

Related to this question is the problem of simultaneous versus serial endosymbiosis. Because no cells are known that contain chloroplasts but no mitochondria, whereas many cells have mitochondria but no chloroplasts, it is widely believed that mitochondria were adopted first and that chloroplasts were taken in subsequently by one or more of the resulting cell lines but not by others. Such a succession of events would explain the main divergence between algae and plants on one hand, animals and fungi on the other. Supporting this view is the fact that chloroplasts seem less ancient than mitochondria; they have a more complex genome. I myself have extended the serial adoption theory to microbodies of the peroxisome/glyoxysome type. Because of their primitive character, complete lack of a genetic system, and widespread distribution, these organelles could have preceded mitochondria (97, 122, 287). The theory of serial endosymbiosis—and, by way of consequence, that of a polyphyletic origin of organelles—is rejected by Cavalier-Smith, who has constructed a reciprocal adaptation script based on simultaneous adoption: purple bacteria lost phototrophy and became aerobes when exposed to the oxygen produced by their cyanobacterial neighbors; the latter lost dark respiration because the purple bacteria were more efficient at it. In support of this model, Cavalier-Smith invokes the enormous evolutionary success a host harboring the two kinds of endosymbionts would have enjoyed (119–121). This hypothesis implies that the eukaryotic precursors of animals lost their chloroplasts, whereas no plant precursor lost its mitochondria. While the second implication would fit with the evolutionary advantage theory, one may well wonder, by virtue of the same theory, why the cells that lost their chloroplasts were not eradicated.

Most likely, this controversy will soon belong to the past, as more sequencing data and other biochemical results are gathered. Present trends reveal increasing support for polyphyletism (219, 229, 230, 232, 399, 401, 403, 409), but the matter is not yet settled.

### *Integration*

If the bacterial ancestors of organelles were first taken up by phagocytosis, what happened to the membrane of the phagocytic vacuole? A popular view is that it was retained to become the outer organellar membrane. Cavalier-Smith has proposed instead that the vacuolar membrane was lost—except from the chloroplasts of euglenoid flagellates and dinoflagellates, which have three membranes—and that the outer membrane of mitochondria and chloroplasts originates from the outer membrane of their gram-negative endosymbiotic forebears (121). Microbodies, he suggests, would similarly have escaped from their containing vacuole but, being derived from gram-positive ancestors, have only a single membrane. Hydrogenosomes are an exception; they have two closely apposed membranes. Cavalier-Smith sees these organelles as originating from mitochondria.

The two theories are compatible with present knowledge. Among known endosymbionts, some are naked in the cytoplasm, others are intravacuolar. Outer membranes of organelles and of gram-negative bacteria show some similarities, for example, the presence of pore-forming proteins (porins) and of focal adhesions with the inner membrane (121). On the other hand, bacterial outer membranes do not contain receptors for protein import, whereas such receptors are present on cytomembranes. I will come back to this question later, but first I must turn briefly to the most remarkable aspect of endosymbiont integration, namely the massive transfer of their genes to the nucleus.

This phenomenon is not simply a matter of endosymbionts simultaneously losing redundant genes and their products, as parasites often do. This may have happened with some genes, those for glycolytic enzymes, for example (except in glycosomes); but it is not the main event. As a rule, only the genes, not their protein products, have been lost by the endosymbionts. What has happened is that the genes have shifted to the nucleus, so that the products are now made in the cytosol and reach their final site in the organelles by translocation across one or more membranes. It is debatable in what measure the shift was a true transfer of endosymbiont genes or consisted simply in their functional replacement by existing nuclear genes. But there is little doubt that authentic transfers of genes from organelle to nucleus did take place, probably on a large scale (125, 227, 308).

In view of the ease with which foreign genes become integrated in the nucleus after microinjection or transfection, the transfer itself does not raise any particular problem. It may not even have required a special mechanism for getting the genes out of the organelles (although such mechanisms have been suggested). Occasional organelle lysis, as was bound to occur from time to time, might have sufficed. On the other hand, correct expression of the transferred genes could have been a difficulty to the extent that organelles and nucleus observed different codes. Translocation of the nuclear gene products into the organelles would have represented a major problem in any event.

In considering these two problems, we must remember that organelles are usually present in the cytoplasm in multiple numbers, up to many thousands in the case of mitochondria. Their growth and multiplication are governed by population dynamics. This means that plenty of time was available for the development of appropriate systems. The endosymbionts could conserve their genes and go on expressing them for as long as the corresponding products could not be correctly synthesized in the cytosol and translocated.

Coding problems would have been encountered only with mitochondrial genes. As known so far, all chloroplasts follow the universal genetic code. In contrast, mitochondria show deviations from the universal code, and these deviations sometimes even vary from one species to another (see Table 3–3). Most likely, these deviations are late acquisitions that occurred after the major gene migration from endosymbionts to nucleus.[22] If this is so, the correct expression of the transferred genes posed no problem. Coding changes would have affected only the expression of the small number of genes that remained in the organelles. The coevolution of tRNA genes and of protein-coding genes becomes more readily conceivable under such conditions.[23] Suppressor mutations in bacteria show how, at least in point fashion, such adjustments can take place.[24]

Translocation raises much greater difficulties. Products of transferred genes can find their correct intraorganellar locations only if they bear the right "addresses," and this problem needs to be solved separately for each product. Much is already known about the mechanisms involved in these translocations (see Footnote 7, page 75). In all cases studied so far, the proteins are made in the cytosol by free ribosomes and transferred posttranslationally across one or more organellar membranes. This transfer depends on the specific recognition of the primary translation products by a

**Table 3–3** Mitochondrial deviations from the universal genetic code.

| | UGA | AUA | AGA | AGG | CUA | CGG |
|---|---|---|---|---|---|---|
| Universal | Stop | Ile | Arg | Arg | Leu | Arg |
| Mitochondrial | | | | | | |
| Human | Trp | Met | Stop | Stop | Leu | Arg |
| *Neurospora* | Trp | Ile | Arg | Arg | Leu | Arg |
| *Saccharomyces* | Trp | Met (Ile) | Arg | Arg | Thr | Arg |
| *Drosophila* | Trp | Met | Ser | Ser | Leu | Arg |
| Plants | Stop | Ile | Arg | Arg | Leu | Trp* |

According to Ebringer and Krajčovič (226).
*This assignment may not reflect a difference in codon usage; it could be the result of a posttranscriptional editing of the genomic CGG codon to the regular UGG tryptophan codon (327).

[22] At first sight, one would be tempted to believe the opposite. Indeed, as emphasized by Syozo Osawa and Thomas Jukes (291), the mitochondrial codes closely resemble the simplest primitive code that can be imagined, much more so than does the so-called universal code. These authors believe, nevertheless, that the primitive character of the mitochondrial codes is the result of a simplifying regression, not of conservation. It is significant in this respect that the mitochondria of *Chlamydomonas* apparently follow the universal code (125). This could also be true of all plant mitochondria, as their only divergence from the universal code (utilization of CGG for tryptophan, see Table 3–3) is probably corrected by a posttranscriptional editing phenomenon that converts CGG to UGG, the normal tryptophan codon (327).

[23] Such a process could even have taken place in complete organisms, both prokaryotic and eukaryotic (for an overview, see Reference 291). The mycoplasmas, which are minuscule eubacteria (see Chapter 2, Footnote 6, page 43), use the stop codon UGA for tryptophan, as do animal mitochondria. A number of ciliated protists designate glutamine by the stop codons UAA and UAG.

[24] A suppressor mutation is one that modifies the anticodon of a tRNA in such a way that a mutated codon in an mRNA will be read correctly.

receptor situated on the organellar outer membrane.[25] The part of the product that is recognized in this way, or topogenic sequence (310), may be N-terminal and either cleaved upon transfer, as is the signal sequence of export proteins made in the rough endoplasmic reticulum, or not cleaved, depending on individual cases. Or it may be situated elsewhere in the protein molecule. In all instances, the transfer requires a source of energy, which may be ATP, protonmotive force, or a membrane potential. Transfer apparently requires the protein to be in an unfolded state.

Bacteria translocate some of their surface and secretion proteins posttranslationally. It thus seems reasonable to assume, as was done for the cotranslational system of the endoplasmic reticulum, that the organellar translocation systems originated from prokaryotic systems. Here the origin of the outer organellar membrane becomes relevant. If it is of endosymbiont origin, there is a polarity problem. Bacteria do not import proteins, they export them.[26] Cavalier-Smith's solution to this problem is to assume a flip-flop (121): genes coding for pieces of the translocation machinery would be transferred to the nucleus, and then their products would be inserted into the organellar outer membrane in the proper orientation for import. A difficulty with this hypothesis is that organelles do not seem to have lost the ancestral ability to export proteins. According to present indications, proteins destined for the intermembrane space or for the outer face of the inner membrane—cytochrome $b_2$ in mitochondria, for example—are first translocated from the cytosol into the organellar matrix through zones of focal adhesion between the two membranes; and they are then exported to their location by a second system situated on the inner organellar membrane and showing clear evidence of originating from a prokaryotic export system (175).[27]

Things would be simpler if the outer organellar membrane were derived from a vacuolar membrane. Such a membrane could, like the rough endoplasmic reticulum, have inherited a correctly oriented translocation system from the ancestral plasma membrane. Extending this hypothesis to microbodies holds the intriguing implication that the single membrane of these particles could have originated from a vacuolar membrane and not, as all proponents of the endosymbiont theory have tacitly assumed, from the plasma membrane of an endosymbiont. Such a possibility is not implausible. With mitochondria and chloroplasts, the main value to the host cell lies in the energy-retrieval systems situated in, or derived from, the plasma membrane of the endosymbionts. The evolutionary retention of this membrane was thus mandatory. No such reason exists for the preservation of the plasma membrane of the primitive bacteria that may have given rise to microbodies.

[25] One such receptor has been identified in pea chloroplasts. It occupies zones where the outer membrane is in contact with the inner membrane (343). A yeast mitochondrial outer-membrane protein that could serve a similar function has also been recognized (362).

[26] This statement may not be entirely true. Bacteria apparently take in some colicins by a receptor-mediated, energy-dependent mechanism (175).

[27] Another kind of flip-flop is conceivable. The translocation receptor could first be inserted into the inner organellar membrane in its ancestral orientation and could then be transferred to the outer membrane by way of a bud that would detach from the inner membrane and join with the outer membrane. It is easily seen that the receptor will now be oriented so as to catch cytosolic proteins and direct them into the organelle.

The two hypotheses are thus defensible in theory. They may no longer be so in practice, at least for chloroplasts. There are so many specific chemical similarities between the outer membrane of these organelles and that of cyanobacteria as to make a phylogenetic relationship between the two membranes very likely (129).

Establishing a translocation system represents only a first step in what must have been a protracted process of evolutionary adjustment that involved matching many different proteins with the same receptor. There are no clues as to how this may have happened. Presumably, the original matching occurred with a single protein. After that, other proteins had to evolve the appropriate topogenic sequence by mutation or insertion. The participation of a common transposon (121) is an attractive possibility. Whatever its mechanism, this kind of matching is perhaps not as improbably intricate as it may appear, considering that it happened at least three times in a row in the same cell line to generate three distinct sets specific for microbodies, mitochondria, and chloroplasts (not counting the targeting of secretory and nuclear proteins).

A striking aspect of the gene transfer phenomenon is its massive scale. Judging from present-day organelles, we get the impression that only major physical hindrances could have stood in its way. In mitochondria and chloroplasts, the remaining genes mostly code for hydrophobic subunits of inner-membrane constituents and for the various rRNAs and tRNAs required for the synthesis of these molecules. Presumably, the subunits proved unamenable to translocation, perhaps because of their insolubility,[28] and import of the RNAs needed for their assembly was either impossible or selected against for some reason.[29] With microbodies, which presumably did not rely on key membrane components for their usefulness—as we have seen, their original membrane could even have been lost—no such obstacle existed and the entire genome was transferred to the nucleus.

These facts are even more surprising because many organelles of the same kind exist in each cell. Thus, the selective pressure favoring the transfer of genes from endosymbiont to nucleus must have been sufficient to cause all the more richly endowed members of the population to be eradicated in favor of those that had lost genes. What evolutionary advantages did the endosymbiont-host cell association derive from gene transfer? In the opinion of Cavalier-Smith, economy was the main factor (121). The more the endosymbionts were relieved of the burden of replicating, transcribing, and translating a large genome, the more of their space and activity became available for the metabolic processes that represented their main asset to the host cells. Whether this explanation, which allows no positive advantage to the transfer of genes to the nucleus, suffices to account for the enormous evolutionary pressure that must have favored this transfer is questionable. Equally deserving of consideration is the argument that single-copy genes, centralized in an organized

[28] Arguing against this explanation is the fact that when appropriately engineered organellar genes are introduced into the nucleus, their products, synthesized in the cytosol, are correctly translocated into the organelles (189).

[29] It is of interest that the intermembrane space of rat-liver mitochondria contains nuclease activities acting on both DNA and RNA (309). Note, however, that some instances of RNA import into mitochondria are known (157).

nucleus accessible to a variety of adaptive controls, must have carried great evolutionary advantages over multiple copies distributed in a large number of individual containers scattered throughout the cytoplasm.

Integration involved much more than a simple exchange of genes for their products between endosymbionts and host cells. Endosymbionts, no doubt, lost a number of gene products irreversibly when they lost the corresponding genes.[30] More important, the endosymbionts probably also gained a number of constituents. Bacteria are not equipped with the complex transport systems through which organelles exchange a number of special substances, including coenzymes and nucleotides, with the surrounding cytoplasm. The organelle ancestors must have acquired these systems in the course of their endosymbiotic integration, while conserving their original electron-transport chains, phosphorylation systems, proton-motive machineries, and, in the case of chloroplasts, photosystems.

### *Genomic Organization*

There are remarkable variations in the organization of organellar genomes, especially in mitochondria (84, 163, 174, 187). In humans and higher animals, the mitochondrial genome is exceedingly compact and includes hardly any base sequence that is not expressed from at least one DNA strand. Each strand is transcribed as a single RNA molecule, which is then cut by nuclease action into the requisite rRNAs, tRNAs, and mRNAs. Intriguing mechanisms control this sectioning, which has to be perfectly precise, as well as the reading of the mRNAs, which are devoid of any 5′-terminal premessage sequence and sometimes even end with a truncated stop codon. In contrast, the mitochondrial DNA of yeasts and other fungi is more than 5 times larger than human mitochondrial DNA and includes a number of split genes. The RNA transcripts of these genes are cut and spliced, as are those of many eukaryotic nuclear genes, but by a different mechanism, which may require the transient expression of intron-encoded messages (see Footnote 14, page 81). Plant cells have an even larger mitochondrial genome, up to more than 100 times bigger than the mitochondrial genome in humans.

In trypanosomatids, the mitochondrial (kinetoplast) DNA consists of a vast network of some 10,000 circles, most of them of small size and probably without coding function. In this tangle, the true genomic DNA is restricted to a few large circles. The trypanosomal kinetoplast has offered molecular biologists what may well be their greatest surprise in the series of unexpected findings that mark the short history of the new field of RNA processing. After RNA splicing, after self-splicing and catalytic RNA, now comes RNA editing (419, 421, 422). This astonishing phenomenon, which now has been observed also in plant mitochondria (318, 324, 327), consists in extensive insertions (and sometimes excisions) of uridine in mRNA precursors. In

[30] This does not mean that the products were necessarily lost by the cell. Some endosymbiont genes integrated in the nucleus could conceivably be expressed into products that are not targeted to the organelles. Such a mechanism could explain the occurrence of some typical eubacterial properties in eukaryotes (see Footnote 2, page 58).

one case, the processing is so considerable as to make the edited RNA unable to hybridize with its parent gene. What guides the process, which is obviously highly specific, is a matter of considerable interest.[31]

In this bewildering variety, even the language is not constant. As we have seen, mitochondria not only use a partly different genetic code but even indulge in the luxury of inter-species variations (see Table 3–3, page 93). Chloroplasts have a more orthodox genomic organization, and they use the universal genetic code, even though they make their own set of tRNAs. Some, like mitochondria, may contain split genes and engage in RNA splicing.

The peculiarities of the genomic organization of mitochondria and chloroplasts have engendered considerable speculation concerning the origin of the organelles. Mitochondrial introns, for example, have provided Mahler with arguments against the endosymbiont hypothesis (233), have confirmed the belief of others that the common ancestor of prokaryotes and eukaryotes had split genes (244, 245, 270, 429–432), and have been explained away by adversaries of the latter theory as being the result of transpositional insertions (121).

The extraordinary diversity in the genomic organization of mitochondria could simply reflect their polyphyletic origin, which seems increasingly likely. But this is probably not the whole explanation, as there is no comparable diversity in the prokaryotic world. It is tempting to assume that the organelles owe at least part of their diversity to the opportunity they have had to evolve for enormous lengths of time within the sheltered environments of their host cells. As already pointed out, endosymbiotic organelles are uniquely open to evolutionary changes: 1) They are numerous and, therefore, can indulge in competitive evolutionary experimentation. 2) Their survival is largely ensured by their host cells, which have appropriated most of their genes. 3) Having few genes of their own, they can more readily adapt to changes that would be highly disruptive to more complex genomes. 4) They can gain certain features from their host cells by gene transfer. As an example of the latter possibility, the primitive splicing mechanisms of certain organelles could be vestigial traits of the host cells (from the time when they were "learning" to splice) conserved by the organelles, rather than heirlooms left by some ancestral endosymbiont.

One last peculiarity inherent to organellar genomes deserves to be mentioned. Their mutations are transmitted cytoplasmically, in non-Mendelian fashion, as illustrated by the historical example of the *petite* mutation in yeast (24). Furthermore, in sexually reproducing organisms, this line of transmission is exclusively maternal, at least for chloroplasts in plants and for mitochondria in animals. This implies that phylogenetic trees constructed from organellar genomic sequences should theoretically lead back to a single female individual, the "mother" of a whole species. A study of this kind made on human mitochondria claims to have identified "Eve" as a woman who lived in East Africa some 200,000 years ago (441).

[31] One mechanism has been identified in *Leishmania tarentolae*. It depends on small "guide" RNA molecules (gRNAs) transcribed from short DNA stretches situated in intergenic regions of the mitochondrial maxicircle DNA (311).

CHAPTER FOUR

# The Ancestral Cell

## THE ROOT OF LIFE

Most phylogenetic classifications of living beings have the tree of life emerging from a single root, even if they may disagree on the exact location of this root (see Figures 2–1, pages 40 and 41, and 3–2, page 59). Thus, by going sufficiently far back in time, we should eventually come to a single ancestral cell or, rather, to a single ancestral population of cells, from which *all* existing living forms, prokaryotes as well as eukaryotes, would have arisen. Is this view really justified?

There are two possible reasons why it might not be: there might not have been a common ancestor; or the common ancestor might have been different from a cell as we understand the word today. The first alternative implies that all the features common to all living cells—the genetic code, for example[1]—owe little or nothing to chance. Otherwise, they could not be the same in two or more independent evolutionary lines.[2] As will be made clear in the second part of this book, I take a highly deterministic view of the origin of living processes, but not to the point of denying a significant participation of chance factors in the development of these processes long before authentic cells emerged.

As to the second alternative, something of the kind is implicit in Woese's identification of the universal ancestor as a progenote, a nebulous entity that he defines as "a theoretical construct, an entity that, by definition, has a rudimentary, imprecise linkage between its genotype and phenotype" (239, page 263). What lies behind this definition is not clear. I have adopted the term "progenote" to designate the first primitive protocell, which must indeed have been very different from the cells we know today (Chapter 9). But this progenote cannot have been the last common ancestor of all living beings. By definition, this last ancestor occupies the origin of the first trifurcation—two successive bifurcations, actually, but of still uncertain structure—that gave rise to the separate lines leading to the archaebacteria, the eubacteria, and

[1] I am taking the view here that the "universal" code was indeed at one time universal and that today's rare deviations are later acquisitions. The reasons supporting this belief have been given in Chapter 3.

[2] This kind of reasoning must always be accompanied by two caveats: one is the possibility of independent development by convergent evolution; the other, that of horizontal, rather than vertical, transmission. However, it is highly unlikely, if not impossible, that such mechanisms could account for all the molecular characteristics shared by all living organisms.

the eukaryotes. It must therefore have possessed all the properties that these three lines have in common and that could not have appeared independently more than once. There is little doubt that we would recognize the common ancestor as a cell, should we meet it today.

The purpose of this brief chapter is to identify those common properties and to draw the most plausible picture of the ancestral cell. Much of the material considered has been covered in the preceding chapters; but bringing it together will be useful in preparation for the second part of the book. If we wish to go back to the very sources of life, the form under which it first manifested itself must be defined as clearly as possible.

## HABITAT

Three properties emerge from sequencing results as possible characteristics of the environment occupied by the ancestral cell: high temperature, low pH, availability of metabolizable sulfur.

We have seen that the closest extant prokaryotic relatives of eukaryotes are the thermoacidophilic archaebacteria, of which *Thermoplasma acidophilum* is the best-studied representative. As it happens, this group is also the most ancient among the archaebacteria. In a review on bacterial evolution, Woese writes: "It would appear that the ancestral archaebacterium was an extremely thermophilic anaerobe that probably derived its energy from the reduction of sulfur" (239, page 258). The eubacterial ancestor may not have been very different, as Woese further states: "It can be compellingly argued that thermophilia is an ancestral eubacterial characteristic" (239, page 260). Although defending an evolutionary tree very different from that advocated by Woese, James Lake agrees with him in this particular respect. "The last common ancestor of extant life," he writes, "probably lacked nuclei, metabolized sulphur and lived at near-boiling temperatures" (400, page 184).

## ENVELOPE

There can be no cell without an envelope. What was this envelope like in the universal ancestor? The most likely answer is that it consisted of a membrane made of lipids and proteins. All living cells have such a membrane, which, we have seen, combines in a particularly advantageous fashion all the properties needed for the multiple functions a cell boundary has to fulfill. It is difficult to visualize a simpler ancestral boundary from which the present membrane structure would have evolved independently more than once.

If the ancestor had such a membrane, were its constituent lipids of the ester or of the ether variety (see Figure 2–4, page 45)? This is a trickier question. We have just seen that the universal ancestor was most likely thermophilic and perhaps acidophilic. If, as suggested by some (120), ether lipids represent an adaptation to a harsh environment, it would seem that the ancestral membrane must have contained ether lipids. However, we have also seen that the ancestral urkaryote probably originated from an archaebacterium. Eukaryotes have ester lipids, as have eubacteria. Hence, the alternative possibility that the ancestral cell had ester lipids and that the archaebacteria acquired ether lipids at a later stage of their development, after the branching of the eukaryotic line from the common trunk. No compelling reasons yet exist for preferring one possibility to the other.

Looking at metabolic pathways could be helpful in this respect. The major components of ether lipids are long-chain alcohols, synthesized from five-carbon isoprene units by the mevalonate pathway, as are carotenes, sterols, terpenes, and many other natural substances. The major components of ester lipids are fatty acids, assembled from two-carbon acetyl groups by way of malonyl-coenzyme A. Both metabolic pathways coexist in archaebacteria and in eubacteria. A review of the subject concludes: "The major difference in archaebacteria seems to be in regulation of the biosynthetic pathways toward increased isoprenoid hydrocarbon production and decreased fatty acid synthesis" (146, page 150). Therefore, metabolism simply tells us that the ancestral cell may well have had the means to make both kinds of lipids or, at least, their building blocks. Which ones it actually made and incorporated into its membranes remains a matter for conjecture.

The membrane proteins must have included a minimum set of transport systems to maintain an adequate intracellular ionic milieu and to mediate the necessary exchanges of matter between the cell and its environment. An intriguing question is whether this set included a proton pump. If the ancestral environment was acidic, this seems likely. Whether the plasma membrane of the ancestral cell also contained proton-extruding electron-transport or photochemical machineries is a more complex question. I shall address it when discussing energy metabolism.

Among the expected components of the ancestral membrane are the systems required for its assembly. These would probably have included some of the enzymes of lipid synthesis, as well as the systems involved in biosynthetic protein insertion. It is particularly significant in this respect that the membranes of bacteria and those of the eukaryotic endoplasmic reticulum share a cotranslational transfer mechanism dependent on cleavable N-terminal signal sequences that are actually so similar as to be interchangeable between the two systems. It seems likely that this mechanism was already present in the common ancestor of the two cell types. Unfortunately, only eubacterial translocation systems have been studied so far. It would be interesting to know how archaebacteria insert their membrane proteins.

Did the ancestral cell have a wall? Did it have a second membrane? It is tempting to answer "no" to both questions, on grounds of economy and because life is possible without a wall or outer membrane, even in a hot, acidic environment (*Thermoplasma,* which has been mentioned above as the closest known prokaryotic relative of eukaryotes, has no wall). However, the evidence from sequencing data indicates that wall-less forms, both eubacterial and archaebacterial, are descended from walled forms and that gram-positive eubacteria are derived from gram-negative forebears (239). Furthermore, cells could have started with a double membrane, rather than with a single membrane, if they arose by folding and closing from what was originally an "inside-out" vesicle (Chapter 9). Such a process would have produced two membranes enclosing an intermembrane space within which structures could have assembled from secreted materials. The problem thus remains open.

Here also, metabolic capabilities could provide a guide. A major difference between eubacteria and archaebacteria is in the chemical structure of the cell wall (Chapter 2). The main constituent of eubacterial cell walls is murein, a peptidoglycan characteristically including D-amino acids as building blocks. Archaebacteria lack murein. Some have a pseudomurein, which, however, does not contain D-amino acids. To the extent that utilization of D-amino acids is an archaic character (Chapter 6), this fact supports Cavalier-Smith's suggestion that the ancestral cell possessed a murein wall and that the loss of the ability to make this compound was a major event in the split of archaebacteria (and their alleged protoeukaryotic relatives) from eubacteria (120).

## METABOLISM

Perhaps some day, an encyclopedic biochemist will accomplish the feat of including in a single map all the metabolic reactions in the biosphere. I have neither the knowledge nor the heroism needed for such an exercise. But I will venture a prediction: it will indeed be a single map on which any point can be reached from any other point. All living organisms use closely similar metabolic processes. Even the methanogens, which are perhaps the most exotic forms of life, have numerous conventional enzymes and coenzymes in addition to their more idiosyncratic armamentarium (146).

What came first and what came later? What was acquired—or lost—and how many times? What evolved from what? What was the ancestral stock from which all metabolic pathways were derived? These questions will perhaps be answered one day, when the universal map can be shaded according to evolutionary age. In the meantime, we may safely assume that the ancestral cell used ATP, NAD, FAD, coenzyme A, and other conventional coenzymes in the same way its contemporary descendants do; that it made and broke down sugars, amino acids, purines, pyrimidines, and other small building blocks by pathways that we would have no difficulty recognizing today; and that it assembled those building blocks into polysaccharides, proteins, nucleic acids, complex lipids, and other macromolecules by mechanisms similar to those that operate now. Some pathways, such as glycolysis or the pentose-phosphate cycle, are so central that they were most likely already functioning in the ancestral cell. Others, such as those leading to the synthesis of complex alkaloids, could have developed later. In between are the problems, evoked above, of exactly when certain characteristic metabolic processes, such as the synthesis of murein, ether lipids, or fatty-acid esters, first appeared. I shall not consider such details further but wish to examine two fundamental questions: Was the ancestral cell heterotrophic or autotrophic? What mechanisms did it use for energy retrieval and ATP regeneration?

A long-popular idea was that the ancestral cell was a clostridium-like anaerobic heterotroph dependent on fermentations associated with substrate-level phosphorylations for its energy supply. This idea was consistent with the view of a prebiotic environment bountifully supplied in fermentable substrates. As I shall argue in Chapter 6, this view implies too rosy a picture of the ingenuity of abiotic chemistry. Substrate-level phosphorylation may well have preceded the carrier-level kind—it almost certainly did—but not as part of a fermentative loop. It was much more likely coupled to a transfer of electrons between exogenous donors and acceptors, from $H_2S$ to $Fe^{3+}$, for example.[3] My contention is that life had few abiotically produced organic substrates to start with and that it had to learn autotrophy almost from the beginning. A fortiori, by the time the ancestral cell population was about to divide into two separate branches, self-sufficiency must have been the order of the day. Therefore, I conclude that the ancestral cell was autotrophic.

Was it phototrophic? Or was it simply chemoautotrophic? At first sight, the answer to this question seems clear. Among bacteria, all the chlorophyll-dependent phototrophs are eubacteria, and their phylogenetic positions indicate strongly that the common eubacterial ancestor was itself phototrophic, perhaps of the green, nonsulfur type. On the other hand, the only known archaebacterial phototroph uses a completely different system based on bacteriorhodopsin. The simplest explanation of this difference is that the ancestral cell was nonphototrophic and that chlorophyll is a eubacterial invention. The present-day world holds many examples of primitive anaerobic chemoautotrophs that could serve as possible models for such an ancestral cell. The methanogens, for example, can grow with the transfer of electrons from molecular hydrogen to carbon dioxide as the sole source of energy (146). Even though the proposal of primitive chemoautotrophy seems simplest, the alternative possibility—that the ancestral cell was actually phototrophic and that an early archaebacterium (before the branching of the urkaryote) lost the ability to make chlorophyll—cannot be excluded. I see no way of distinguishing between the two alternatives.

Even if only chemoautotrophic, the ancestral cell almost certainly contained membrane-associated, proton-extruding electron-transport chains and was able to perform carrier-level phosphorylation. Components of such chains are widely distributed among both eubacteria and archaebacteria. The only possible uncertainty concerns the eukaryotic line. Present-day eukaryotes owe all their membrane-bound energy-retrieval systems to their endosymbiotic organelles, but this does not mean that their earliest ancestor lacked such systems. This could be so only if this early ancestor had detached from the main prokaryotic line before the development of protonmotive carrier-level phosphorylation. This is contrary to the widely held belief that eukaryotes branched from archaebacteria after these had already diverged from eubacteria. It seems more likely that the ancestral cell possessed protonmotive systems and that these were lost at some stage in the evolution of eukaryotes. This loss could have occurred on the way from archaebacterium to urkaryote or at some later stage. It is even possible that proton-extruding electron-transport chains were lost

[3] It is also significant that fermentative anaerobes such as clostridia do not occupy a very low position in phylogenetic trees reconstructed from sequencing results (239).

from the eukaryotic membrane system only after the adoption of endosymbionts (or, in certain cases, as a result of anaerobic parasitism).

## GENETIC ORGANIZATION

The ancestral cell most likely possessed DNA genes, had available accurate replication and transcription systems, used prokaryotic-type (70S) ribosomes, and followed the universal genetic code. The properties of its contemporary descendants leave little doubt on these accounts.

The genes of the ancestral cell must have been comparable in length to those of contemporary organisms. The many sequence similarities existing among homologous genes and among their products throughout the biosphere would be unexplainable otherwise. These similarities enforce the conclusion that many proteins and functional RNAs most likely had reached a level of refinement close to what it is today by the time the first major split in the tree of life occurred. By that time, therefore, the whole process of gene assembly from minigenes coding for miniproteins must have been largely completed. It must have taken place earlier, during the transition from progenote to ancestral cell. Whether it left traces in the form of introns in the ancestral cell's genes or whether these genes were already "streamlined" remains a moot point (Chapters 8 and 9). What can be stated with some confidence, though, is that if the ancestral cell had many genes split by introns, it must also have possessed splicing mechanisms comparable in precision and in scope to those of higher eukaryotes. It would not have been viable otherwise. On the other hand, with only a few split genes, coding, for instance, for one or the other functional RNAs, it could have made do with a rudimentary splicing mechanism, perhaps self-catalyzed, such as is found in cytoplasmic organelles and in lower eukaryotes.

## PASSENGER GENES

If the ancestral cell was in any way similar to present-day prokaryotes, we must keep in mind the possibility that it served as a vehicle for the transport of various passenger genes, borne by resident plasmids or by phagelike viruses able to move in and out of cells among the primeval population. This point is mentioned because early genes that were subsequently discarded in the course of the evolutionary development of the ancestral cell could possibly have been carried along in this fashion. Genes coding for some enzymes of RNA metabolism, such as RNA replicase or reverse transcriptase, are possible examples.

## CONCLUSION

The main conclusion arising from this brief discussion is that the ancestral cell, defined as the last common ancestor of all living beings, must have looked like a fairly typical prokaryote, except, possibly, for the possession of introns. It was probably a chemoautotrophic, anaerobic respirer (using electron acceptors other than oxygen) or, perhaps, an anaerobic phototroph. If it resembled its closest extant relatives, it was thermophilic, probably acidophilic, and metabolized sulfur. It may have had a murein wall and, perhaps, an outer membrane, resembling present-day gram-negative bacteria in these respects. Its membrane lipids were probably made of ether-linked isoprenoid alcohols but could have been built from fatty acid esters.

This general conclusion entails an important corollary: speculations on minigenes, miniproteins, RNA worlds, shadowy progenotes, hypothetical ribo-organisms, and the like do not pertain to the ancestral cell, but to its history. Such speculations concern the long transition from progenote to ancestral cell, which is even more shrouded in mystery than is the subsequent evolution of the ancestral cell itself. For this reason, we must be extremely cautious in viewing certain seemingly archaic peculiarities as vestiges from a remote past when cells were very different from what they are today or, perhaps, did not even exist. The ancestral cell served as a bottleneck that stopped most vestiges from getting through. It is simply not possible, for example, that the last common ancestor of all living beings still depended on small RNA genes and on highly imprecise replication and expression mechanisms, as is sometimes surmised. Steven Benner has similarly argued that all that life can have retained from its remote origins must have come through a single "breakthrough organism" (115). In the second part of this book, I shall try to guess what happened before the breakthrough.

PART TWO

# BIRTH OF THE CELL

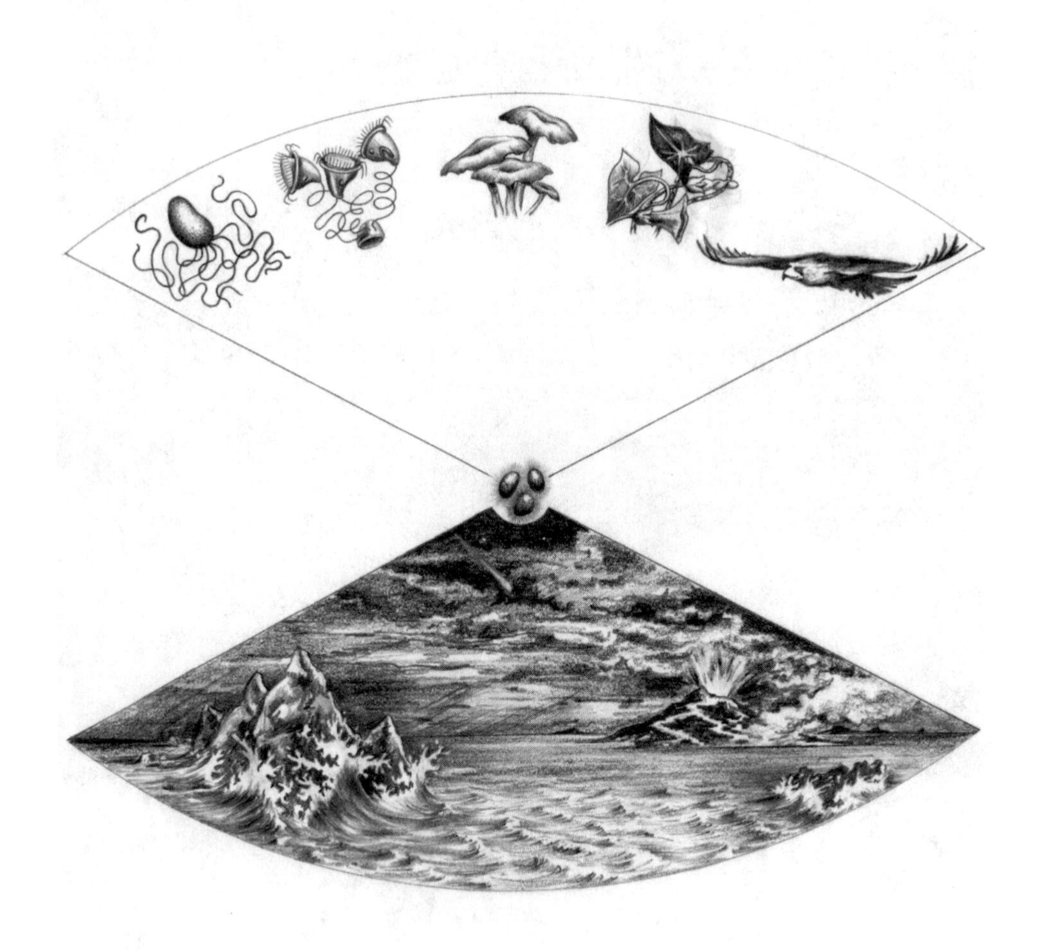

CHAPTER FIVE

# THE ORIGIN OF LIFE

## AN AGE-OLD MYSTERY

The origin of life has always been an object of wonder and fascination, though not of rational inquiry. There were two ready answers to the mystery: creation, as exemplified by the Book of Genesis; and spontaneous generation, essentially an accepted fact until the end of the 17th century (61, 253). Then, starting with the experiments of the Italian biologist Francesco Redi, the debate about spontaneous generation was launched. It raged for two centuries, until Louis Pasteur finally settled the matter. After that, until the early 1950s, the problem of the origin of life was relegated to the realm of the unknowable and viewed by most scientists as unworthy of serious pursuit.

There were some notable exceptions: the Soviet biochemist Alexander Oparin, the acknowledged father of the field, who first published a booklet on the origin of life in 1924 and later expanded it to a full-size book that went through numerous editions (61); the British geneticist J. B. S. Haldane, who addressed the topic in a brief, but influential, essay published in 1929 by *The Rationalist Annual* (28); and the British physical chemist and crystallographer John D. Bernal (48), one of the founders of biophysics. It is of possible historical relevance that all three were confirmed Marxists, militant defenders of dialectic materialism. One may wonder to what extent ideology had something to do with their desire to explain life as a naturally emerging phenomenon in the evolution of the Earth. It certainly had with Oparin, whose book is peppered with references to the philosopher Friedrich Engels. Rumor even has it that he was set on the problem by the Party.

Attitudes began to change by the end of the Second World War. Biochemistry had made important advances and opened the possibility of formulating the problem of the origin of life in concrete terms. But hardly any biochemist saw the subject as appropriate for experimental investigation. Then the bombshell fell. In 1953, Stanley Miller, a young chemist working in the laboratory of physicist-turned-cosmologist Harold Urey, attempted to reproduce conditions that Urey, an authority on the origin of planets (41), believed might have prevailed on the surface of the primitive Earth. Miller took a simple reducing gas mixture ($CH_4$, $H_2$, $H_2O$, and $NH_3$), assumed to simulate the prebiotic atmosphere, and exposed it to repeated electric discharges, taken to mimic prebiotic lightning flashes. After a few days, several amino acids and other typically biological organic compounds had accumulated in the water phase (382). Not since Friedrich Wöhler synthesized urea in 1828 had a chemical experiment been hailed as a comparable milestone. After the organic world, the prebiotic world had been freed from the vital spirit and had entered the laboratory.

Since this historical experiment, the field has veritably exploded. In the last three decades, the origin of life has been the subject of dozens of books, scores of essays, thousands of articles, relating an enormous amount of experimental and theoretical work. Periodicals devoted exclusively to the subject have been founded. Textbooks dedicate whole chapters to it. The reason for this upsurge of interest is simple. As I have attempted to show in the first part of this book, we have come to know enough about life to draw the basic blueprint according to which all extant living organisms are constructed. Scientists faced with the blueprint (or, rather, with their own version of the blueprint, because they tend to see life though different glasses, depending on their fields of specialization) find the problem of how the plan materialized almost inescapable. This turned out to be my case as well.

But I must add a warning. If not considered totally outlandish any more, the field still remains largely confined to speculation. When it comes to events that happened several billion years ago, hard data are scarce and, perforce, are supplemented by reasoning and imagination, if not blind faith. Yet, life did start somewhere, sometime, somehow. Trying to reconstruct the events that led to its birth holds almost irresistible fascination, especially now that we have available so much new knowledge on the nature of life and so many new tools for digging into the past and approaching the problem.

## DEFINING THE PROBLEM

The origin of life raises four questions: Where? When? How? and Why? Of the four, the last one, no doubt, is of greatest concern to most of us. Unfortunately, the answer to it, if there is one, is beyond the reach of science, although the kind of answers given to the other questions may have some bearing on our attitude to the metaphysical problem, as I shall mention in Chapter 10. As to the first three questions, science is beginning to offer us, if not definitive answers, at least a number of suggestive indications.

Concerning, first, the site where life originated, we have the choice between planet Earth, some other planet or celestial body, and outer space. The theory that life came to us from outer space had a number of adherents in the 19th century. The Swedish physical chemist Svante Arrhenius, the discoverer of electrolytic dissociation and of the law, which bears his name, relating the velocity of chemical reactions to temperature, defended the theory of a spatial origin of life with almost mystical fervor. He believed that the whole universe was filled with seeds of life and coined the term "panspermia." The theory of an extraterrestrial origin of life has been revived by the British astronomer Fred Hoyle, the father, with Hermann Bondi and Thomas Gold, of the "steady state" theory of the universe (involving continual creation of matter). In recent years, Hoyle, in collaboration with the Sri-Lankan scientist Chandra Wickramasinghe, has advanced increasingly controversial theories of the origin of life, to the point of claiming that viruses and bacteria arose in outer space and were brought to the Earth by comets (37–39). A modified form of the theory, termed "directed panspermia," has been proposed by Francis Crick. Together with Orgel, Crick has argued that there may not have been enough time in the early history of our planet for the development of something as complex as a living cell; he has suggested that the first germs of life reached the Earth in a spaceship sent from "some distant planet," by "a form of higher creature who, like ourselves, had discovered science and technology, developing them far beyond anything we had accomplished" (27, page 117).

The main objection to such theories is, of course, that unless life is eternal, it must have originated somewhere. Until proof to the contrary is obtained, judicious use of Occam's razor counsels the assumption that life started where we know it actually is.[1] Furthermore, as I hope to show, life could have arisen much faster than is often assumed.

Considering next the time at which life appeared on Earth, it is placed with a fair amount of certainty at between 3.5 and 4.0 billion years ago. This estimate is based on three presumptive geological indicators of bacterial life that can be traced uninterruptedly from present time until about 2.5 billion years ago and, more spottily, back to 3.5–3.6 billion years ago, perhaps even as far back as 3.8 billion years ago (388). According to a comprehensive survey of geological data pertaining to the origin of life put together by William Schopf (89), these indicators are:

1. Organic carbon significantly enriched in the lighter $^{12}C$ isotope as opposed to the heavier $^{13}C$ isotope, a typical property of bioorganic compounds.
2. Microfossil traces believed to originate from authentic cells resembling prokaryotes.
3. Layered rocks, called stromatolites, taken to represent the fossilized remains of superimposed mats of bacterial colonies, topped and fed by phototrophic organisms. Live formations of this sort still exist. They are highly organized. If their ancient counterparts were in any way similar, they must have been preceded by more primitive forms of life.

The age of the Earth is put at about 4.5 billion years. Our newborn planet was certainly not immediately conducive to a successful biogenetic process. In particular, it is estimated that impacts by large asteroids may have precluded the continued

[1] Manfred Eigen and coworkers (377) have recently reported the results of a sophisticated comparative analysis of nearly 1,000 tRNA sequences by "the method of statistical geometry in sequence space." The authors have compared "horizontal" with "vertical" divergences—i.e., structural differences among the different tRNAs of a given species with the differences among the same tRNAs of different species—restricting the analysis to positions that change sufficiently slowly to provide valid information. Their conclusion may be summarized as follows: on the assumption of a constant mutational rate, the time separating the primeval ancestor of all tRNA molecules (Chapter 8) from the last common ancestral cell (Chapter 4) would be about half the age of this last common ancestral cell. For example, if archaebacteria, eubacteria, and eukaryotes diverged 2.5 billion years ago, the genetic code would be about 3.8 billion years old. The authors conclude that "the genetic code is not older than, but almost as old as our planet," which implies that life may have arisen on Earth. We shall see that the critical divergence probably occurred much earlier, around 3.5 billion years ago, which would make the genetic code 5.25 billion years old, or older than the Earth. It is likely, however, that mutations were faster in prebiotic times than later, so the author's conclusion probably remains valid.

existence of life on the Earth's surface for several hundred million years (381, 387, 391). If authentic cells were present on Earth as early as 3.8 billion years ago, their generation cannot have taken more than a few hundred million years. This is still a very long time span: about what it took the first mammals to develop into their present-day descendants, including us; 100 times as long as it took apes to become humans.

As to the third question—how did life originate?—this will be the main topic of discussion in the following chapters. As a general a priori answer to the question, it may be taken as virtually certain, unless one adopts a creationist view, that life arose through the succession of an enormous number of small steps, almost each of which, given the conditions at the time, had a very high probability of happening. This assumption simply amounts to a rejection of improbabilities so incommensurably high that they can only be called miracles, phenomena that fall outside the scope of scientific inquiry. The calculations have been made so often as to need no repetition. Hoyle's image of a Boeing 747 assembling spontaneously from a tornado-swept junkyard is as good as any to illustrate the impossibility of a living cell's coming together in a single shot from whatever materials may have been present on our young planet. Clearly, there must have been many successive steps, each leading to a slightly more complex and organized system. Viewing each step as highly likely, if not bound, to happen under the conditions that prevailed follows from the fact that the number of individual steps must have been very large. Let the probability of each step be even moderately low—say 50%—and the combined probability, which is the product of the individual probabilities, soon reaches levels that border on the miraculous ($10^{-300}$ for as few as 1,000 steps). I shall come back to this important point in Chapter 10.

## A PREVIEW OF THE SCRIPT

In the coming chapters, I shall retrace the main steps of what, in the light of present knowledge, I see as a plausible course for the natural development of life on Earth. The story is necessarily long, involved, and further complicated by numerous uncertainties. In many ways, it is the story of a personal voyage of discovery, and it is written in a correspondingly discursive style. Readers who enjoy a suspense story may wish to skip the rest of the present chapter and accompany me on my journey with the kind of open mind that I had when I set forth. For those who would like an easier thread to follow, I offer here a synoptic preview of my script. The tale is told in simple historical style, without any of the probability weighings, plausibility assessments, and other precautionary periphrases that it requires.[2] These will come in due course.

According to my reconstruction, emerging life went through four main successive stages—or "worlds," to use a popular expression: the primeval prebiotic world, the thioester world, the RNA world, and the DNA world. This version of the script differs from the current favorite mainly by the insertion of a thioester world. I consider this

[2] The readers' attention is called to this point, lest they be misled by the apparently dogmatic style of the script. All statements should be read as conditional and hypothetical.

insertion essential because I cannot accept the view of an RNA world arising through purely random chemistry.

### *The Primeval Prebiotic World*

**Generation of reducing equivalents.** When our newly condensed planet cooled, it lost all of its molecular hydrogen and thereby became unfit for the production of the organic building blocks of life. Fortunately, rescue came in the form of an abundant source of electrons—$Fe^{2+}$ ions—and of an equally abundant source of energy—ultraviolet light. The ready photochemical oxidation of ferrous iron, both catalyzed and energized by UV light, produced vast amounts of high-energy reducing equivalents that enriched the atmosphere with such gases as $H_2$, $CH_4$, $NH_3$, HCN, $H_2O$, $H_2S$, and others. The $Fe^{3+}$ ions thus generated precipitated as insoluble oxides and other complexes, which gave rise to the first banded-iron formations.

**Synthesis of organic building blocks.** As reduced gases accumulated in the atmosphere, synthetic reactions, sparked by electric discharges and other physical forms of energy, converted them into small organic molecules. Among these were a variety of carboxylic acids, including difunctional acids, such as amino acids, hydroxy acids, and dicarboxylic acids, as well as thiols wherever $H_2S$ was present in large amounts. These substances accumulated in oceans and lakes, where they reached high concentrations, especially in sheltered, shallow areas. Organic material formed in outer space and brought to Earth by comets and meteorites may also have contributed to these stores.

### *The Thioester World*

**The central postulate.** Without additional help of both catalytic and energetic nature, the "prebiotic broth" would have remained sterile. Fertilization of the broth was accomplished by thioesters arising through the condensation of carboxylic acids with thiols. These are the main contentions on which my proposal of a thioester world rests, a proposal supported by a wealth of biochemical data that attest to the remarkable power and versatility of thioesters. Let us look first at what thioesters were capable of doing, before addressing the key problem of their primeval formation.

**Multimerization.** The first function of thioesters was to support the assembly of multimers.[3] Amino acids and other difunctional acids, activated as thioesters, were able to participate in a variety of transacylation reactions of the kind that still occur today on a large scale. This process derived its energy from the consumed thioester bonds and may have occurred without the help of a catalyst. Although essentially random and undirected, the multimerization process, nevertheless, had a highly selective yield. Many combinations failed to be made in significant amounts because of energetic or kinetic impediments of various kinds or were too unstable to reach

[3] The choice of the term "multimer" is justified in Chapter 6 (Footnote 4, page 137).

appreciable steady-state levels or fell out of solution owing to their low solubility in water. Because of the stringent screening effect of these factors, the resulting water-soluble mixture contained at significant concentrations only a tiny fraction of the statistically possible combinations, mostly relatively long chains, compactly folded or cyclic, many of them chirally homogeneous or nearly so, sometimes associated into complexes of two or more molecules. In addition, an abundance of insoluble multimers and aggregates also arose to form films, coatings, and sediments of various sorts. The soluble and insoluble multimers and higher-order structures that originated in this way played an essential role in setting the biogenic process on course by providing incipient life with structural support and, especially, catalysis.

**Catalysis and the emergence of protometabolism.** The minute subset of multimers that survived included a number of crude catalysts foreshadowing, at least qualitatively, the activities of all main present-day enzymes. This fact had three crucial consequences:

1. Thioesterification and multimerization both became catalytically self-supporting and increasingly efficient thanks to the multimeric catalysts they generated.
2. A protometabolic network, resembling in rudimentary fashion the metabolic network of present-day organisms, was laid down progressively, thanks to the encounters of primary building blocks—and, later, of their products and of their products' products and so on—with catalysts capable of acting upon them. This development was initially limited to simple, reversible reactions, until mechanisms arose for overcoming energy barriers.
3. Catalysts that had become functional—i.e., had acquired a substrate—could, if they had a sufficiently high affinity for their substrates, be protected by substrate binding. Catalysts thus protected emerged above the nonfunctional multimers to form an interrelated multiprotoenzyme system. They also served as preferential objects for elongation and other modifications that converted them to more stable, or otherwise improved, catalysts subject to stabilization by their substrates.

**Thioester-linked oxidation-reductions.** Characteristic of the thioester world were a number of electron-transfer reactions. Still of major importance today, these reactions served crucial energetic functions in the early stages of developing life. Depending on the kind of electron donors and acceptors offered by the environment, two kinds of processes could take place.

1. Given donors of low enough oxidation-reduction potential, thioesters could be subjected to reductive splitting, leading to the formation of aldehydes, or to carboxylating reductive splitting, leading to the formation of $\alpha$-keto acids. Thus, the highly endergonic reduction of carboxyl to carbonyl groups, a reaction of absolutely primordial importance, was energetically supported, as it still is today, by the concomitant exergonic splitting of thioester bonds. $\alpha$-Keto acids formed in this manner could be further reduced to $\alpha$-hydroxy acids, or reductively carboxylated to hydroxy dicarboxylic acids, or reductively aminated to $\alpha$-amino acids. The foundations of autotrophy were thereby laid.

2. In the presence of appropriate electron acceptors, the same reactions could operate in reverse, at the expense of the compounds (hemithioacetals) formed by the addition of thiols to aldehydes or $\alpha$-keto acids, initiating the oxidative synthesis of thioesters, the basis of substrate-level phosphorylation and of heterotrophy.

**The iron cycle.** As in organisms living today, these reactions took place in one direction or the other, depending on local conditions. The two directions were linked together by an iron cycle. The electrons needed for the reductive syntheses were supplied, at the required high level of energy, by $Fe^{2+}$ ions with the help of UV-light energy, whereas the $Fe^{3+}$ ions generated by this reaction served as electron acceptors in the oxidative processes. Thus, a UV-light-powered $Fe^{2+}/Fe^{3+}$ cycle supported burgeoning life in the same way that the visible-light-powered $H_2O/O_2$ cycle supports the major part of the biosphere today, with the crucial difference that the iron cycle could operate efficiently without any membrane, chromophore, or photocatalyst. Simple iron-sulfur complexes, the precursors of today's iron-sulfur proteins, sufficed to catalyze these early electron transfers.

The banded-iron formations, which were deposited in many widespread areas of our planet between 3.8 and 1.5 billion years ago, are a lasting testimony to the immensely important role of this iron cycle, not only in the development of life, but also in its further evolution up to the time when oxygen appeared in the atmosphere. The thioester world is really a thioester-iron world.

**Thioester-dependent phosphorylations.** Just as occurs today in the major substrate-level phosphorylation processes, inorganic phosphate entered protometabolism by attacking thioesters phosphorolytically. The resulting acyl-phosphates were again attacked by inorganic phosphate (instead of ADP, not yet invented) to yield inorganic pyrophosphate, the precursor of ATP as bearer of high-energy phosphate bonds. The first substrate-level phosphorylations were thereby initiated. In turn, pyrophosphate could transfer phosphoryl groups to a variety of acceptor molecules. In this way, phosphate progressively infiltrated the whole of protometabolism and gained its primary importance as a structural and functional constituent of all forms of life. Synthetic reactions started to take place by sequential group transfer. Many energy barriers to the development of protometabolism were overcome.

**The blossoming of protometabolism.** Having inaugurated electron transfer, phosphorylation, group transfer, and energy coupling, all thanks to the thioester bond, protometabolism could now exploit the full catalytic potential of the multimer population. The thioester world had reached its acme. Among its major achievements were AMP and the other mononucleotides, which provided the thioester bond with yet another opportunity to play a star role. Attacked by AMP, thioesters yielded acyl adenylates, which, under the attack of inorganic pyrophosphate, gave rise to ATP. From then on, ATP progressively took over the functions of pyrophosphate. Adenylyl transfer became a major mechanism of activation for synthesis. Coenzymes began to be made. Other mononucleoside triphosphates appeared. Finally, the first oligonucleotides were assembled. Emerging life had arrived at the threshold of the RNA world.

**Birth of the first thioesters.** There is a condition to the origin of the thioester world. Somewhere in the prebiotic world, thiols and carboxylic acids had to join spontaneously to form the combinations that were to fertilize life. At least two distinct mechanisms may be invoked to explain this highly endergonic reaction: simple dehydrating condensation, thermodynamically favored by a very hot and acidic medium; oxidative synthesis at the expense of thiols and carbonylic substances. Other reactions can be considered also. Unfortunately, we do not have the means to estimate the relative roles these various processes could have played to prime the thioester world. Their roles, in any case, were transient; the thioester world soon became self-sufficient, thanks to the iron cycle.

### *The RNA World*

**Molecular replication, competition, and selection.** With the appearance of the first oligonucleotides, replication by complementary base pairing entered the scene. Inevitable replication errors generated variant molecules, which were subjected to a molecular form of Darwinian competition that selected those that combined replicatability and stability in the most effective way. The winner in this competition has been identified by Manfred Eigen as the common ancestor of all transfer RNAs or, rather, as the "quasi-species" of molecular siblings that deserves this name.

This population included certain molecules capable of appropriating L-$\alpha$-aminoacyl groups by attacking the corresponding thioesters or adenylates by means of their 3′-terminal hydroxyl group. These molecules gained, in aminoacylated form, an additional selective value. Molecules that, in addition, possessed a site capable of specifically binding a given kind or family of amino acids were particularly favored by selection. In this way, a first small set of proto-tRNAs emerged through simple molecular selection. Reciprocally, the amino acids that were recognized by one or the other proto-tRNAs were themselves selected for any kind of metabolic utilization that depended on RNA binding (eventually including protein synthesis). This mutually selective marriage between proto-tRNAs and amino acids was a gradual process. It started with the most abundant amino acids and then extended progressively to others as they arose metabolically.

**The beginning of protein synthesis.** Thanks to interactions with other RNA species (proto-rRNAs and/or proto-mRNAs), helped perhaps by some multimeric catalysts, aminoacylated proto-tRNAs came to be immobilized side by side and to transfer aminoacyl or peptidyl groups from one to the other. The central process of protein synthesis was born, though, at this stage, in an essentially undirected form. Coding was still to come. Even this random assembly process was an appreciable asset because it offered an alternative to the thioester-dependent mechanism for making catalytic multimers. The new process was perhaps more efficient than the old one. It may have made better or more stable catalysts. Most important, because of its dependence on replicatable molecules, it had the crucial advantage of being perfectible by selection. But, for selection to work, the prebiotic system had to be subdivided into a large number of supramolecular units capable of competing with each other.

**Confinement.** Right from the start, hydrophobic and amphiphilic compounds were abundantly present among the products of multimerization. Later, with advancing metabolic sophistication, amphiphilic lipids were also formed. Conditions suitable for the spontaneous assembly of these molecules into vesicular structures were common. Under most circumstances, however, confinement was unfavorable to the survival of the imprisoned systems because of the many hindrances it opposed to the exchanges between the systems and their surroundings. As long as was possible, life developed in some sort of extended protocytosol, largely unstructured except for the eventual participation of catalytic or sheltering surfaces. At some stage, however, compartmentation became an indispensable condition of further progress. The latest limit for this was when progress came to depend on Darwinian selection among complex competing systems.

Lipid bilayers and hydrophobic polypeptides participated—in what order remains to be established—in the formation of the first membranes. Whatever its mechanism, this construction took place in progressive steps, so as to allow insertion of the minimum of transporters and channels needed for the exchanges with the outside before completion of a sealed envelope. After ion-tight membranes had developed, they served as fabric for the assembly of all sorts of systems—among them ATPases, electron-transport chains, and photosystems—sharing the ability to create a proton-motive force with the help of the energy they used up. But all these developments came later. At the time under consideration, the rudimentary envelopes that surrounded nascent living systems were still porous. They allowed small molecules to pass through freely, but they were impermeable to polynucleotides and to polypeptides, which property sufficed to allow Darwinian competition to take place among the enveloped systems.

**The emergence of translation.** The first benefit of competition was improvement of the RNA-dependent system of polypeptide assembly. This system was useful as an alternative to the thioester-dependent mechanism for the maintenance of the multimer population—still randomly assembled—in which were to be found the protoenzymes required by protometabolism. Any change—reproducible by replication—in the proto-tRNAs, proto-rRNAs, and proto-mRNAs involved in polypeptide assembly that increased the efficiency of the process carried a selective advantage. This is how information entered the system.

An important step in this information acquisition process was the appearance of a mechanism aligning aminoacylated proto-tRNAs along a proto-mRNA strand according to a triplet periodicity. This mechanism was selected because it proved particularly favorable to an efficient transfer of aminoacyl and peptidyl groups. At first, the interactions between proto-tRNAs and proto-mRNAs were haphazard, but selection soon favored molecules with complementary relationships of increasing regularity and specificity. RNA sets that avoided jamming caused by the presence of overlapping binding triplets along the proto-mRNA were particularly favored. In this way, a rudimentary, comma-less code emerged, initially as an asset to the efficiency of a peptide-assembly process that was still essentially random at this stage. Henceforth, however, some sort of primitive relationships progressively became established between the nucleotide sequences of proto-mRNAs and the amino acid sequences of the polypeptides assembled along these RNAs. Not yet full-fledged translation, these relationships sufficed to allow the minimum of feedback necessary for the quality of the polypeptides to affect the selective value of the corresponding proto-mRNAs. The way to authentic translation was opened. Progress in this direction was inevitable.

An alternative scenario has aminoacylated proto-tRNAs first joining on a polypeptide template to assemble a copy of the template and the first proto-mRNAs arising subsequently by reverse translation of the anticodons displayed in register by the template-immobilized aminoacylated proto-tRNAs. Contemporary organisms hold not a shred of evidence supporting such a mechanism, which contradicts the "Central Dogma" of molecular biology. But who can know what orthodoxy was like four billion years ago?

**The rescue of protometabolism.** The "heretical" hypothesis just cited has the advantage of offering a ready explanation for the metabolic transition from the thioester to the RNA world. In this scheme, the sequences of the original thioester-dependent protoenzymes were simply copied and later reverse translated into proto-mRNAs; the first protoenzymes were the authentic precursors of today's enzymes. Even in the framework of the more orthodox hypothesis, the metabolic information of the thioester world—if not necessarily its structural information—could be rescued by the network of existing protometabolites. This network formed a powerful screen that automatically selected those catalytic peptides that it provided with substrates. Such peptides occupied a functional niche in the protometabolic network and thus offered a selective advantage. In this way, a second multiprotoenzymatic system, functionally similar to the first one, was progressively installed. This second system had the enormous advantage of being encoded into replicatable messages and, thereby, of being perfectible by selection. New reactions may have arisen during this long developmental process. On the whole, however, the need for continuity limited considerably the possibilities of metabolic innovation. The new protometabolism, catalyzed by polypeptide protoenzymes synthesized under the control of RNA, was not very different from the first one catalyzed by multimers randomly assembled at the expense of thioesters. Indeed, the two synthetic mechanisms continued to operate side by side until all the required thioester-derived catalysts had been made redundant by the appearance of functionally equivalent catalysts made by the RNA machinery.

**RNA-dependent evolution.** Thanks to compartmentation into multiple competing units provided with an elementary translation system, incipient life had reached a stage where it existed as a population of autotrophic protocells equipped with rudimentary protoenzymes encoded by a primitive RNA genome. It had acquired some properties of cellular life as we know it today but was still a long distance away from it. The genes were very short—they were minigenes, 50 to 100 nucleotides long—and they were present in numerous copies of their two complementary forms. Replication and translation were very imprecise, with the consequence that each macromolecular population offered many variants for screening by natural selection.

This stage in the development of life may be taken to correspond to Woese's progenote (Chapter 3), the distant precursor of the universal ancestral cell.

Especially important for the further evolution of this progenote was the development of RNA-splicing mechanisms (perhaps RNA-catalyzed at first) of both *trans* and *cis* kinds, which allowed genetic experimentation to take place by the modular assembly of minigenes. Longer genes made for better catalysts, and these made for more efficient protocells, provided the precious new information did not succumb to replication and translation errors. Thus, at this stage, the commanding steps that both set the pace of evolution and limited it were marked by the appearance of more reliable replication and translation catalysts. These allowed for longer genes, which, in turn, allowed for further catalyst improvement. Genes and information-transfer catalysts thus reinforced each other until authentic enzymes, encoded by strings of hundreds of nucleotides, appeared. Protometabolism turned progressively into metabolism. Major innovations arose, including the assembly of membrane-associated energy-retrieval systems. At some stage in this evolution, transition from the RNA world to the DNA world took place.

## *The DNA World*

**From RNA to DNA.** Passage from an RNA genome to a DNA genome was relatively simple. All it required was a reductive system capable of removing the 2′-oxygen of ribose in nucleotides, a system for methylating uracil (perhaps not essential in the beginning), and a set of enzymes—most likely mutants of the primeval RNA replicase—for reverse transcribing RNA into DNA, replicating the DNA, and transcribing the DNA back into RNA. Transfer of the genetic information to DNA entailed several advantages. These may have included a greater chemical stability of the genetic material, a lesser propensity to participate in all sorts of molecular rearrangements, perhaps greater fidelity of replication, which, in turn, made it possible for genes to be linked progressively into strings of increasing length, until they all came to be accommodated in a single chromosome. Probably the most important advantage was the separation between replication and expression of the genetic information. This separation allowed the information to be stored in single copies, or in the minimum number of copies needed for adequate expression, and to be replicated in coordinate fashion at a specific time. The expression and amplification of the information were left to transcription, which could be controlled indepen-dently of replication and individually for each separate gene or group of genes. These advantages were, no doubt, of paramount importance. Evolution has taken good care to avoid a return to the RNA world. Except for RNA viruses, believed to be vestiges of the RNA world, RNA replicase and reverse transcriptase have largely disappeared from the biosphere.

**From protocells to cells.** In the course of the numerous evolutionary trials in which the first protocells were engaged, certain decisive events progressively narrowed the range of possibilities, finally letting through a single population of cells, which appeared between 3.8 and 3.5 billion years ago. The members of this population resembled present-day bacteria in their structures and main metabolic attributes. A Gram test would probably have revealed them as negative. They occupied hot and acidic sulfurous waters. They were anaerobic and almost certainly chemoautotrophic (or possibly phototrophic). All the forms of life that compose the biosphere today have arisen from the further evolution of this primeval cell population.

CHAPTER SIX

# BUILDING WITHOUT A BLUEPRINT

## THE FIRST BIOMOLECULES

Having, in the preceding chapter, defined the problem raised by the origin of life, we may now look at the blueprint and ask, first, whether the seven characteristics that I singled out in Chapter 1 as essential to life appeared together, or whether some may have—or perhaps even must have—preceded others. Clearly, incipient life needed *constituents* in the first place; which means *raw materials, energy,* and, perhaps, *catalysis*. Obviously also, the first steps in this construction process had to be strictly abiotic until some sort of organization began to emerge and the lifeward bound system had acquired a dynamic structure of its own.

### *Prebiotic Syntheses in the Laboratory*

Although organic chemistry has succeeded in making a large number of biological substances, the conditions needed for such processes were judged too artificial to be relevant to the spontaneous origin of these molecules on Earth. The first deliberate—and spectacularly successful—attempt at reproducing a prebiotic synthesis is generally credited to Miller. His experiments, mentioned in the preceding chapter, have yielded a variety of organic compounds under surprisingly simple conditions (see Table 6–1).

In the wake of Miller's success, a wide variety of experiments have been carried out under allegedly prebiotic conditions. They have yielded an impressive collection of biochemical compounds, including many amino acids, purines, pyrimidines, and sugars. The list has become so long that emphasis is now put on those substances that still await "prebiotic" generation. Among these, Miller mentions arginine, lysine, histidine, straight-chain fatty acids, porphyrins, pyridoxal, thiamine, riboflavin, folic acid, lipoic acid, and biotin (110, 111). A crucial factor in the interpretation of these experiments is the distinction between "prebiotic" (before life) and "abiotic" (without life). Classical organic chemistry may be termed abiotic (except that the chemist himself is very much alive), but this does not make every one of its achievements into a prebiotic event. Miller acknowledges this. "In some cases," he writes, "the conditions are so forced (e.g., the use of anhydrous solvents) or the concentrations so high (e.g., 10 millimolar formaldehyde) that such conditions could not have occurred extensively on the primitive earth" (111, page 23). There is also a lot of discussion

**Table 6–1** Prebiotic building blocks. Yields from sparking $CH_4 + NH_3 + H_2O + H_2$.

| Compound | % of $CH_4$ carbon |
|---|---|
| Formic acid | 4.0 |
| Glycine | 2.1 |
| Glycolic acid | 1.9 |
| Alanine | 1.7 |
| Lactic acid | 1.6 |
| $\beta$-Alanine | 0.76 |
| Propionic acid | 0.66 |
| Acetic acid | 0.51 |
| Iminodiacetic acid | 0.37 |
| $\alpha$-Aminobutyric acid | 0.34 |
| $\alpha$-Hydroxybutyric acid | 0.34 |
| Succinic acid | 0.27 |
| Others | 0.62 |

According to Miller (110, 111).

as to what the real conditions may have been. In particular, the Urey-Miller assumption of a highly reducing prebiotic atmosphere is now seriously questioned. According to the latest geochemical reconstructions, our planet, as it condensed, must have lost most of its hydrogen, leaving $CO_2$, rather than $CH_4$, as the main source of carbon (89). Under such conditions, the yield of amino acids by electric discharges would be drastically reduced.

## *Syntheses in Outer Space*

Recent advances in the exploration of space by means of spectroscopic techniques of various kinds have opened the alternative possibility that the primary biogenic building blocks were the products, not of atmospheric, but of interstellar chemistry. It has been found that outer space contains, in extremely dilute form but in huge amounts, a variety of potentially biogenic organic molecules (221). They are mostly "exotic" ions and radicals, made of a small number of atoms of carbon, hydrogen, nitrogen, oxygen, and, sometimes, sulfur or silicon. It is speculated that these molecules, if concentrated on the surface of dust grains, could well combine into more conventional substances, even polymers. As put by Joshua Lederberg, "any process that permits the large-scale production of oligomeric carbon could hardly avoid its polymerization." And he asks further: "Does the cosmic condensation already furnish enormous quantities of matter preadapted to the requirements of biological macromolecules? Will we find the raw material of DNA already present there?" (108, page vii).

The analysis of comets (371, 379, 384) by infrared spectroscopy and, more recently, by spacecraft-mediated mass spectrography, has confirmed Lederberg's main assertion. The message from both comet Halley and comet Wilson is that these celestial conglomerates of "dirty ice" contain organic polymers. The identification of these molecules is still uncertain. It ranges from a fairly well-documented polyoxymethylene to a highly problematic cellulose, even to frozen viruses and bacteria, as proposed by the enthusiastic defenders of panspermia, Hoyle and Wickramasinghe (37–39). DNA is yet to be identified.

More solid messages from outer space have been delivered by meteorites. These do indeed contain organic material, which includes truly biogenic building blocks. Most impressive is the Murchison meteorite, a carbonaceous chondrite that fell in 1969 in Murchison, Australia. This body has been found to contain a number of authentic amino acids, remarkably similar, in both quality and relative quantity, to the yields of electric discharge experiments (see Table 6–2).

The possibility of an extraterrestrial origin of biogenic building blocks is thus clearly substantiated, but there is at present no quantitative estimate of what this contribution might have been. On the other hand, there are strong reasons for believing that Miller's experiments are significant. Correlations of the kind shown in Table 6–2 seem particularly convincing in this respect. There remains, however, the problem of where the reducing equivalents needed for atmospheric syntheses could have come from if, as is now admitted by most authors, degassing of our cooling planet had left it virtually devoid of hydrogen.

**Table 6–2** Extraterrestrial amino acids. Relative abundances of amino acids in the Murchison meterorite and in the products of an electric discharge synthesis.

| Amino acid | Murchison meteorite | Electric discharge |
|---|---|---|
| Glycine | ++++ | ++++ |
| Alanine | ++++ | ++++ |
| α-Amino-*n*-butyric acid | +++ | ++++ |
| α-Aminoisobutyric acid | ++++ | ++ |
| Valine | +++ | ++ |
| Norvaline | +++ | +++ |
| Isovaline | ++ | ++ |
| Proline | +++ | + |
| Pipecolic acid | + | <+ |
| Aspartic acid | +++ | +++ |
| Glutamic acid | +++ | ++ |
| β-Alanine | ++ | ++ |
| β-Amino-*n*-butyric acid | + | + |
| β-Aminoisobutyric acid | + | + |
| γ-Aminobutyric acid | + | ++ |
| Sarcosine | ++ | +++ |
| *N*-Ethylglycine | ++ | +++ |
| *N*-Methylalanine | ++ | ++ |

From Miller (110, 111).

### *The Source of Abiotic Reducing Equivalents*

Recent experiments have provided a likely answer to this question, pointing to ferrous iron ($Fe^{2+}$) as the source of the required electrons and to UV light as the means whereby they were both released and brought to a high enough energy level to reduce hydrogen ions and other potential acceptors.[1] It has long been known that hydrogen is produced upon irradiation of aqueous acid solutions of ferrous salts with UV light of short wavelength (lower than 250 nanometers). It has now been found that this reaction also takes place at neutral or alkaline pH and, in fact, does so much more readily under these conditions. Even UV light of high wavelength (to about 400 nanometers) is effective, and the quantum yield is remarkable: at least 30% (see Figure 6–1).

The discovery of this process has thrown a revealing light on a major geochemical puzzle, namely the origin of the so-called banded-iron formations. These are layered, sedimentary deposits rich in iron (at least 15%), found in Precambrian strata in many parts of the world. Their age ranges between 1.5 billion years and the earliest available geological record (3.8 billion years). The iron in them is partly ferrous and partly ferric, with the proportion of ferric iron ranging between 30% and 60%. It is generally accepted that these deposits sedimented in ferruginous bodies of water as a result of the oxidation of ferrous ions and the consequent flocculation and precipitation of insoluble ferric or ferro-ferric oxides and other complexes. For want of another explanation, traces of oxygen were held responsible for the oxidative process; but this left unsolved the question of where the oxygen came from. With the discovery of the photochemical process, a strong oxidant is no longer needed to explain the phenomenon. Even hydrogen ions, nitrogen, or $CO_2$ could have served as electron acceptors, with UV light providing the required energy.

The discovery of the photochemical oxidation of ferrous iron at neutral pH not only explains the banded-iron mystery (373), it also provides a possible solution to the riddle of the primeval reducing equivalents (372). According to the oldest banded-iron deposits, the process was already going on 3.8 billion years ago, possibly earlier, certainly early enough to have supplied abiotic syntheses with reduced raw materials such as $H_2$, $CH_4$, $NH_3$, HCN, and others. It was plentifully supplied with energy. UV irradiation, especially with the longer wavelengths included, was a major source of energy in the prebiotic world, estimated to have been about 1,000 times more abundant than electric discharges (110, 111).

[1] The photooxidation of ferrous iron at neutral pH was discovered a few years ago by Paul Braterman, Graham Cairns-Smith, and Robert Sloper, who pointed out that it could be responsible for the generation of banded-iron formations (373). The reaction was studied further by Zofia Borowska and David Mauzerall, who have emphasized its possible importance in providing the reducing equivalents needed for prebiotic syntheses (372). Detailed information on banded-iron formations is found in Reference 89.

**6–1** *Photochemical generation of reducing equivalents.*
UV light of 400-nanometer wavelength (72 kilocalories per einstein) supports the transfer of electrons from $Fe^{2+}$ to protons with the generation of $Fe^{3+}$ and molecular hydrogen ($\Delta G°$ at pH 7.0 = +54.6 kilocalories per pair of electron-equivalents). Ferric ions coprecipitate with ferrous ions as an insoluble mixed oxide (magnetite), which sediments and gives rise to a banded-iron formation. (Energy data from Reference 154).

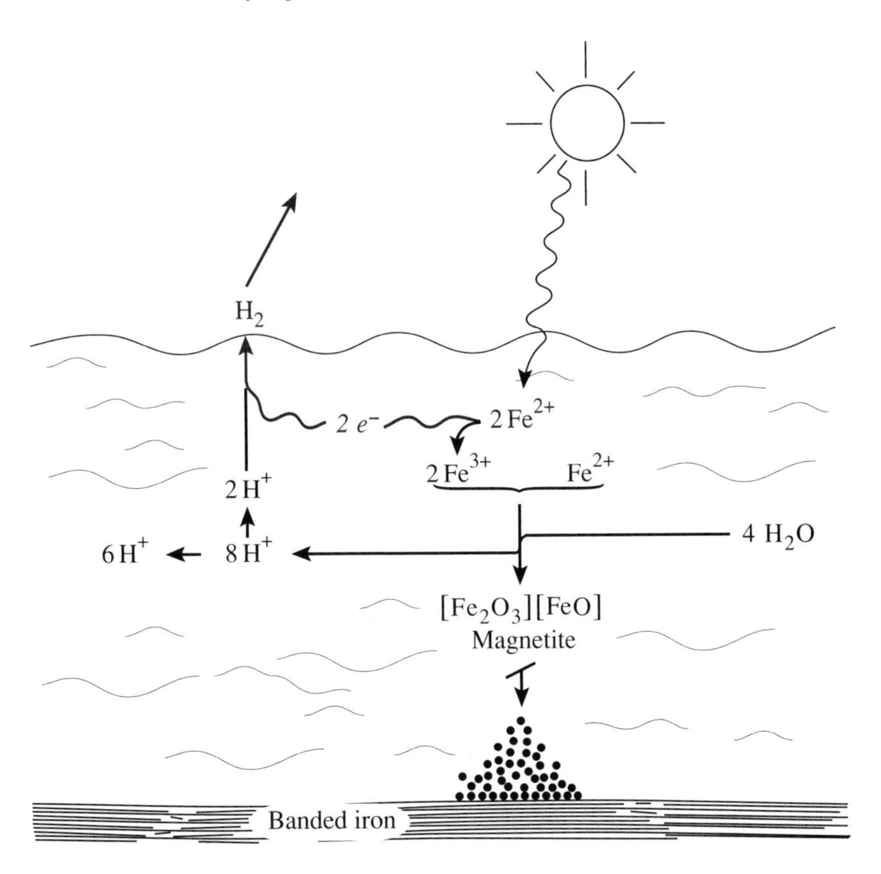


## *The Primeval Soup*

In consideration of all the available evidence, we may reasonably assume that a number of biogenic building blocks were produced abiotically in large quantities through random chemical processes, both in the Earth's atmosphere and in outer space, and were collected in unknown proportion from these two sources by the oceans and lakes of the primitive Earth. There, thanks to very favorable partition coefficients (306), they could have accumulated up to high concentrations, especially in shallow areas sheltered from the more destructive forms of radiation. The so-called prebiotic, or primeval, "soup" is believed to have formed in this manner, although estimates of the thickness of the soup are widely divergent, as we shall see in Chapter 9.

The nature of the soup's ingredients is also a matter of debate. Results such as those shown in Tables 6–1, page 124, and 6–2, page 125, make it very likely that carboxylic acids of various kinds, including amino and hydroxy acids and, perhaps, other small molecules, figured prominently among the ingredients. Much more questionable, however, is the possibility that the soup contained significant amounts of the macromolecules, especially RNA-like polynucleotides, considered by many to be prerequisites for the launching of a biogenic evolutionary process. In most instances, attempts at simulating the prebiotic formation of macromolecules have succeeded only under highly artificial conditions and, even then, have yielded complex mixtures in which the "right" compounds were present in minute amounts. Yet, it is virtually taken for granted by numerous authors today, and even stated as accepted fact in several textbooks, that primitive Earth conditions were such as to yield, not only the main building blocks of life, but also a variety of complex combinations of these building blocks, including replicatable, information-containing polynucleotides.

## THE LIMITS OF ABIOTIC CHEMISTRY

### *The RNA World*

The following quotation out of the latest edition of a popular textbook illustrates the current creed: "Thus, as the prebiotic cauldron cooled, the resulting soup of organic chemicals probably contained oligonucleotides, peptides, and oligosaccharides" (13, page 1,103). This conclusion follows a statement proposing the participation of some "water-hungry condensing agent like a polyphosphate" in the required assemblies. No indication is given as to how the appropriate building blocks might have been made, selected, and assembled in the correct fashion, nor as to where the water-hungry polyphosphate came from and how it acted. A detailed account of the origin of life based on the concept of a protein-less "RNA world" then follows, under the subhead: "The first 'living' molecule was almost certainly a nucleic acid."

The possibility that RNA may have preceded proteins in the development of life was first evoked more than 20 years ago, after the key functions that RNA molecules fulfill in the synthesis of proteins, as informants (mRNAs), adaptors (tRNAs), and catalytic assembly sites (rRNAs), had become clear. The knowledge gained made it, in the words of Crick, "not impossible to imagine that the primitive machinery had no protein at all and consisted entirely of RNA" (242, page 50). Another strong point in favor of RNA in what is often referred to as the "chicken-or-egg" problem—the "protein-or-nucleic-acid-first" controversy—is the fact that RNA can be replicated directly and, therefore, can store genetic information, undergo mutations, and be subject to Darwinian selection, whereas proteins can do so only by way of nucleic acids. Also pointing to the importance of RNA in the development of life is the dominant role of nucleotides and related molecules as metabolic cofactors, which has been particularly emphasized by Harold White III, who proposed the view that the nucleotide coenzymes are "fossils of an earlier metabolic state," vestigial remnants of RNA enzymes (237).

These ideas gained little success among biochemists, who were not ready to accept that RNA could have preceded proteins, obviously needed in the first place as enzymes. Then, in 1981, came the discovery by Cech that a ribosomal RNA from the mitochondria of the ciliate *Tetrahymena pyriformis* could catalyze its own splicing, in the complete absence of proteins (314). Other instances of RNA catalysis quickly followed, leading to the concept of "RNA as an enzyme," vigorously advocated by Cech (161, 162). In the eyes of many molecular biologists, these findings clinched the matter: in the beginning was a protein-less "RNA World," in which, according to Walter Gilbert, its foremost proponent, "RNA molecules and cofactors [were] a sufficient set of enzymes to carry out all the chemical reactions necessary for the first cellular structures" (98, page 903). As we have seen, this doctrine has gained so much credence as to become textbook material in a matter of only a few years.

There can be no doubt that the discovery of RNA catalysis was of major importance. As we saw in Chapter 3, it goes a long way in explaining the origin of RNA splicing. Yet, it must be pointed out that catalytic RNAs have so far been found to act only on RNA substrates, often the catalytic molecule itself, in which case the term "catalyst" in its accepted sense cannot even be said to apply. In all known cases, the catalyzed reaction, whether it leads to splicing, hydrolysis, or replication, is limited to a single kind: transesterification from a phosphodiester bond between nucleotides. Finally, reaction rates are very slow, of the order of one molecule of substrate modified per minute per molecule of catalyst. These facts hardly support the image of an RNA-catalyzed protometabolism conjured up by the term "RNA as an enzyme."

Even if RNA catalysts should one day turn out to be more versatile and active than those presently known, their abiotic origin would still pose a major problem. Witness the following statements made at a 1987 symposium grouping most experts in the field (80): "it is still hard to believe that a series of well-defined 3′,5′-linked oligo-β-D-ribonucleotides could have formed de novo" (112, page 14); "many sticky problems remained to be solved concerning the prebiotic synthesis of RNA" (106, page 31); "the general point remains that RNA is exquisitely difficult to make" (103, page 934). The fact is that no one has as yet succeeded in synthesizing an oligonucleotide from scratch under even remotely plausible prebiotic conditions—or under any conditions, for that matter. Even such dedicated proponents of prebiotic chemistry as Miller and Orgel (60) admit this. A recent joint paper from their groups states that "the accumulation of relatively pure mononucleotides on the primitive Earth is highly implausible" (380, page 4398). The workers propose instead the possibility that some "simpler"—though still enormously difficult to make—oligonucleotide analogues may have been the first genetic material. Gerald Joyce, one of the coauthors of this proposal, concludes a 1989 review with the admonition: "It is time to go beyond talking about an RNA world and begin to put the evolution of RNA in the context of the chemistry that came before it and the biology that followed" (222, page 223). And Orgel adds: "There is at present no convincing theory that can account for the origin of replicating RNAs" (133, page 223). Summing up these doubts, Mitchell Waldrop heads a recent commentary with the question: "Did life really start out in an RNA world?" (435a). His answer is another question, a noncommital "Who knows?"; mine, also interrogative but more straightforward: "Did God make RNA?" (428).

Yet, judging from the recent literature (241, 244, 245, 256, 429–432), the widespread faith in a primeval, protein-less RNA world remains unshakable and continues to be popularized (420, 435). Perhaps, this attitude is to be explained by the lack of a plausible alternative, short of special creation. "It must have happened that way," the underlying reasoning seems to go, "because it could not have happened otherwise."

## *Alternatives to the RNA World*

Not all authors are ready to accept the popular view of prebiotic chemical wizardry. Its most persuasive critic is Graham Cairns-Smith, who has written what amounts to an indictment of the theory (50). Especially as concerns the nucleic acids, he has detailed the host of unlikely operations the abiotic systems would have had to accomplish in order to achieve what even the most contrived laboratory attempts have failed to reproduce. For Cairns-Smith, the inescapable conclusion could be written: "It did not happen that way because it could not have happened that way." Other authors have reached the same conclusion, sometimes pushing the scientific argument to the point of seriously considering the alternative of special creation. *The Mystery of Life's Origin,* by Charles Thaxton, Walter Bradley, and Roger Olsen, a chemist, an engineer, and a geochemist, is a case in point (65).

Cairns-Smith does not go so far. Searching after another way for life to have started, he has proposed that the main constituents of present-day life were produced by mineral organisms, probably made of clays, capable of storing information in the form of crystal defects, of propagating the information by replication and division, of undergoing and propagating "mutations," and, therefore, of being subject to natural selection. Evolution of these mineral bionts would have been in the direction of increasing organic synthetic sophistication, until they succumbed to what the author calls "genetic takeover" by the products of their own industry. In support of his scripturally appealing hypothesis, he offers many facts attesting to the versatility of clays and examines in considerable chemical and physical detail the possible evolutionary history of our putative mineral ancestors.

This theory has the merit of drawing attention to the potentialities of clays as catalysts, adsorbents, shelters, concentrators, separators, and other physical adjuvants of prebiotic syntheses, in line with a proposal made earlier by Bernal, one of the first scientists to become interested in the origin of life (Chapter 5). A weakness of the theory, apart from its lack of factual support, is that it does not convincingly link phenotype to genotype in a way likely to favor the selection of clays more proficient at organic chemistry. Granting clays the status of "selfish genes,"[2] I fail to see what reproductive advantage a clay mutant would gain from having acquired the capacity to catalyze oxidative phosphorylation, for example, or to synthesize some polysaccharide.

[2] *The Selfish Gene* is the title of a book by Richard Dawkins (67) in which the old saying "A chicken is an egg's way of making another egg" is modernized into the statement that "A phenotype is a genotype's way of making another genotype." That is, the gene is the fundamental unit of natural selection.

Another doubter of the nucleic-acid-first theory is Robert Shapiro, who has cast a skeptic's eye over the various models that compete to account for the origin of life (64). Like Cairns-Smith, he has underlined the many improbable steps needed for prebiotic nucleic acid synthesis (389). Even D-ribose alone, he contends, could not have arisen spontaneously in significant amounts on the prebiotic Earth (390). Opting for proteins as prerequisites of the metabolic ingenuity needed to produce nucleic acids, while unwilling to abandon a Darwinian view of chemical evolution, Shapiro resurrects protein copying to account for the progressive development of the necessary machinery by natural selection. This hypothesis was much in favor before the mechanism of protein synthesis was elucidated. Felix Haurowitz, for example, proposed that like-for-like affinities, of the sort at work in crystallization, could explain the alignment of amino acids along a protein template before assembly (44). Shapiro rejects such a mechanism. He postulates instead some sort of adaptor molecules that would align activated amino acids on their assembly line, as do tRNAs on ribosomes, but that would recognize amino acid residues in a polypeptide template rather than nucleotide triplets in an mRNA template. The possibility of protein copying, he adds, whatever its mechanism, is suggested by the existence of prions, infective agents that cause such diseases as scrapie in sheep and, in man, Creutzfeldt-Jakob disease and, perhaps, kuru (198). Prions are made of protein and, unlike viruses, seem to contain no nucleic acids or, at least, too little for the coding of their protein component. Later evidence, however, has shown that the prion proteins are encoded by the host genome and reproduced in orthodox fashion (316, 341, 346). Nevertheless, as I shall show in Chapter 8, Shapiro's proposal deserves consideration.

An even more radical hypothesis has been put forward by Günter Wächtershäuser. This investigator, who refers to the authority of the philosopher of science Karl Popper, goes so far as to deny any contribution of Miller-type abiotic syntheses—even of the most simple kind—to the origin of life. He categorically rejects the "prebiotic broth theory" as having "received devastating criticism, for being logically paradoxical, incompatible with thermodynamics, chemically and geochemically implausible, discontinuous with biology and biochemistry, and experimentally refuted" (274, page 453). He proposes a bidimensional organism, made of a monomolecular layer of anionic organic compounds electrostatically bonded to a positively charged mineral surface (of pyrite, for example). The author calls up ingenious concepts to account for the genesis, the development, and the evolution of this system, including its transformation into a tridimensional entity. He does not explain, however, in a way that I consider satisfactory, the origin of the first constituents of the "organism," the nature of the catalytic actions that determined its "metabolism," nor, especially, the source of the energy flux indispensable to its dynamic maintenance and progress. We shall see in Chapter 9 that his condemnation without appeal of the "prebiotic soup" is not justified in as peremptory a manner as the author would have it. Certain particular aspects of his model remain interesting, nevertheless, and will be mentioned later.

Freeman Dyson (54) has found an ingenious way out of the information quandary by applying to molecular populations a mathematical apparatus first developed by Motoo Kimura (69) to explain the evolution of biological populations by random drift, without natural selection. Dyson's "toy model" comprises a fenced-in population of catalytically active polymers (e.g., peptides), adequately supplied with monomeric building blocks and energy and busily reshuffling each other's structures by substituting one monomer for another. By introducing a few additional assumptions, he estimates the probability of such a system's jumping from one stable configuration (a basin in the multidimensional space of molecular populations) to another stable configuration of a higher degree of organization. For the jump to have a chance of occurring within acceptable limits of time and space, Dyson finds that the number of distinct monomer species must be at least 9 (possible for amino acids, but not for nucleotides), whereas the discriminating factor of the catalysts need be only rudimentary (hardly higher than that of simple inorganic catalysts). Under such conditions, the model will include a few hundred molecules of polymers made up of 5 to 10 monomeric units each (i.e., a few hundred penta- to decapeptides if the monomers are amino acids). Such a system is theoretically capable of further improvement, conditional on a parallel rise in the number of distinct monomer species available and in the catalytic discrimination of the polymers.

The model, as Dyson points out, leads to plausible predictions. It is generally believed that the initial number of amino acids used by prebiotic systems was fewer than the 20 used by contemporary organisms, and Dyson's predicted value of 9 is close to 8, the number of amino acids that are readily derived from intermediates of the core carbon metabolism (50). Furthermore, populations of penta- to decapeptides could conceivably form spontaneously under abiotic conditions, and the fact that it is enough for these molecules to have rudimentary catalytic activities is reassuring. However, the nature of these catalytic activities raises a problem. The present-day world holds nothing remotely comparable to enzymes that spend their time remodeling each other's primary structures. In addition, the postulated system needs a boundary and a very cooperative environment in order to function.

A common weakness of origin-of-life theories based on natural selection is their requirement for almost faultless copying. This, for example, is a cornerstone of the theory proposed by Manfred Eigen (Chapter 8). There must be enough error to give natural selection something to work on but not enough to trigger a catastrophe. Between the Charybdis of slavishness and the Scylla of laxity, the strait is very narrow: too narrow, some believe, for a primitive prebiotic craft to have negotiated successfully. Dyson's model, on the other hand, tolerates a high error rate, of the order of 20%, but, as pointed out, seems unrealistic in other respects.

### *Something Is Missing*

All in all, the impression one gains from the abundant literature devoted to the origin of life is that there exists an uncomfortably wide gap between, on one side, what random prebiotic chemistry may reasonably be expected to deliver without magic or miracle and, on the other, the minimum set of self-replicating, information-conserving molecules required to initiate the sort of selective evolutionary process envisioned by most investigators. Shapiro sums up the situation by pointing to a "missing fragment in our picture of the origin of life . . . a principle that governs the gradual evolution of simple chemical systems into more sophisticated ones capable of replication and Darwinian natural selection" (64, page 205). In my opinion, there is nothing mysterious about this principle. It belongs to the basic blueprint: it is an intrinsic part of the metabolic organization of all living beings.

## PROTOMETABOLISM

### *The Indispensable Catalysts*

What random chemistry cannot do by itself, only some sort of channeled metabolism can accomplish. This necessarily implies catalysis. Most speculations on the origin of life call for catalysts. For example, the possible involvement of catalytic clays is often mentioned and has become one of the foundations of Cairns-Smith's theory of mineral genes (50). A catalytic role has also been attributed to some of the polymers that could have arisen from the random condensation of amino acids. This theory has long been defended by Sidney Fox (57, 58), who has indeed reported that "proteinoids," generated by heating dry mixtures of amino acids under certain specified conditions, display a number of rudimentary catalytic activities, reminiscent of enzymes. As mentioned, much emphasis is also now put on the catalytic properties of RNA molecules, but without an explanation as to how these could have arisen in the first place.

Proposals of this sort seldom state with any precision the nature of the required catalytic activities. This, however, is an essential point. In present-day living organisms, the construction of an RNA molecule from the kinds of building blocks that abiotic mechanisms may reasonably be expected to supply requires a minimum of 30 to 40 enzymes, several coenzymes, which themselves need a number of enzymes for their synthesis, and an ATP-generating system. Altogether, close to one hundred or more enzymes may well be necessary. Other metabolic pathways likewise proceed by a number of discrete steps catalyzed by distinct enzymes and guided by their specificities. This organization is part of the basic blueprint of life, and one must therefore ask when and how it originated. Does it go back to prebiotic times? Or was it developed later and preceded by cruder mechanisms that took fewer steps but did so at the expense of yield and specificity? This question is rarely discussed explicitly, but it is my impression that the general belief favors the second alternative. For the prebiotic synthesis of purines, for example, most authors attach considerable significance to the fact that adenine is obtained in up to 0.5% yield when concentrated solutions of ammonium cyanide are refluxed for a few days (385, 386). Mention is rarely made of a possible pathway that would start from glycine, formate, $CO_2$, and ammonia (as provided by aspartate and glutamine), which are the actual biological precursors of the purine ring and which eminently qualify as prebiotic building blocks.

To me, the latter possibility seems more credible than that of a natural phenomenon somehow imitating the refluxing of concentrated ammonium cyanide. This, I admit, is a matter of personal bias. Miller and Orgel, for example, strongly defend the opposite view. Referring to the possibility that "metabolic pathways parallel the corresponding prebiotic syntheses that occurred on the primitive earth," they wrote: "It is not difficult to show that this hypothesis cannot be correct in the majority of cases. Perhaps the strongest evidence comes from a direct comparison of known contemporary biosynthetic pathways with reasonable prebiotic pathways—in general, they do not correspond at all" (60, page 185). Much hinges on what is meant by "reasonable." To Miller and Orgel, the term obviously refers to the kind of chemistry that can be done with some of the simpler tools of organic chemistry, whereas I would prefer to apply it to what can be done with some of the simpler tools of biological chemistry. My suggestion, accordingly, is that the prebiotic environment contained a variety of catalysts displaying, be it in crude form, activities and specificities similar to those of the main present-day enzymes.

## *Protometabolic Pathways*

As I hope to show later, the presence of a diverse collection of catalytic activities in the prebiotic milieu is not as implausible as it may first appear. Once this possibility is accepted, prebiotic chemistry loses much of its mystery. Only simple substances, of the kind that are readily produced in Miller-type experiments, need be made by random abiotic mechanisms. After that, the catalysts take over. One or more of the primary products, upon encountering appropriate catalysts, are converted into new molecules that were not initially present. These, in turn, if they meet suitable catalysts, generate other molecules, which themselves give rise to others, and so on. A network of "protometabolic" pathways thereby progressively extends, signposted, as it were, by the string of "protometabolites" that appeared one after the other and by the catalysts that produced them (see Figure 6–2).

A major advantage of such a script over the random phenomena posited by most other theories is that it selects a coherent set of interrelated intermediates and makes them emerge way above the chemical "noise" created by the primitive cauldron, up to concentrations equivalent, within the limits set by chemical equilibria, to those of the most abundant abiotic products. To appreciate the full value of this advantage, one must remember that the concentration problem plagues every attempt at reconstructing the abiotic synthesis of complex biochemical compounds. Consider, for instance, the synthesis of RNA. Even the most optimistic models have to postulate multiple steps, each afflicted by low yields and many side reactions. By the end of the sequence, only extremely minute amounts (if any at all) of correct polynucleotides can realistically be expected to be present, swamped by hosts of molecular "misfits" from which, according to prevailing theories, they would somehow be rescued by some rudimentary process of selective replication. This possibility strains credibility. Indeed, a variety of mechanisms, including evaporation, adsorption, compartmentation, even freezing, have been invoked, in essentially ad hoc fashion, as possible solutions to the concentration problem. With catalysts to guide the reactions, this problem largely vanishes.

Another advantage of the proposed model is that the selected intermediates could well have been the kind found today in living organisms. Emerging life could have invented "real" metabolism right from the start, without first resorting to strange

**6–2** *The unfolding of protometabolism.* Only substances A, B, C, D are assumed to be products of truly random abiotic syntheses, for example, substances such as are listed in Tables 6–1, page 124, and 6–2, page 125. The other protometabolites (M, N, O, P, Q, R, S, T, U, W) are taken to arise through the intervention of catalysts (*Cat*), as in biological metabolism.

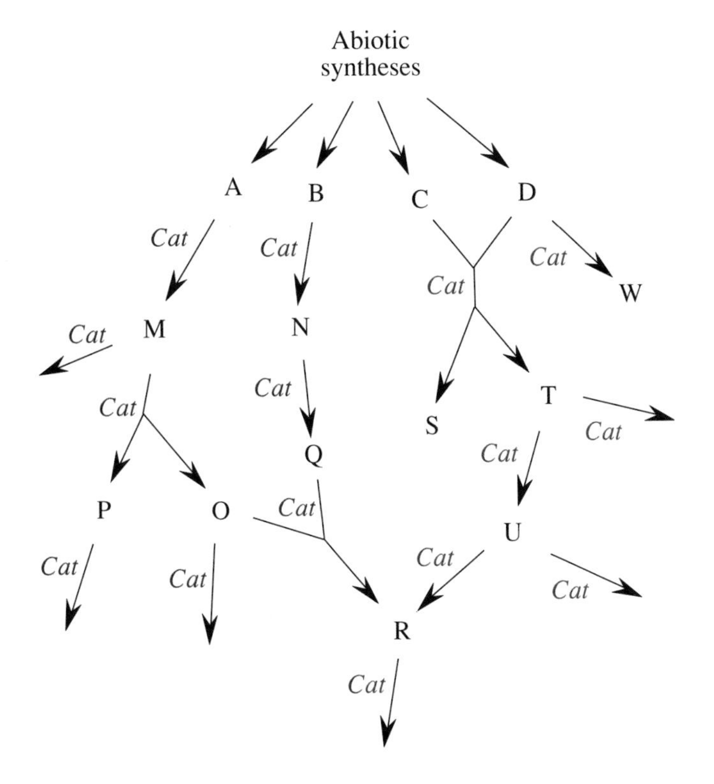


intermediates or contrived reaction conditions. Indeed, the substrates that were probably available in the largest quantity were such as to favor such a process. That this is so for the synthesis of the purine ring has already been mentioned. It is also true of the pyrimidine ring, which is derived biologically from aspartate, $CO_2$, and ammonia, and for citric acid-cycle intermediates, from which, in turn, a number of major metabolic pathways radiate.

The notion that metabolism was discovered progressively is an old one in biochemistry. Marcel Florkin, one of the pioneers of comparative biochemistry and of molecular evolution, called it "biochemical orthogenesis" (55). Sam Granick, discussing the biosynthesis of porphyrins, summed it up in a beautifully terse statement, which recalls Haeckel's aphorism (Chapter 3, Footnote 8, page 77): "Biosynthesis recapitulates biogenesis" (145). Robert Eakin took the notion one step further with his concept of autocatalytic cycles in which intermediary metabolites acted as primitive cofactors, eventually to be replaced by present-day coenzymes (248). This idea has been elaborated in great detail by G. A. M. King (289). As emphasized by Hyman Hartman (257), Lawrence Dillon (53), and Cairns-Smith (50), among others, the structure of metabolism supports the view that metabolic pathways unfolded, in "onion" fashion, from a core made up of the more common products of simulated abiotic syntheses. However, the possibility that the unfolding was *guided* by catalysts (105, 247) has not, to my knowledge, been considered before, except in a brief letter to the journal *Nature* by Narendra Mehta, who proposes a model very similar to mine. He envisages a "miniprotein" somehow assembling spontaneously around a prebiotically formed stable molecule and writes further: "If the miniprotein had catalytic activity for the stable molecule, this 'substrate' would be converted to a new molecule, setting in motion the formation of a pathway" (434). I have not seen this proposal elaborated further.

Granting that protometabolism developed in the way envisaged, it remains to be seen to what extent the physical conditions of the prebiotic environment—its temperature or its acidity, for example—would have made protometabolism different from metabolism. Energy is another problem. Without a source of usable energy, the network would have been restricted to simple reversible reactions and its development soon blocked by energy barriers. This condition will be addressed in the following chapter. But first, let us look at the catalysts.

## PROTOENZYMES

### *The Catalysts of Protometabolism*

The first question that arises concerns the activities and specificities of the alleged prebiotic catalysts. Obviously, if protometabolism prefigured metabolism and accomplished feats as elaborate as coupled phosphorylation and the assembly and replication of RNA-like molecules, a diverse set of catalysts, foreshadowing all major present-day enzymes, would have been needed. But they could have been less efficient than enzymes, and their specificities could have been broader. The international classification of enzymes lists only six categories, which may even be reduced to four: oxido-reductases, or electron transferases; group transferases; hydrolases, which are really group transferases that use water as an acceptor; lyases, catalyzing additions to double bonds; isomerases; and ligases, which catalyze assemblies coupled to ATP cleavage (by sequential group transfer). It is likely that the prebiotic world could have functioned with fewer and less specific catalysts than do living organisms today. But there is a limit to this kind of economy, as different substrates often require different catalytic mechanisms, even though the reactions the substrates undergo fall into the same formal category. Also, as we shall see, substrate specificity could have played a significant role in the selection and evolution of early catalysts.

It is conceivable that the postulated catalysts were of mineral nature, but this does not seem very likely. Even with due regard to the versatility of clays and other minerals, one can hardly expect such materials to provide the variety of catalytic activities and specificities the model demands, although, to be sure, metal ions may have been involved as cofactors in a number of reactions, as they are today. As a more plausible hypothesis, I suggest that the prebiotic catalysts were organic multimers[3] formed by the random assembly of small abiotic building blocks. Leaving aside for the time being the mechanism of such a process, which will be considered in Chapter 7, let us look first at its possible products.

It is striking that all the more abundant substances obtained in abiotic simulation experiments (see Tables 6–1, page 124, and 6–2, page 125) are carboxylic acids, many of them difunctional ones, to wit, amino acids, hydroxy acids, and dicarboxylic acids. If prebiotic building blocks were in any way similar to these substances, they most likely combined by acylation. Such a process, if essentially random, would have given rise to a motley collection of linear and cyclic molecules of various sizes, in which available building blocks would be linked together by amide, ester, and, perhaps, other bonds. A few authentic, chirally homogeneous (either L or D) oligopeptides would be included, but, for the most part, less regular assemblies would be expected on statistical grounds. Many such molecules exist today, especially in the bacterial world, but also in eukaryotes. Examples are glutathione, amino acid conjugates, the lactyl-peptide of the prokaryote cell-wall murein, a number of bacterial peptides, such as the cyclic antibiotic gramicidin S, and the ubiquitous pantetheine (see Figure 6–3).

If we assume that a random multimerization of difunctional acids could have taken place as I envisage, the question then arises whether the various catalysts required by my model could have been present in the mixture of multimers thus produced. If present, would they have reached an efficient concentration? In order to address these problems, we must first find out what is the minimum level of complexity needed for a multimer—an oligopeptide, for example—to display a sufficient catalytic activity to serve as a protoenzyme. We shall then examine the probability of finding molecular assemblages possessing the required properties among the products of multimerization.

## *The Conditions of Catalytic Activity*

Among the favorite probabilistic calculations of origin-of-life specialists is the demonstration that a full-blown enzyme—even a small protein the size of cytochrome *c*—is no more likely to come together by random assembly of amino acids than is a monkey to rattle off "To be or not to be" on a typewriter. The enzyme is, in fact, immeasurably less likely to come together in this way than is the monkey to emulate Shakespeare.[4] There simply was not enough time, space, and matter on the

[3] I have adopted the term "multimer" in spite of its unfortunate hybrid structure—it combines the Latin "multus" (many) with the Greek "meros" (part)—because the more orthodox "polymer" conveys the image of a "giant molecule formed from smaller molecules of the same kind" (Webster). On the other hand, "oligomers" (from "oligos," few) somehow seem too short for my purpose.

[4] The probability of human cytochrome *c* (104 amino acids) coming together by chance, assuming all possible combinations of 104 amino acids to have the same probability, is $20^{-104}$, or about $10^{-135}$. That of a monkey rattling off "To be or not to be" on a typewriter with 27 keys is of the order of $10^{-26}$. Even if the typewriter has 100 keys, the probability is $10^{-36}$. To equal the feat of getting cytochrome *c* by chance, our monkey would have to continue almost to the end of the second verse of Hamlet's monologue.

**6–3** *Some examples of natural heterogeneous multimers.*

Murein peptide

D-Lactyl → L-Ala → γ-D-Glu → L-Lys → D-Ala

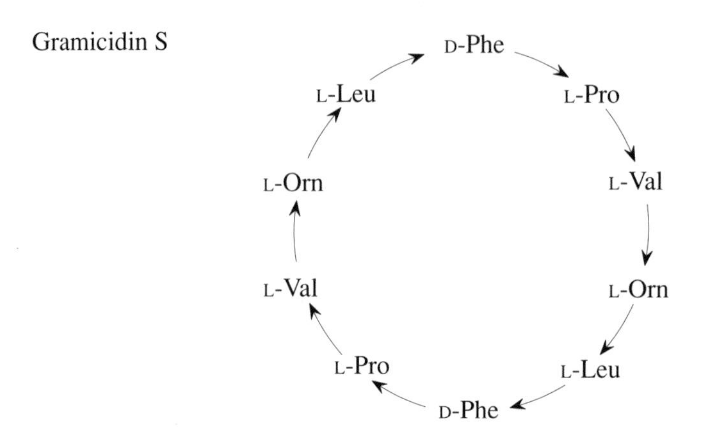


Pantetheine

$$HO\text{-}CH_2\text{-}\underset{CH_3}{\overset{CH_3}{C}}\text{ - }\overset{OH}{CH}\text{ - }\overset{O}{\overset{\|}{C}}\text{-}NH\text{-}CH_2\text{-}CH_2\text{-}\overset{O}{\overset{\|}{C}}\text{-}NH\text{-}CH_2\text{-}CH_2\text{-}SH$$

prebiotic Earth, even in the entire universe, to try more than an infinitesimal fraction of the possible amino acid combinations. Proof of this can be found in the values of Table 6–3 and in Figure 6–4. Yet, enzymes exist. To explain this, one could imagine

**Table 6–3** The protein lottery. Listed are the numbers of distinct polypeptides of different lengths that can be made with a set of 4 or 8 amino acids, and with the standard set of 20 (see also Figure 6–4).

| | Number of different polypeptides | | |
|---|---|---|---|
| Number of residues | 4 amino acids | 8 amino acids | 20 amino acids |
| 5 | $1.0 \times 10^{3}$ | $3.3 \times 10^{4}$ | $3.2 \times 10^{6}$ |
| 8 | $6.6 \times 10^{4}$ | $1.7 \times 10^{7}$ | $2.6 \times 10^{10}$ |
| 10 | $1.0 \times 10^{6}$ | $1.1 \times 10^{9}$ | $1.0 \times 10^{13}$ |
| 20 | $1.1 \times 10^{12}$ | $1.2 \times 10^{18}$ | $1.0 \times 10^{26}$ |
| 50 | $1.3 \times 10^{30}$ | $1.4 \times 10^{45}$ | $1.1 \times 10^{65}$ |
| 100 | $1.6 \times 10^{60}$ | $2.0 \times 10^{90}$ | $1.0 \times 10^{130}$ |

**6–4** *Relationship between chain length and average concentration of individual multimers in a 10% solution containing all possible multimers in equal amounts.*
The curves represent the following formula:
Molar concentration of individual multimers = $\frac{C}{n\,M\,N}$ moles per liter in which:
$m$ = number of distinct monomers
$M$ = molecular weight of monomers
$n$ = number of residues in multimers
$C$ = total concentration of multimers, in grams per liter
$N$ = number of distinct multimers = $m^n$, (see Table 6–3)
Assumed for the calculation: $M$ = 110
$C$ = 100 grams per liter

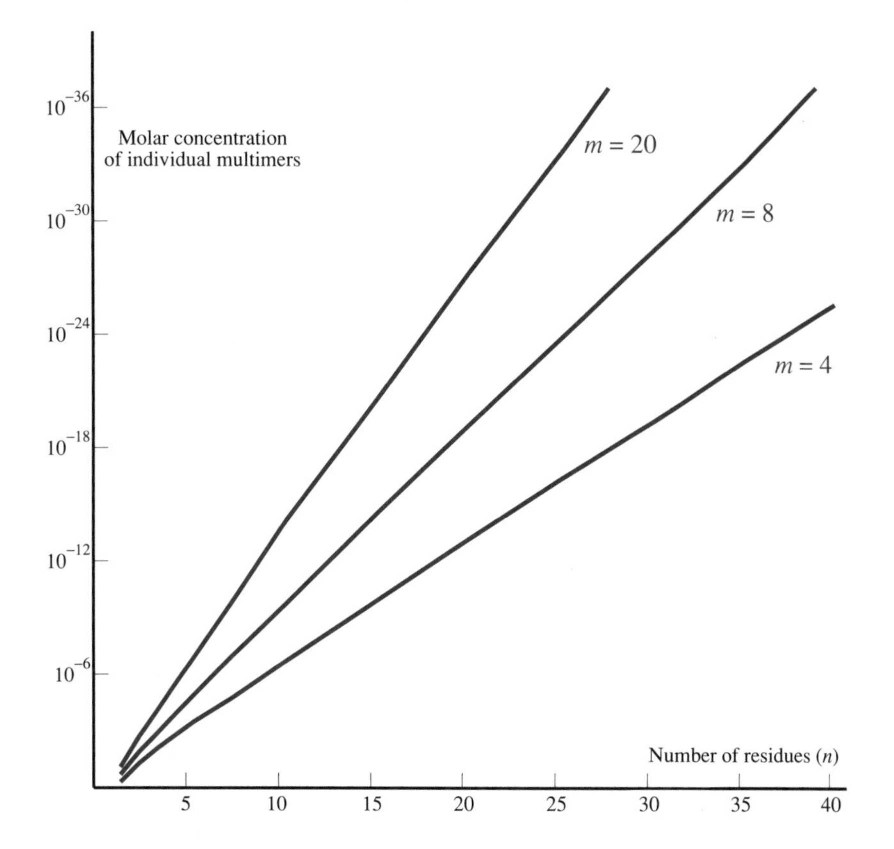


that the number of amino acid combinations whereby a given activity can be realized is so astronomically high that enzymes could, in fact, arise full blown after only a limited number of random trials. Enzymes do indeed tolerate numerous amino acid substitutions without loss of activity (141), but not enough, by far, to allow this hypothesis to be entertained seriously.[5] A more likely explanation is that enzymes had

[5] Two recent investigations, one theoretical (263), the other experimental (326), suggest that this view may need to be reconsidered. One concludes that "there is a significantly nonzero probability of the origin of an enzyme from a random sequence of amino acids" (263). According to the other, "the notion that enzymes are the extremely rare products of an evolutionary process that cannot be simulated is unnecessarily defeatist" (326).

humble beginnings and evolved progressively from short precursors with rudimentary catalytic activities, which could themselves have arisen by a random process. This theory has been ably defended by Jacques Ninio (73).

The view that enzymes developed from small-size precursors does not rest simply on theoretical arguments. It is supported by the existence of split genes. As pointed out in Chapter 3, there is considerable evidence that exons are not random cuts through polypeptide molecules but often represent distinct structural or functional domains, "miniproteins," as they have been called. A theory of modular construction of proteins by exon shuffling has been developed on the strength of these findings. Gilbert, the main advocate of this theory, has pointed out that a process of this sort would solve the "numerology" problem of evolution (98). In terms of the monkey analogy, our simian Shakespeare would be given words to put together, rather than letters, and would have a chance of succeeding in a reasonable time if his choice of words were not too great.

Exons, with an average length of around 40–50 amino acid residues (98), are probably not the smallest modules that served in the construction of proteins. This role could have belonged instead to units of the order of 20 residues, which, according to Gō (99, 322), represent entities sufficiently compact to have had an autonomous existence in remote times (Chapter 3). That even smaller elements may have been used originally is not excluded. For example, the reconstructed evolutionary history of a bacterial ferredoxin indicates that this small protein started as a simple tetrapeptide, 7 times repeated in the ancestral polypeptide (see Figure 7–4, page 165). We shall see in Chapter 8 that considerations based on the maximal permissible length of the first genes independently imposes an upper limit of about 20 residues on the length of the peptides issued from the translation of the genes.

We may conclude from these various theoretical and experimental data that proteins must have arisen in the form of small polypeptides, no more than 20 residues long, perhaps shorter. To be retained, and then further improved, by natural selection, these molecules must have had a function. This function may have been structural for some; but for many others, it was probably catalytic. There is widespread agreement on these points, though not on the origin of the ancestral polypeptides themselves. According to the partisans of the RNA world, these polypeptides arose from the translation of RNA minigenes. For my part, since I am unwilling to accept that such minigenes and their translation apparatus could have come into existence without the help of the peptides they are supposed to have produced, I propose for the latter an origin by random assembly. Before adopting this hypothesis, we must examine its implications more closely. Is it realistic and quantitatively plausible? Or doesn't it rather amount to invoking a miracle as incredible as that of a spontaneous generation of RNA?

## *The Likelihood of Catalytic Multimers*

The values of Table 6–3, page 138, and the curves of Figure 6–4, page 139, provide a framework for our discussion. It has been assumed, for the sake of simplicity, that only L-$\alpha$-amino acids participate in the multimerization process and that only linear peptides are formed. What must be confronted with the values shown is the mean chain length required for all the needed catalytic activities to be present. Should pentapeptides suffice, there is no serious problem, as the following estimate shows. Let us assume that eight different amino acid species were available for peptide synthesis—no doubt a maximum in an early prebiotic system. Let us assume next that all possible pentapeptides were made and that they were all present in solution at the same steady-state concentration—manifestly gross oversimplifications, but they should suffice provisionally. Then, the concentration of each individual pentapeptide would represent 1/33,000 of the total polypeptide concentration, that is, of the order of 6 micromolar for a total concentration of 100 grams per liter. At such a concentration, even a rudimentary catalyst should exert some activity. Admittedly, the value of 100 grams per liter for the total multimer concentration represents a generous estimate. But it gives a ballpark. We are not in the realm of total implausibility. On the other hand, if the required average chain length is 20 residues, the concentration of each individual peptide falls, under the same hypothetical conditions, to $4 \times 10^{-14}$ micromolar, or only some 24 molecules per cubic centimeter, which seems hardly compatible with the manifestation of even a crude catalytic activity.

Faced with such figures, we could suppose first that the earliest catalytic multimers were no more than five residues long, on average. Indeed, certain present-day peptides (see Figure 6–3, page136), as well as the ancestor of bacterial ferredoxin, mentioned earlier, have such a size. We could suppose further that longer chains arose mostly by modular assembly, of which traces are found in bacterial ferredoxin (see Figure 7–4, page 165) and in gramicidin S (see Figure 6–3). Finally, there is the possibility of cyclization, also illustrated by gramicidin S. In the event of cyclic molecules, the concentration estimates of Figure 6–4, page 139, would have to be raised by almost one order of magnitude (the same cyclic decamer can be made from up to 10 distinct linear decamers). We shall see in the following chapter that the thioester-dependent mechanism I propose for multimerization, indeed, readily allows for iterative chain elongation, for the combination of small modules, and for chain cyclization. Molecular closure by cyclization could even have been an important mechanism of chain-elongation arrest.

These considerations show that the random-assembly hypothesis is plausible, even though its degree of probability could be a matter for discussion. But an additional factor that could be particularly important remains to be taken into account. The conditions that I have posited for estimating the multimer concentrations are totally unrealistic.

First, it must be emphasized that "random" multimerization in no way means that the composition of the resulting mixture would be a simple statistical function of the relative abundances of the different building blocks. It is clear that significant differences in bond energies and, especially, in activation energies are likely to exist between different associations, thereby affecting the probability of their formation. Indeed, as shown by Fox, when mixtures of amino acids are polymerized by heating, the composition of the products shows striking and reproducible deviations from simple statistical randomness (57, 58). The same holds true, according to Dean Kenyon and Gary Steinman, when polymerization is induced by dicyanamide (59).

Fox attaches great significance to this fact, which he refers to by the term "molecular selection." The involvement of catalysts in multimerization could bias the process even further. Consequently, assembly will definitely not yield identical amounts of each possible multimer. Most likely, only a small fraction of the statistically possible combinations will actually be made in significant amounts.

Second, with the expected kind of building blocks (see Tables 6–1, page 124, and 6–2, page 125), a large number of the multimers that are made should be very hydrophobic and form insoluble aggregates. Some of these insoluble multimers could have catalytic activities or exert some other biogenic role, as will be seen in Chapter 9 (see also Reference 240). In any event, their precipitation should give the more soluble multimers, among which the protoenzymes required by the model are more likely to be found, the opportunity to reach higher concentrations.

Finally—and this is probably the most important factor—a certain proportion, perhaps a majority, of the synthesized soluble molecules will never reach an appreciable concentration because they are degraded almost as fast as they are made. A property most likely to enhance the resistance of multimers to breakdown would be the ability of the molecules to fold into some sort of stable conformation. This ability depends on the nature and sequence of the monomeric building blocks and increases with chain length. It is possible that the rare, chirally homogeneous peptides predicted statistically could have been highly favored in this respect because of their ability to adopt particularly stable conformations. A cyclic structure is also likely to contribute to stability. Another stability factor could be the joining of two or more multimers into plurimolecular complexes. Therefore, only a small subset of molecules of sufficient length and appropriate sequence, some perhaps of cyclic structure or endowed with associative properties, enter into account as candidates for long-term survival. Even this small lot must be further reduced inasmuch as the synthesis of such molecules would require numerous steps, each beset by many hazards due to the instability of an intermediate or to other kinetic impediments. Consequently, many of the potential long-term survivors will never make it in reality.

The properties of the rare survivors of this multiple selection agree well with those that one would be inclined to associate with the manifestation of a catalytic activity. Indeed, a certain conformational stability may be considered a cardinal quality of a potential catalytic multimer. Without such stability, the binding of substrates, that, perhaps, of mineral ions, within a precise catalytic environment is likely to occur too infrequently and fleetingly to lend itself to an effective chemical interaction. Compactness, it will be remembered, is a key criterion in Gō's identification of modules in protein molecules (99, 322). We have seen that these modules, or their equivalents, may well originate from small ancestral polypeptides with catalytic function. A cyclic structure could also favor a catalytic function. So would the association into plurimolecular complexes, a formula widely exploited by present-day enzymes. Furthermore, because of the many severe constraints affecting the synthesis, solubility, and survival of the multimers, the selected molecules could have been few enough to reach concentrations compatible with the display of an eventual catalytic activity. Finally, because of these very constraints, the composition of the population would have been under strong deterministic control. This composition would have been very constant and reproducible, in spite of the random character of its mode of generation. Consequently, if the required catalysts were present at the start, they had every chance of being present indefinitely, as long as the conditions were not perturbed too drastically.

The model that I propose is thus quantitatively plausible as well as qualitatively realistic. Provided that the multimerization of bifunctional acids were possible by the mechanism I shall describe in the following chapter, or by some analogous mechanism, the resulting mixture had serious chances of containing, at efficient concentrations, rudimentary catalysts prefiguring the principal enzymes. As a main drawback, the active multimers would have been in the company of a large number of inactive congeners, whose presence constitutes the unavoidable price imposed by the random character of the assembly. There are no a priori reasons for assuming that this dilution, possibly bothersome in certain respects, could have been so to the extent of preventing the catalysts from exerting their actions. It is, however, not excluded that the catalytic multimers or, at least, some of them could have benefited from an additional selection phenomenon, due to stabilization by their substrates.

## *Selection by Substrates*

Enzymes owe their specificity to their ability to bind their substrates at sites strategically situated in relation to the catalytic sites. It is known that this binding often protects the enzymes against degradation. Should the same hold true for the crude catalysts under consideration, those that are actually working (i.e., are supplied with substrates) would tend to accumulate selectively (247).

This mechanism is interesting because it does not just rely on favorable coincidences. It is truly selective, in that it contains a built-in feedback loop that actually makes it save what is useful. In contrast to Darwinian selection, however, it selects by differential survival, not by differential replication; it can operate with a random system of production and does not require the elaborate machinery needed for replicative synthesis. In this respect, this mechanism could have played an important role in the development of this machinery from simple abiotic beginnings. But is it plausible? Some calculations are needed to answer this question.

If it is assumed, for the sake of simplicity, that the synthesis of catalytic multimer $j$ takes place at a rate $R_j$ independent of the amount of $j$ present and that its breakdown obeys first-order kinetics with rate constant $k_j$, then the steady-state level of $j$, $N_j$, is given by:

$$N_j = \frac{R_j}{k_j} \tag{23}$$

If it is assumed further—this is the most favorable assumption—that the catalyst-substrate complex is totally resistant to degradation, we find that the ratio of the steady-state amount $N'_j$ of compound $j$ in the presence of substrate to the amount $N_j$ in the absence of substrate is given by:

$$\frac{N'_j}{N_j} = 1 + \frac{[S_j]}{K_j} \tag{24}$$

in which $[S_j]$ is the substrate concentration and $K_j$ the dissociation constant of the catalyst-substrate complex (the equivalent of the Michaelis constant of an enzyme, see Equation 22, Footnote 15, page 35).

It is seen that for significant protection to occur in the absence of any change in either the synthesis rate or the breakdown rate-constant, the substrate concentration should be at least equivalent to, and preferably higher than, the Michaelis constant of the catalyst. Many enzymes have Michaelis constants in the millimolar range, and it does not seem very likely that crude catalysts of the kind considered would bind their substrates more tightly than do enzymes. It follows that high substrate concentrations, on the order of centimolar or higher, may be needed for the selection mechanism to function as postulated.

This requirement is not as unrealistic as it may appear at first sight, since the potential stabilizers are either common abiotic products or substances formed from them by catalytic action and thereby raised to comparable concentrations within the limits set by the equilibrium constants of the catalyzed reactions. There is no reason why abiotic products synthesized steadily over millions of years could not have accumulated up to high concentrations, if not in the oceans, at least in shallow waters.[6] Indeed, the requirement would be even more stringent in a system relying only on random chemistry. As pointed out above, all the models that attempt to explain the origin of life by natural chemical phenomena encounter the same problem of how intermediates reach effective concentrations. Within this context, the proposed mechanism thus seems far from implausible.

Once the possibility of such a mechanism is accepted, a remarkable interplay between catalysts and substrates suggests itself. As substrates are recognized by catalysts and give rise to new products, which themselves propagate the process further, the catalysts involved are simultaneously protected and made to accumulate selectively. Thus, the progressive emergence of protometabolites over the "chemical noise" would be accompanied by a similar emergence of protoenzymes over the "multimeric noise." The network would stabilize itself as it is laid down. But this would be only a first step. By mass-action effect, emerging catalysts would themselves become favored substrates for chemical modification, including elongation. Improved or novel catalysts arising in this way would, in turn, be selected if they were supplied with substrates that stabilized them. Some catalyst evolution would thereby become possible. The model allows a progressive exploration of the catalytic, as well as of the metabolic, landscape.

[6] Miller (111) has attempted to estimate the concentration that amino acids could have reached in the primitive ocean. His maximal value is 0.3 millimolar for an average ocean depth of 3,000 meters. Substantially higher concentrations could be expected in shallow waters.

Interestingly, the system considered is inherently self-supporting. All it needs is priming: the first multimers must form spontaneously or, at least, without the help of a multimeric catalyst. After that, the system would generate its own catalysts, and these would be among the first to be selected by substrate stabilization, since the building blocks of multimerization would be among the most abundant abiotic products. At a later stage, one could also expect the selection of iterative assembly catalysts—that is, chain elongation catalysts—whose products are also substrates and, therefore, potential stabilizers. Consequently, even if it were inefficient at the start, multimerization would be self-improving in some sort of autocatalytic fashion.

The model also has defects. The ability to bind—and be protected—obviously does not necessarily imply the possession of catalytic activity. There would be a lot of deadwood among the stabilized multimers, although components of structural value could emerge in this manner. Waning catalyst protection due to substrate dilution as the system evolves toward greater diversification is another flaw, responsible for a built-in limitation to progress. As the number of intermediates increases, their maximal permissible average concentration must perforce decrease. If we assume, for example, a maximal total metabolite concentration of 100 grams per liter and an average molecular weight of 100, the maximal individual average concentration would be 0.1 molar with 10 intermediates, centimolar with 100, and millimolar with 1,000. The ability to protect catalysts would decrease in parallel, at least statistically. Certain catalysts would benefit from a relatively efficient protection because their substrates were present at high concentrations in the steady-state system. Others, on the other hand, would not be significantly protected. Although it is impossible to appreciate the advantages of such a partial and incomplete protection process, the balance, in any case, can only be positive. Even without any stabilization by substrates, the model that I propose remains plausible.

CHAPTER SEVEN

# HARNESSING ENERGY

## THE FIRST SOURCE OF PREBIOTIC ENERGY

In the preceding chapter, we have seen the possible role of UV light, electric discharges, and other physical forms of energy in supporting the abiotic reductive and synthetic reactions that may have led to the appearance of the primary organic building blocks of life. For the biogenic process to advance further, whether along the lines I have proposed or otherwise, a more specific harnessing of energy eventually became necessary. Whatever source of energy was tapped, it had to be funneled into the system by some sort of coupling. As put by Lipmann, "coupling by crossover of two forms of chemical energy might have been the first event on the way to life" (109, in Reference 22, page 215). And he goes on to speculate that substrate-level phosphorylation may have been the first such coupling, with pyrophosphate, rather than ATP, acting as the conveyer of "energy-rich phosphate bonds." Others, like Clair Folsome (56) and Harold Morowitz (32, 264, 265), visualize some sort of primitive photophosphorylation as the first coupling, again with pyrophosphate as the initial phosphoryl carrier. Accordingly, they give priority to the development of a phototransducing membrane capable of supporting protonmotive force or some other charge displacement usable for the performance of work. As stated by Morowitz, "the first crucial event in the origin of cellular life is the formation of a plasma membrane" (32, page 234). Gerald Weissmann expresses the same idea. "In the beginning," he writes, "there must have been a membrane" (86, page xiv). Franklin Harold is even more affirmative. In a book devoted to bioenergetics, he does not hesitate to proclaim: "In the Beginning was the Membrane" (43, page 168).

Such speculations proceed from what may be called a "biomorphic" type of reasoning. That is, they take their cue from the existing biological order of things, simply substituting pyrophosphate for ATP. Primitive biosynthesis is seen as proceeding like its modern counterpart but with pyrophosphate as the condensing agent; and the primary energy coupling is viewed as serving to regenerate the consumed pyrophosphate with the help of catabolism or light, that is, to accomplish what energy coupling does for ATP in present-day organisms.

There are, indeed, good reasons to suspect that pyrophosphate may have preceded ATP as the central energy currency. Even in the world today, pyrophosphate serves as phosphoryl donor for certain reactions in several bacterial species (95, 155, 156), in some eukaryotic microorganisms (152), and in the cytosol of plant cells (366).[1] There are, however, serious difficulties in making the pyrophosphate bond the *first* bearer of utilizable energy in the history of life. In fact, the unique biological importance of phosphate, while readily justified on the strength of theoretical considerations (278), nevertheless poses a major problem that has puzzled many origin-of-life specialists. The concentration of inorganic phosphate in oceans and freshwaters is low—around 0.1 nanomolar (126), or one hundred-thousandth of what it is in living cells—largely because of the low solubility of calcium phosphate at neutral pH. Furthermore, pyrophosphates and polyphosphates, the putative primary energy donors, are rare minerals, at least on the Earth today. It is difficult to explain the spontaneous emergence, under such circumstances, of organisms in which phosphate plays as central a structural and functional role as it does in all forms of life. For these reasons, I am inclined to question all models that derive their primary energy supply from the pyrophosphate bond, whether provided from natural stores or produced from inorganic phosphate by some transducing system. I believe we must look for an earlier form of energy that preceded the pyrophosphate bond and facilitated the entry of phosphate into metabolism. We need not look far for such a form. It stands out in glaring red letters in all textbooks of biochemistry. It is the thioester bond.

Many workers have emphasized the biochemical importance of the thioester bond. Already in 1951, Theodor Wieland, impressed by the discovery of acetyl-coenzyme A by his mentor, Feodor Lynen (300), started a long series of investigations into the synthesis and chemical properties of thioesters of amino acids (114). Wieland discovered that amino acid thioesters can assemble spontaneously into multimers in the absence of a catalyst, long before a similar process was found by Lipmann and associates to be involved in the biological synthesis of gramicidin S and other bacterial peptides (296). The importance of thioesters in substrate-level phosphorylations prompted Efraim Racker to write in 1965: "Thiol esters may possibly represent the evolutionary ancestors of 'high-energy' ATP" (45, page 7). Similar considerations have guided experiments by Arthur Weber in search of conditions under which phosphoanhydrides could form at the expense of thioesters (392, 393). Lars Onsager, the founder of nonequilibrium thermodynamics who developed a late interest in the origin of life, is quoted by Lipmann as having suggested "thioesterification of carboxylic acids for activation as an early event in prebiology" (149b, page 31).

These views all make eminent biochemical sense (see Figure 7–1). The thioester bond is at the root of substrate-level phosphorylation, presumably the oldest form of metabolic ATP regeneration, in both the glycolytic chain and the citric acid cycle. It is the precursor of virtually all acyl-ester bonds in nature and of many carbon-carbon bonds—for example, in fatty acids (through malonyl-coenzyme A), in citric acid-

[1] The significance of these facts has been discussed in Chapter 1 (page 11). Contrary to what is sometimes assumed, however, they are not likely to reflect legacies from a distant past when ATP had not yet been invented. The scattered phylogenetic distribution of metabolic pyrophosphate utilization precludes this possibility. On the other hand, the existence of such utilization in a number of present-day living organisms is clear proof that pyrophosphate can substitute for ATP in the absence of pyrophosphatase activity and, therefore, that it could have played the primitive role assigned to it.

**7–1** *The central role of the thioester bond.* Carboxylic acids are linked to thiols with the help of energy to give rise to thioesters (1). These, in turn, act as donors of activated acyl groups in many major biosynthetic pathways (2). In all these reactions, the phosphate ester of pantetheine (see Figure 6–3, page 138) serves as the acyl carrier, either as such or as part of coenzyme A. Several important substrate-level phosphorylation reactions also operate at the expense of thioesters, with acyl-phosphates as central intermediates. The typical two-step process (see Figure 1–5, page 15) involves phosphorolysis of the thioester, giving rise to an acyl-phosphate (3), followed by phosphoryl transfer from the acyl-phosphate to ADP, giving ATP (4).

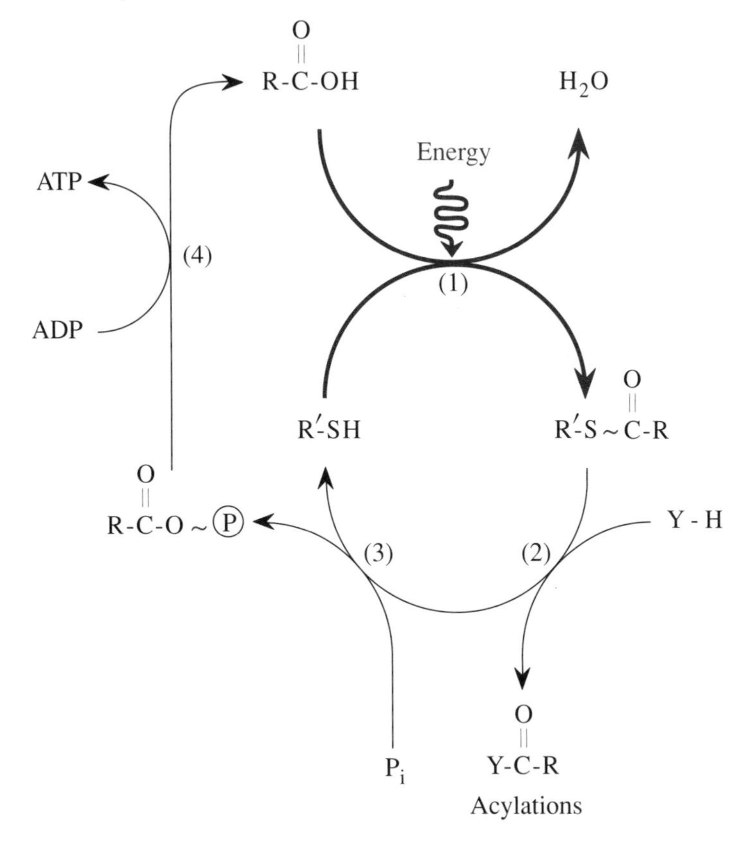


cycle and glyoxylate-cycle intermediates (through acetyl-coenzyme A, via citrate or malate), in most bacterial and animal porphyrins (through $\delta$-aminolevulinate), and in the vast terpenoid-steroid family (through mevalonate). Even more suggestive, the thioester bond is also the precursor of some amide bonds, in amino acid conjugates, for instance, and, especially, in those possibly vestigial products, the stereochemically mixed bacterial peptides that include the antibiotics gramicidin S and tyrocidin and several other products of biological interest (148, 149b, 296). In the synthesis of all these compounds, the phosphate ester of pantetheine, itself an interesting hybrid that could have an ancient history (see Figure 6–3, page 138), serves as the universal acyl carrier. It does so as such (fatty acids, peptides) or as part of coenzyme A, possibly a later invention developed when ATP came on the scene. The suggestion by

Lipmann (149a) that pantetheine-dependent peptide synthesis may have preceded the RNA-dependent mechanism certainly deserves serious consideration. As intimated here, the more primitive process could even go back to prebiotic times.

## THE THIOESTER WORLD

### *The Central Postulate*

Thioesterification presupposes the presence of thiols and of carboxylic acids. This poses no problem. Thiols may be expected to form readily under prebiotic conditions in the presence of $H_2S$. Indeed, several sulfur products have been obtained from $H_2S$ in simulation experiments (60, 374), and it has even been found that $H_2S$ can act as an efficient catalyst in reactions powered by UV irradiation (60). As to $H_2S$ itself, its presence in the prebiotic environment raises no difficulty. In fact, there are many reasons for believing that life started in a sulfur-rich milieu. As concerns carboxylic acids, we have seen that they are expected to have been plentiful in prebiotic times.

Thus, the building blocks for making thioesters were likely to be there. But the question remains as to how thiols and acids could have condensed, with loss of water, to give rise to thioesters:[2]

$$R'\text{—SH} + R\text{—}\overset{\overset{\displaystyle O}{\|}}{C}\text{—OH} \longrightarrow R'\text{—S—}\overset{\overset{\displaystyle O}{\|}}{C}\text{—R} + H_2O \qquad \textbf{(25)}$$

Under physiological conditions, such a reaction is strongly endergonic—the free energy of hydrolysis of the thioester bond is equivalent to that of the terminal pyrophosphate bond of ATP—and can yield only traces of thioester. We must therefore ask how this energy hurdle could have been overcome. This question will be more easily addressed after the main features of a hypothetical thioester world have been outlined. Let us provisionally assume that, wherever life started, the various organic acids that arose abiotically were present, at least partly, in the form of thioesters, and let us ask what could have happened under such conditions. To answer this question, we need only look at present-day organisms. They tell us that, fueled by no more than thioesters, a whole metabolic network could have developed along the lines that I have postulated.

### *Thioester-Dependent Multimerization*

The discovery by Lipmann and coworkers (148, 149a, 296) that a number of bacterial peptides are assembled from amino acid thioesters provides us with a ready mechanism for the synthesis of the multimers among which the catalysts required by my

[2] All chemical equations in this chapter assume an acidic medium (pH 2.0 or lower), in conformity with one of the models proposed for the primary generation of thioesters.

model are to be found. This mechanism proceeds by what Lipmann has called head growth. With amino acids, for example, the succession would be as follows:

$$R'\text{-S-}\overset{\overset{\displaystyle O}{\|}}{C}\text{-}\underset{\underset{\displaystyle NH_3^+}{|}}{CH}\text{-}R_1 \quad + \quad R'\text{-S-}\overset{\overset{\displaystyle O}{\|}}{C}\text{-}\underset{\underset{\displaystyle NH_3^+}{|}}{CH}\text{-}R_2$$

$$\longrightarrow R'\text{-SH} + H^+ + R'\text{-S-}\overset{\overset{\displaystyle O}{\|}}{C}\text{-}CH\text{-}R_2\;(\text{-NH-}\overset{\overset{\displaystyle O}{\|}}{C}\text{-}\underset{\underset{\displaystyle NH_3^+}{|}}{CH}\text{-}R_1)$$

$$+ \quad R'\text{-S-}\overset{\overset{\displaystyle O}{\|}}{C}\text{-}\underset{\underset{\displaystyle NH_3^+}{|}}{CH}\text{-}R_3 \longrightarrow R'\text{-SH} + H^+ + R'\text{-S-}\overset{\overset{\displaystyle O}{\|}}{C}\text{-}CH\text{-}R_3\;(\text{-NH-}\overset{\overset{\displaystyle O}{\|}}{C}\text{-}CH\text{-}R_2\;(\text{-NH-}\overset{\overset{\displaystyle O}{\|}}{C}\text{-}\underset{\underset{\displaystyle NH_3^+}{|}}{CH}\text{-}R_1)) \qquad \textbf{(26)}$$

A characteristic of this mechanism is that the carboxyl end of the growing chain remains continually engaged in a thioester linkage (just as the carboxyl end of a growing peptide chain in conventional protein synthesis remains continually engaged in an ester linkage with the corresponding tRNA). In Scheme 26, this end is shown as reacting with the amino group of a monomeric aminoacyl thioester (elongation). But it could equally well react with the amino group of an oligomeric chain (modular assembly) or with the terminal amino group of its own chain, if it were long enough (cyclization). The latter reaction is the only one, besides hydrolysis of the head thioester bond, to bring the growth of the multimer to an end. As pointed out in the preceding chapter, such an assembly process is particularly well adapted to the production of the small subset of stable multimers among which active catalysts with specific substrate-binding properties are most likely to be found. Note that a similar process could also incorporate other monomers besides amino acids—hydroxy acids, for example—into the growing chain. Transacylation from a thioester (acyl-coenzyme A) onto hydroxyl groups is the universal biological mechanism of ester assembly.

An interesting aspect of the reaction considered is that it could have occurred with the help of some crude mineral catalyst, or even without a catalyst. Wieland (114) has found that peptides form spontaneously from aminoacyl thioesters simply upon incubation in a weakly alkaline aqueous medium. Once primed, the reaction would be expected to generate its own catalysts, thereby becoming self-supporting. It could even be self-improving thanks to substrate stabilization, as suggested in the preceding chapter.

Thus, a first, and crucially important, benefit to be expected from thioesters is the generation of the multimeric protoenzymes required for the unfolding of protometabolism. Let us now look at how thioesters could support this unfolding energetically.

### *Thioester-Dependent Reductions*

We have seen (Chapter 6) that the reducing equivalents needed for the abiotic synthesis of primary building blocks were probably supplied in plentiful amounts by the UV-supported photochemical oxidation of $Fe^{2+}$ ions. A possible alternative is the oxidation of ferrous sulfide to ferrous disulfide (pyrite), as proposed by Wächtershäuser (275). Although of great importance for the generation of the first biogenic materials, these primeval reactions are not likely to have included among their products the trioses and the other carbonyl derivatives (aldehydes and ketones) on which autotrophic syntheses are critically dependent. Such substances are characterized by particularly low oxidation-reduction potentials (i.e., have electrons of a particularly high energy level); they are typically produced in nature at the expense of low-potential electron donors (NADH or NADPH) with the additional help of thioester energy. Essentially the same could have happened in prebiotic times.

We can thus visualize the reduction of the abiotically produced carboxylic acids to the corresponding aldehydes as taking place by the reductive splitting of their thioesters:

$$R'-S-\overset{\overset{\displaystyle O}{\|}}{C}-R + 2\,e^- + 2\,H^+ \rightleftharpoons R'-SH + R-\overset{\overset{\displaystyle O}{\|}}{C}-H \qquad \textbf{(27)}$$

A variant of this reaction is the carboxylating reductive splitting of thioesters, giving rise to $\alpha$-keto acids:

$$R'-S-\overset{\overset{\displaystyle O}{\|}}{C}-R + 2\,e^- + 2\,H^+ + CO_2 \rightleftharpoons R'-SH + R-\overset{\overset{\displaystyle O}{\|}}{C}-COOH \qquad \textbf{(28)}$$

This reaction could be followed by further reduction of the $\alpha$-keto acid to an $\alpha$-hydroxy acid:

$$R-\overset{\overset{\displaystyle O}{\|}}{C}-COOH + 2\,e^- + 2\,H^+ \rightleftharpoons R-CHOH-COOH \qquad \textbf{(29)}$$

or by its reductive carboxylation to a hydroxy dicarboxylic acid:

$$(H)R-\overset{\overset{\displaystyle O}{\|}}{C}-COOH + 2\,e^- + 2\,H^+ + CO_2 \rightleftharpoons HOOC-R-CHOH-COOH \qquad \textbf{(30)}$$

or by its reductive amination to an $\alpha$-amino acid:

$$R-\overset{\overset{\displaystyle O}{\|}}{C}-COOH + 2e^- + 2H^+ + NH_4^+ \rightleftharpoons R-\overset{\overset{\displaystyle NH_3^+}{|}}{CH}-COOH + H_2O \qquad \textbf{(31)}$$

Starting from an acetyl-thiol, for example ($R = CH_3$), the products would be acetaldehyde, pyruvic acid, lactic acid, malic acid, and alanine in Reactions 27, 28, 29, 30, and 31, respectively.

All these reactions are important components of metabolism in most organisms today. In the direction shown from left to right and with suitable electron donors, they play major roles in autotrophic reductions, carbon dioxide fixation, and amination. Reaction 27 runs reversibly with NADH or NADPH as the electron donor. Reaction 28 is more endergonic by about 4.5 kilocalories per mole (the standard free energy of carboxylation of an aldehyde). To run in the reductive direction, it needs an electron donor of correspondingly lower oxidation-reduction potential, a ferredoxin as a rule. Or it could be driven by reductive trapping of the $\alpha$-keto acid by Reactions 29 or 30, which are strongly exergonic with NADH or NADPH as the electron donor.

For the reactions to proceed prebiotically as assumed, appropriate electron donors would have been needed. Possible candidates for this role would not have been lacking if, as assumed above, the prebiotic world was flooded with high-energy reducing equivalents provided by the photochemical oxidation of $Fe^{2+}$ ions.

Reactions 27 to 31 are all readily reversible; they can easily run in the oxidative direction, from right to left, if supplied with an appropriate electron acceptor. This is their preferred direction in heterotrophic organisms. In particular, Reactions 27 and 28, mostly with $NAD^+$ as the electron acceptor, accomplish the oxidative synthesis of thioesters that drives the main substrate-level phosphorylations (the oxidative decarboxylation of pyruvate depicted in Figure 1–5, page 15, is an example of such a reaction). I shall come back to this important point when discussing the primary generation of thioesters.

## *Thioester-Dependent Phosphorylations*

In substrate-level phosphorylations, the thioesters synthesized oxidatively are attacked phosphorolytically by inorganic phosphate, giving rise to the corresponding acyl-phosphates:

$$R'-S-\overset{\overset{\displaystyle O}{\|}}{C}-R + HO-\underset{\underset{\displaystyle OH}{|}}{\overset{\overset{\displaystyle O}{\|}}{P}}-O^- \rightleftharpoons R'-SH + R-\overset{\overset{\displaystyle O}{\|}}{C}-O-\underset{\underset{\displaystyle OH}{|}}{\overset{\overset{\displaystyle O}{\|}}{P}}-O^- \qquad \textbf{(32)}$$

In the example of Figure 1–5, the phosphorolysis of acetyl-coenzyme A to acetyl-phosphate represents this reaction.

I submit that a reaction of this type may have signaled the primeval entry of inorganic phosphate into the fabric of life. Utilizing the energy stored in the thioester bond, it would have served to activate the phosphate ion into a highly reactive, energy-rich, acyl-bound phosphoryl group.

In present-day organisms, the phosphoryl group of acyl-phosphates is exploited metabolically by way of its transfer to ADP (or, sometimes, GDP) to form the universal energy donor ATP (or GTP):

$$R-\overset{\overset{\displaystyle O}{\|}}{C}-O-\underset{\underset{\displaystyle OH}{|}}{\overset{\overset{\displaystyle O}{\|}}{P}}-O^- + ADP \rightleftharpoons R-\overset{\overset{\displaystyle O}{\|}}{C}-OH + ATP \qquad \mathbf{(33)}$$

It is conceivable that the same reaction occurred prebiotically with inorganic phosphate as the attacking agent and thereby gave rise to inorganic pyrophosphate, which, we have seen, very likely acted as the primitive carrier of high-energy phosphoryl groups before ATP became available:

$$\begin{array}{c} R-\overset{\overset{\displaystyle O}{\|}}{C}-O-\underset{\underset{\displaystyle OH}{|}}{\overset{\overset{\displaystyle O}{\|}}{P}}-O^- + HO-\underset{\underset{\displaystyle OH}{|}}{\overset{\overset{\displaystyle O}{\|}}{P}}-O^- \\ \Updownarrow \\ R-\overset{\overset{\displaystyle O}{\|}}{C}-OH + {}^-O-\underset{\underset{\displaystyle OH}{|}}{\overset{\overset{\displaystyle O}{\|}}{P}}-O-\underset{\underset{\displaystyle OH}{|}}{\overset{\overset{\displaystyle O}{\|}}{P}}-O^- \end{array} \qquad \mathbf{(34)}$$

As Weber (392, 393) has shown, both Reactions 32 and 34 can take place in the absence of enzymes. The sequence of these reactions would have been of immense importance in the prebiotic world by creating the first sequential group-transfer system. This would have coupled the dehydrating joining of two phosphate ions into inorganic pyrophosphate to the hydrolytic splitting of a thioester, with an acyl-phosphate as the double-headed intermediate (for details of such a mechanism, see Chapter 1, pages 14, 15). From here on, thanks to phosphoryl-transfer reactions from pyrophosphate (in lieu of ATP), phosphate could invade protometabolism and come to serve the functions that it accomplishes in all living organisms today—as energy conveyer, as mediator of synthetic reactions, as metabolic "handle," and as structural component. Through pyrophosphate, the thioester bond could thus have supported all the energy needs of protometabolism.

## *The Unfolding of Metabolism and the Origin of ATP*

Consider the situation we have reached. Thioester-generated multimers provide the necessary catalysts. Thioester-supported endergonic electron transfers allow the most demanding biosynthetic reductions to take place. Thioester-fueled phosphorylations supply phosphate-bond energy for the support of dehydrating assembly processes occurring by sequential group transfer. Powered only by thioester bonds, emerging life has gained the freedom to explore the entire metabolic landscape. The glycolytic chain would be developed, leading to the hexoses and, thanks to a side branch, to the pentoses. The core reactions of the citric acid and glyoxylate cycles would begin to function, connecting with a number of amino acids by transaminations. Fatty acids would be formed, and the first lipid esters would appear, eventually leading to phospholipids. In other words, the metabolic "onion" (Chapter 6) would develop, layer upon layer. Assembly of the purine ring is not inconceivable—the ingredients were there—and the pyrimidine ring could also be made if carbamoyl phosphate was among the early phosphorylated products. Thus, the first mononucleotides, including AMP, would eventually be generated, paving the way for the entrance of ATP on the protometabolic scene.

To explain this cardinal event, it is tempting to picture the thioester bond as once again playing a central role. From knowledge of present-day metabolism, one can readily imagine a primary attack of a thioester by AMP, with formation of an acyladenylate:

$$R'{-}S{-}\overset{\overset{\displaystyle O}{\|}}{C}{-}R + AMP{-}OH \rightleftharpoons R'{-}SH + AMP{-}O{-}\overset{\overset{\displaystyle O}{\|}}{C}{-}R \qquad \textbf{(35)}$$

In turn, this molecule could be attacked by pyrophosphate, giving rise to ATP:

$$AMP{-}O{-}\overset{\overset{\displaystyle O}{\|}}{C}{-}R + PP_i \rightleftharpoons ATP + R{-}\overset{\overset{\displaystyle O}{\|}}{C}{-}OH \qquad \textbf{(36)}$$

Note the primordial importance of these two reactions. In the direction from left to right, the sequence would be a new mechanism of substrate-level phosphorylation at the expense of thioesters. As far as is known, such a sequence does not fulfill this function in energy metabolism today because the hydrolysis of pyrophosphate displaces the equilibrium of Reaction 36 far to the left. But things would have been different in a pyrophosphate-based economy.[3]

Running backwards, the sequence Reaction 36 + Reaction 35 represents the main mechanism of activation of carboxylic acids, including amino acids in the synthesis of certain bacterial peptides. The same reverse sequence, but with tRNA replacing a

[3] It has been pointed out to me that the proposed mechanism meets with a thermodynamic difficulty. The synthesis of ATP (under physiological conditions) costs some 2.4 kilocalories per gram-molecule more at the expense of AMP and $PP_i$ than at the expense of ADP and $P_i$ (154). It follows that ATP formation is thermodynamically more difficult to accomplish by Reactions 35 and 36 than by Reactions 32 and 33. The difficulty, however, is only apparent in the present case, because the pyrophosphate is produced by Reaction 34, which is thermodynamically favored in similar proportion. It may therefore be assumed that the energy barrier will be overcome by a high pyrophosphate concentration.

The following reasoning demonstrates this point in another way: the overall balance is obviously the same whether one spends two molecules of thioester to phosphorylate two molecules of ADP by Reactions 32 and 33 or first uses the two molecules of ADP to make ATP and AMP (see Reaction 13, page 10) and then spends two molecules of thioester to convert the AMP to ATP by Reactions 32, 34, 35, and 36. The first mechanism is perfectly possible thermodynamically. Hence, the second one must be equally possible.

thiol as the acyl acceptor in Reaction 35 (see Reaction 42, page 183), accounts for the activation of amino acids in the synthesis of proteins, presumably a later evolutionary development.

Thus, starting from the thioester bond, Reaction 35 introduces the all-important double-headed acyl-adenylates, whereas Reaction 36 inaugurates the central energy carrier, ATP, with its predecessor in this capacity, inorganic pyrophosphate, actually providing the original terminal pyrophosphoryl group of the molecule. After ATP became available from another source, Reaction 36, reversed, would offer an alternative mechanism for making acyl-adenylates, destined to dominate biosynthesis eventually.

Once introduced, ATP could spread throughout metabolism and progressively assume its pivotal function. It could replace pyrophosphate as phosphoryl donor in phosphorylation reactions and thereby give rise to ADP, which could itself substitute for inorganic phosphate in Reaction 34 to regenerate ATP (Reaction 33). Or two molecules of ADP could react together to form one molecule of ATP and one of AMP (see Reaction 13, page 10). AMP itself could be rephosphorylated to ATP via Reactions 35 and 36. Figure 7–2 provides a synoptic view of the main reactions of the thioester world.

## *The Dawning of the RNA World*

Not only ATP, but other nucleoside triphosphates, including GTP, UTP, and CTP, would have appeared, exchanging phosphoryl groups with ATP, sometimes substituting for it in certain reactions. Coenzymes could even have formed. Nicotinamide, for example, could have formed from simple prebiotic building blocks by the quinolinic-acid pathway (302) and then entered the nucleotide circuit. Pantetheine, a possible early oligomer (see Figure 6–3, page 138), could have been phosphorylated and subsequently converted to coenzyme A. Nucleoside diphosphates could have become available for the transport of activated biosynthetic carbohydrate and lipid groups. Thus, the era of coenzymes, which is often viewed as an important gateway to the RNA world (237, 289), would have begun.

Most important of all, nucleoside triphosphates could have started transferring nucleotidyl groups to each other and to the resulting products, thus inaugurating the RNA world. Not the one generally pictured, however. An RNA world requiring no miracle for its creation. An RNA world emerging naturally from the thioester world.

One crucial point, however, remains to be examined, namely the manner in which the primordial thioesters, which make up the very foundation of the thioester world, first could have appeared on the prebiotic Earth.

## THE PRIMARY GENERATION OF THIOESTERS

Two relatively simple mechanisms of thioesterification can be envisaged (see Figure 7–3, page 158): one spontaneous and supported by favorable environmental conditions, the other oxidative. Each has advantages and drawbacks, making their relative merits difficult to assess.

**7–2** *Synoptic view of the thioester world.* (1) Thioester pool, including aminoacyl thioesters. (2) Assembly of multimers, including a variety of protoenzymes (Reaction 26). (3) Phosphorolysis of thioesters generates acyl-phosphates (Reaction 32). (4) Phosphorolysis of acyl-phosphates generates pyrophosphate, the first purveyor of phosphate-bond energy (Reaction 34). (5) Reductive splitting of thioesters (with or without carboxylation) generates carbonyl groups and supports synthetic reductions (Reaction 28 or 27). (6) Adenylolysis of thioesters generates acyl-adenylates (Reaction 35). (7) Pyrophosphorolysis of acyl-adenylates generates ATP, which takes over as purveyor of phosphate-bond energy (Reaction 36). (8) RNA-lysis of (amino)acyl-adenylates generates (amino)acyl-RNA complexes (Reaction 42, page 183). (9) Peptides are formed at the expense of aminoacyl-RNA complexes.

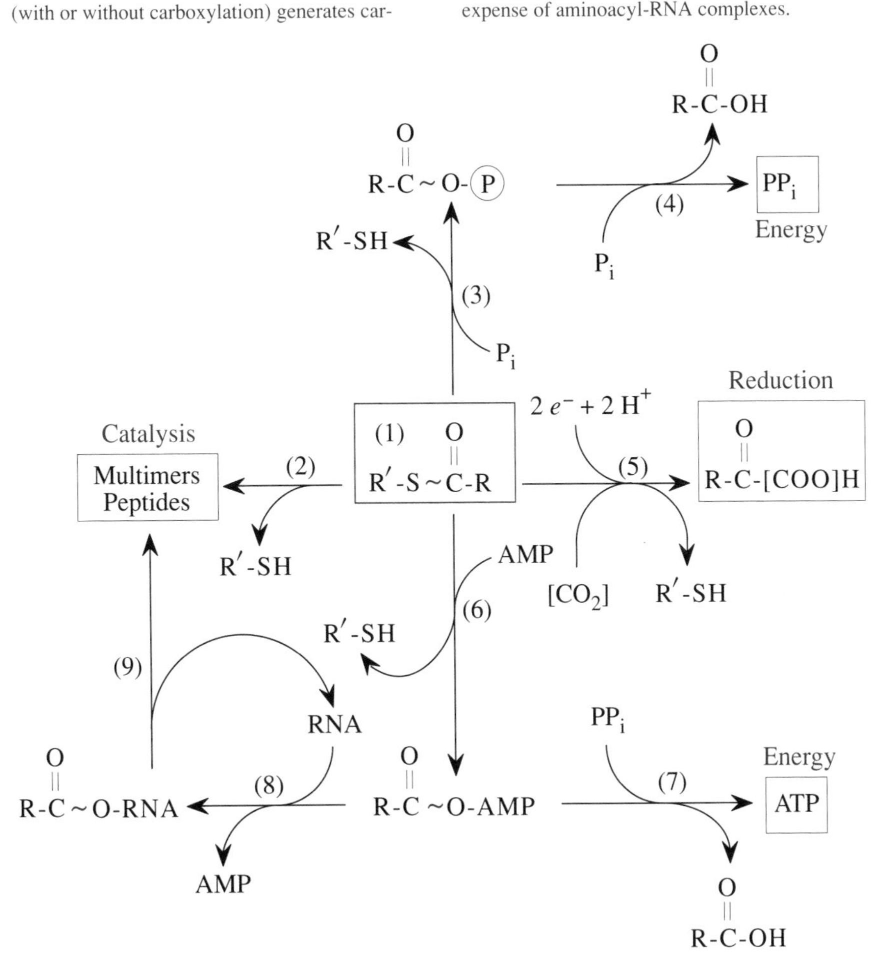

**7–3** *Generation of thioesters.*
(1) Spontaneous assembly of thioesters as favored by hot, acidic environment. (2) Oxidative synthesis of thioesters (alternative to 1, or later development). Reverse of Reaction 27 or 28, with $Fe^{3+}$ as the electron acceptor (see the discussion of the iron cycle on pages 166–168). (3) Thioester pool (see Figure 7–2), possibly maintained by transthiolation reactions (Reaction 40, page 165).

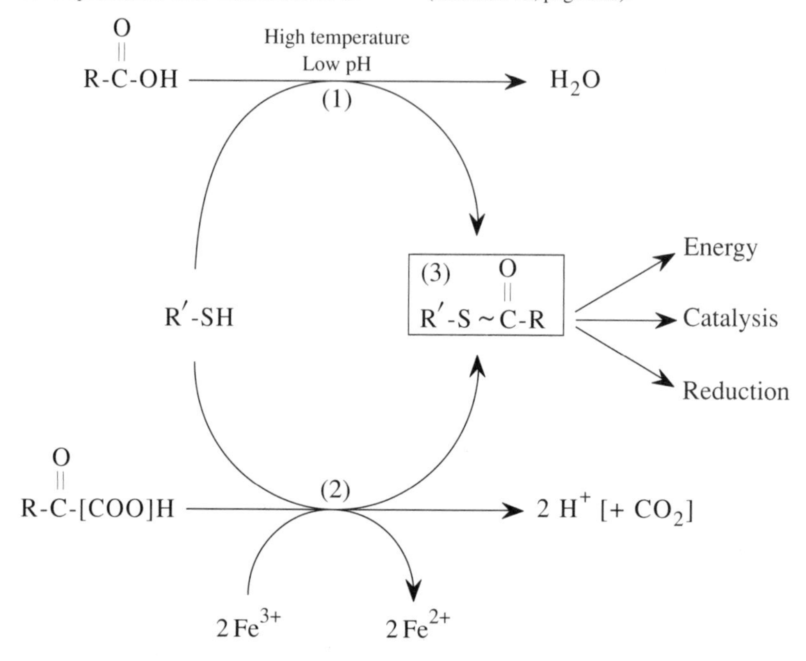


## *Spontaneous Formation of Thioesters*

Thioesters are classified by biochemists as energy-rich compounds, with standard free energies of hydrolysis at pH 7.0 (group potentials) of the order of –7.5 to –8.5 kilocalories per gram-molecule (154, 297). This precludes their formation in more than trace amounts by simple spontaneous assembly according to Reaction 25. Wieland has overcome this thermodynamic hurdle by the classical means of using activated acids, such as acyl chlorides and other mixed anhydrides, as reactants (114). This, of course, does not solve the prebiotic problem; it simply pushes it back to the genesis of some other primordial energy-rich bond. Weber (394) has observed an interesting spontaneous formation of thioesters from glyceraldehyde and a thiol. This reaction, which proceeds in neutral medium and at room temperature with good yield and at an appreciable rate, actually represents a special case of the oxidative mechanism. It will be considered in the next section.

When I started reflecting on a possible thioester world, the thought struck me that a sizable fraction of the free energy of hydrolysis of thioester bonds under physiological conditions is contributed by the free energy of dissociation of the acid. At a sufficiently low pH (one pH unit or more below the pK of the acid), the major part of this contribution would be suppressed, and reversal of the reaction would be correspondingly less difficult. If, as seemed likely, thioester hydrolysis were exothermic, increasing the temperature would give an additional boost to reversal, perhaps to the point that significant amounts of thioesters could arise spontaneously, without an additional supply of energy. Hence the proposal that life arose in a hot, acidic, and sulfur-rich environment (105).

A quantitative assessment of this hypothesis appears in Table 7–1 below. As predicted, the gain is considerable. Changing from pH 7.0 and 25°C to pH 2.0 and 100°C produces a 600-fold increase in the equilibrium concentration of thioester. With high enough (near-molar) concentrations of thiol and acid, a thioester concentration in the millimolar range would be reached, which should be more than sufficient to "start the ball rolling." Some sort of catalyst would probably be needed, but this should not concern us too much. Even a very crude and inefficient catalyst would do to prime the system. After that, better catalysts would arise among the thioester-generated multimers.

Several facts make this hypothesis attractive. First, it so happens that the postulated milieu corresponds to the most likely habitat of the common ancestral cell (Chapter 4) and therefore appears as a very plausible setting for the origin of life. Hot, acidic, sulfur-rich environments exist today in a number of places and could have been more widespread on a young planet subject to intense volcanic activity. They could have been found in hot sulfurous springs, shallow pools fed by such springs, volcanic lakes, or, alternatively, deep-sea hydrothermal vents. The latter formations have evoked considerable interest since their discovery some years ago. They have been found to harbor several strange forms of life, including extremely thermophilic chemoautotrophic bacteria. The possibility that life originated in such an environment is now receiving serious consideration (375).

Another attractive feature of the model is that it provides a ready solution to the phosphate paradox mentioned at the beginning of this chapter. Phosphate is a constituent of many important biological substances, including nucleic acids and phospholipids. It is an essential participant of metabolism, in which it is involved in numerous capacities. Life is truly built around phosphate. How this could have happened if soluble phosphate had been as rare in the prebiotic world as it is today is one of the most mysterious aspects of the origin of life. In the postulated acidic medium, calcium phosphate trapped in rocks should readily dissolve to give appreciable

**Table 7–1** Energetics of thioester formation. Equilibrium constants of Reaction 25, under the conditions and with the reactants indicated. Values at 25°C are derived from $\Delta G°$ values reported by William Jencks and Mary Gilchrist (297). The value of $\Delta H°$ at 25°C given by Ingmar Wadsö (305) for the hydrolysis of *S,N*-diacetyl-mercaptoethylamine has been used for the estimate of the effect of temperature on the equilibrium of both reactions (by van 't Hoff's equation).

| pH | *T*, °C | $10^6 \times K$ (gram-molecules per liter)$^{-1}$ Acetyl thioester of | |
|---|---|---|---|
| | | Mercaptoacetic acid | *N*-Acetyl-mercaptoethylamine |
| 7.0 | 25 | 5.2 | 3.0 |
| 2.0 | 25 | 920 | 536 |
| 2.0 | 100 | (2,800) | 1,640 |

concentrations of inorganic phosphate, which could enter protometabolism in the manner suggested.

A third argument supporting an acidic prebiotic medium is represented by the important role played by proton movement in bioenergetic processes. This point will be elaborated on below.

There are also serious objections against a hot, acidic medium having provided the birthplace of life. The instability of organic substances at high temperature is one. This point was underscored more than 15 years ago by Miller and Orgel, who wrote: "The instability of various organic compounds and polymers makes a compelling argument that life could not have arisen in the ocean unless the temperature was below 25°C. A temperature of 0°C would have helped greatly and –21°C would have been even better" (60, page 127). More recently, this argument has been reemphasized and bolstered experimentally by Miller and Jeffrey Bada, who conclude that the hydrothermal-vent origin-of-life theory is untenable. They state: "Any origin-of-life theory that proposes conditions of temperature and time inconsistent with the stability of the compounds involved can be dismissed solely on this basis, unless some protective mechanism exists, but no such mechanisms are known at present" (383, page 610). However, the measurements by these investigators refer mostly to very high temperatures (250–350°C) and pressures (265 atmospheres) at neutral or slightly alkaline pH values, not to the more moderate conditions in which thermoacidophilic bacteria do, in fact, thrive. Furthermore, as I shall point out later, my model allows for a much more rapid development from simple building blocks to the first cell than is generally considered (Chapter 10). Nevertheless, the problem cannot be evaded. Some of the intermediates considered, such as acyl-phosphates and acyl-adenylates, are exquisitely sensitive to hydrolysis in hot acid. For such molecules to have survived and played their role under the conditions I postulate, some sort of protection, for example, by a binding catalyst, would have been mandatory.

Another objection to the proposed model is that, even under the most favorable conditions, it still needs very high concentrations of reactants, preferably in the molar range or higher, to provide appreciable concentrations of thioester. As shown in Table 7–1, it requires molar acid and thiol for the equilibrium concentration of thioester to reach the millimolar range. Because there are two reactants and only one product (discounting water, assumed to be present at high and essentially constant concentration), the yield at equilibrium is a function of the product of the reactant concentrations. Thus, with one-tenth molar thiol and acid, the equilibrium concentration of thioester would fall to some 20 micromolar; it would drop into the nanomolar range with millimolar reactant concentrations. This brings us back to the concentration problem, which has already been raised several times.

Also to be considered very carefully is the influence of the postulated environment, especially of the low pH, on the equilibrium of protometabolic reactions. In examining this question, I shall consider only gross effects, neglecting, for obvious reasons, minor effects due, for instance, to modifications in the ionization constant of a given protonated group as it occurs in a reactant and in a reaction product. Right from the beginning, a major hurdle faces us, namely, the key multimerization reaction with amino acids as substrates. As shown by Reaction 26, this process releases one hydrogen ion at each transacylation step. This means that transacylation is considerably less exergonic at pH 2.0 than it is at pH 7.0. At 25°C, the decrease amounts to about 7 kilocalories per gram-molecule; it would reach almost 9 kilocalories per gram-molecule at 100°C (leaving out the effect of temperature on the equilibrium constant, which is likely to be small with respect to the pH effect). On the basis of what we know of the thermodynamic properties of thioester and peptide bonds (see, for example, References 140, 297), we may state that these effects should be more than sufficient to obliterate the free-energy difference that drives thioester-dependent peptide synthesis under physiological conditions. Because of this fact, the concentrations of peptides at equilibrium are likely to be lower than the concentrations of the reacting thioesters—which we have seen are themselves expected to be quite low—*and* to decrease further at each successive elongation step. It is possible that these drawbacks could have been overcome to some extent by folding of the products into a different conformation, by their adsorption to solid surfaces, or by their insolubilization (240). Since we are concerned with the synthesis of substances that are needed in catalytic amounts, it remains possible that the reaction could have played its role under the postulated conditions. But the handicap is a serious one and must be kept in mind. Note that the difficulty affects only the synthesis of peptide bonds. The energetics of the synthesis of esters, of acyl-phosphates (Reaction 32), and of acyl-adenylates (Reaction 35) by transacylation should not be greatly affected by the change in environmental conditions.

The electron-transfer reactions (Reactions 27, 28, 29, 30, and 31) are all thermodynamically very sensitive to pH. Because of the presence of two hydrogen ions in the left-hand member of the chemical equations, lowering the pH from 7.0 to 2.0 boosts reduction by almost 14 kilocalories per pair of electron-equivalents at 25°C and by some 18 kilocalories per pair of electron-equivalents at 100°C. This means, in practice, that the oxidation-reduction potential of the reaction partner must be higher by 300–400 millivolts at low pH than in a neutral medium for the system to achieve the same equilibrium state. Note, however, that this is true only if the electrons participate as such in the reaction, as they would with a mineral partner. If they are exchanged as hydrogen atoms with an organic partner, the pH effect would be cancelled out. The energy difference would be half of the values mentioned above with a partner, such as NAD or NADP, that exchanges one electron and one hydrogen atom.

Thus, thioester-dependent reductions would not be hindered, or would be favored, under the hot, acidic conditions I postulate. But the problems of stability, concentration, and multimerization are serious. A possible way out of the quandary would be to assume the existence of steep pH and temperature gradients, as indeed are present in certain hydrothermal vents. The thioesters would be formed at high temperature and low pH and would then be conveyed to more temperate regions, where they could better play their metabolic role. Another possibility is that the thioesters that spawned the prebiotic thioester world were made by a mechanism different from the one I postulated originally and adapted to milder environmental conditions. A possible mechanism of this sort exists. It relies on oxidation.

### *Oxidative Synthesis of Thioesters*

In the discussion of thioester-dependent reductions, it was pointed out that the key reactions (Reactions 27 and 28) are readily reversible. In heterotrophic organisms, they do, in fact, run mostly from right to left and accomplish the oxidative syntheses of thioesters that support substrate-level phosphorylations through Reactions 32 and 33.

The possibility thus exists that Reactions 27 and 28 developed first in the oxidative direction, helped in this respect by the ability of thiol groups to join with carbonyl groups into addition compounds (hemithioacetals), which could, themselves, have served as substrates for oxidation reactions:

$$\begin{array}{c} R'\text{—}SH + R\text{—}\overset{\overset{\displaystyle O}{\|}}{C}\text{—}H \\ \upharpoonleft\downharpoonright \\ R'\text{—}S\text{—}\underset{\underset{\displaystyle H}{|}}{\overset{\overset{\displaystyle OH}{|}}{C}}\text{—}R \\ \upharpoonleft\downharpoonright \\ R'\text{—}S\text{—}\overset{\overset{\displaystyle O}{\|}}{C}\text{—}R + 2\,e^- + 2\,H^+ \end{array} \qquad \textbf{(37)}$$

$$\begin{array}{c} R'\text{—}SH + R\text{—}\overset{\overset{\displaystyle O}{\|}}{C}\text{—}COOH \\ \upharpoonleft\downharpoonright \\ R'\text{—}S\text{—}\underset{\underset{\displaystyle COOH}{|}}{\overset{\overset{\displaystyle OH}{|}}{C}}\text{—}R \\ \upharpoonleft\downharpoonright \\ R'\text{—}S\text{—}\overset{\overset{\displaystyle O}{\|}}{C}\text{—}R + 2\,e^- + 2\,H^+ + CO_2 \end{array} \qquad \textbf{(38)}$$

Such an assumption would be consistent with the often-voiced opinion that the first living organisms were heterotrophs living off abiotically produced foodstuffs. It remains worthy of serious consideration, even if one takes, as I do, a minimalist view of the proficiency of abiotic synthetic mechanisms. No more than some of the simple building blocks that I envisage as starting material would have sufficed to support the process. A possible sequence of events would be the oxidation of lactic acid, presumably an abundant prebiotic component (see Table 6–1, page 124), to pyruvic acid by the reversal of Reaction 29, followed by the thioester-building oxidative decarboxylation of pyruvic acid by Reaction 38:

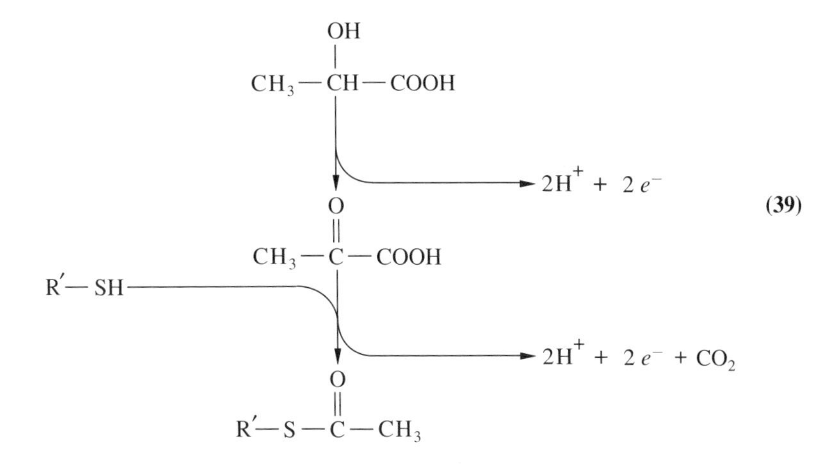


The oxidative deamination of the equally abundant alanine to ammonia and pyruvic acid by the reversal of Reaction 31 could lead to a similar result.

For such reactions to take place, an appropriate electron acceptor would be needed, but this does not seem to pose any special difficulty. Oxygen, which was almost certainly absent, or present only in trace amounts, in the prebiotic atmosphere, is ruled out in this capacity; but examples from the bacterial world offer many suitable alternatives, including ferric iron, nitrate, nitrite, sulfate, sulfur, hydrogen ions, or nitrogen. As will be shown, ferric iron is a particularly attractive candidate for the job of prebiotic electron acceptor.

A second condition would be the availability of appropriate catalysts (before the appearance of catalytic multimers, if the reaction served in the primary generation of thioesters). It is suggestive, and perhaps significant, in this respect that the primitive bacterial pyruvate oxidase and its electron acceptor (a ferredoxin) both belong to the group of iron-sulfur proteins (147). The catalytically active part of these substances consists of iron-sulfide complexes organized into clusters of four [4Fe-4S], sometimes two [2Fe-2S] or three [3Fe-3S], cradled by cysteine residues. Such complexes could have formed spontaneously in the prebiotic world, and they could, by themselves, have displayed enough catalytic activity to allow the reactions considered to take place. This is the opinion of Richard Cammack, who states in a review: "The iron-sulfur clusters are among the simplest electron-transfer groups found in biological systems, and have a good claim to have been the first to be produced during chemical evolution. [4Fe-4S] clusters can readily be formed from iron, sulfide, and thiol or cysteine peptide, under anaerobic conditions such as are supposed to have existed on the primeval Earth" (143, page 87). Supporting the possibility of an early role of iron-sulfur complexes in electron transfer is the fact that the protein part of iron-sulfur proteins may itself go back to very early times. The small ferredoxin from *Clostridium pasteurianum*, one of the most ancient members of the group (228), has

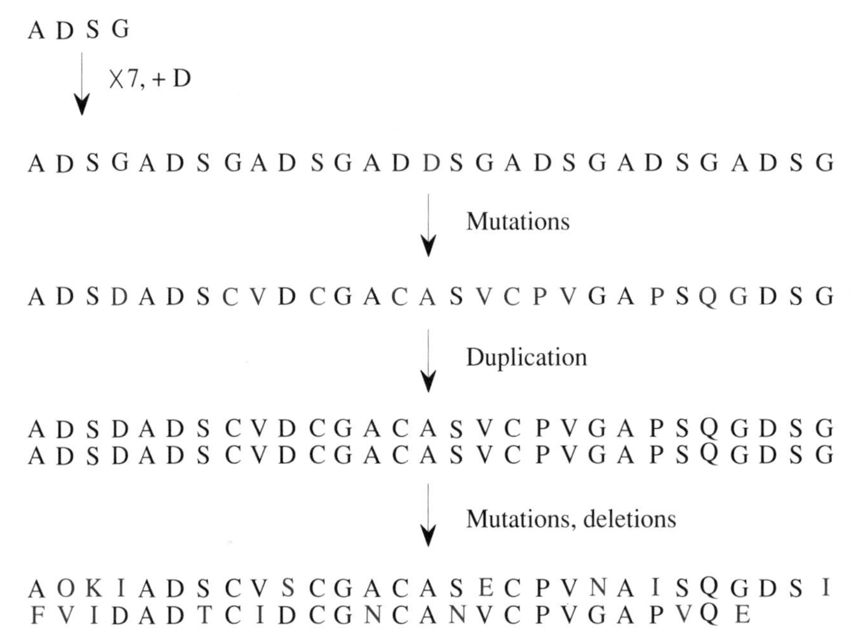


**7–4** *Evolution of a bacterial ferredoxin.* According to Eck and Dayhoff (397). Amino acid sequences are given in one-letter symbols. According to the proposed model, the original ancestor was the tetrapeptide alanyl-aspartyl-seryl-glycine. Combination of this tetrapeptide into a seven-fold repeat chain, with insertion of an aspartyl group in the middle, led to the next step. A number of mutations, followed by duplication, gave rise to a 58-residue polypeptide, which, after additional mutations and deletion of three residues at the C-terminal end, resulted in the ferredoxin of *Clostridium pasteurianum.*

been sequenced by Richard Eck and Dayhoff (397), who see it as derived, through iteration and mutation, from the simple tetrapeptide alanyl-aspartyl-seryl-glycine (see Figure 7–4). Not an unlikely product for the kind of primitive multimerization process that I visualize.

Such facts are very impressive. But they in no way prove that the reactions considered *started* in the oxidative direction. They are equally compatible with the reactions having served first in thioester-dependent reductions. In agreement with this possibility, the older type of pyruvate-oxidase reaction—in contrast with the more recent NAD-linked reaction—is freely reversible, thanks to the participation of a low-potential ferredoxin as redox partner (147). Also to be kept in mind is the fact that both types of reactions require thiamine pyrophosphate, which seems to be an essential catalyst of $\alpha$-keto-acid decarboxylations. To have served in the primary assembly of thioesters, such reactions would have had to proceed without thiamine, which is not a likely product of abiotic syntheses (110). Interestingly, thiamine, through its thiazole moiety, is a sulfur-containing substance.

We may conclude from this brief analysis that the generation of primordial thioesters by oxidative synthesis represents an attractive alternative to the spontaneous assembly process favored by acidity and heat. It does not require the high reactant concentrations that are needed by the spontaneous process. Nor does it rely on harsh environmental conditions likely to endanger the stability of fragile intermediates. It does have some drawbacks. Because of its specificity, such a reaction could produce only a single kind of thioester (acetyl-thiol in Scheme 39) or, at best, a very small number of such compounds. Should this process have served to support multimerization, thiol transfer from the primary thioesterification product(s) to other acids, especially amino acids, would have been necessary:

$$R'-S-\overset{O}{\overset{\|}{C}}-R_1 + R_2-\overset{O}{\overset{\|}{C}}-OH \rightleftharpoons R'-S-\overset{O}{\overset{\|}{C}}-R_2 + R_1-\overset{O}{\overset{\|}{C}}-OH \qquad \textbf{(40)}$$

Whether transthiolation, which is not a simple group-transfer reaction, could have taken place without a catalyst or with the help of some crude mineral catalyst is debatable. Also, the arguments in support of an acidic prebiotic medium should not be forgotten, especially the phosphate problem. If the medium was neutral, would the phosphate concentration have sufficed to drive the key phosphorolytic reactions (Reactions 32 and 34) on which the entrance of phosphate into protometabolism presumably depended? The possible role of an acidic medium in the generation of the first proton potentials, which will be considered later in this chapter, must also be kept in mind.

The matter remains open. The reactions that couple endergonic electron transfers to the breakdown of thioesters and, in the reverse direction, thioester assembly to exergonic electron transfers are of central importance in metabolism. It is very likely that they both played an early role in the development of life and in the provision of bioenergy. Which of the two directions was adopted first, assuming that one preceded the other, cannot be decided on the basis of the theoretical arguments that I have considered.

Before closing this discussion, I must mention an alternative oxidative model, proposed by Weber (277) under the name "triose model." As this name indicates, the central components of the model are the two trioses, glyceraldehyde and dihydroxyacetone, assumed to arise abiotically from formaldehyde. Oxidation of glyceraldehyde hemiacetals (or thiohemiacetals) yields glyceroyl esters (or thioesters), as in Reaction 37. These products, in turn, polymerize to polyglyceric acid, which is taken to play a catalytic role in the polymerization process and, perhaps, a crude informational one favoring self-replication. The esters (or thioesters) are also subject to phosphorolysis, yielding glyceroyl phosphate, as in Reaction 32, pyrophosphate, as in Reaction 34, and other high-energy phosphate compounds. The electron acceptor in the oxidative reaction is itself a triose, as is the donor (yielding glycerol), or some analogous molecule. This core reaction system, which allows a number of variants, is believed to be at the origin of glycolysis, with which it has a number of features in common. Further unfolding of this primitive metabolism would occur through the subsequent involvement of nitrogen and sulfur compounds (Weber accepts the possibility of an early participation of thiols in the system, as noted above, but does not consider this very likely).

Weber's triose model relies, like my own oxidative model, on the oxidative formation of activated acyl groups that further serve for multimerization and for the generation of activated phosphoryl groups. It differs from my model mainly by the nature of the starting materials, which do not contain nitrogen nor, perhaps, sulfur;

by that of the electron acceptor, which is endogenous to the system and, thus, of the fermentative kind; and by that of the multimerization product, which can hardly possess the catalytic versatility of the multimers postulated in my model. The most important difference concerns the required initial conditions. The formation of formaldehyde under prebiotic conditions (for example, by the photochemical reduction of $CO_2$) is an obvious possibility. But whether this highly reactive substance ever could have reached a high enough concentration to permit its effective polymerization to sugars is a question that has given rise to considerable debate. Even if the reaction could have taken place, it would have yielded a wide variety of different aldoses and ketoses in addition to the expected trioses (see, for example, Reference 390). The prerequisites of the thioester world are less exacting and closer to what the prebiotic environment, as presently conceived, is likely to have offered. Weber's model does, nevertheless, have interesting aspects and remains worthy of further consideration.

## THE IRON CYCLE

We have seen in the preceding chapter that UV-activated $Fe^{2+}$ ions were probably the major source of biosynthetic electrons in prebiotic days. As the amount of electrons stored in the prebiosphere increased, a corresponding quantity of $Fe^{2+}$ ions would have been replaced by $Fe^{3+}$ ions in the lithosphere. It is tempting to view $Fe^{3+}$ ions as having served as the main electron acceptor in the first biooxidations. Such reactions might have been needed very early if thioester assembly depended upon them. Even in the opposite eventuality, a variety of oxidative reactions would soon have been launched as part of protometabolism.

It is interesting to examine briefly the properties of such a $Fe^{2+}/Fe^{3+}$ cycle (see Figure 7–5). The standard oxidation-reduction potential of the iron couple at pH 7.0 is +772 millivolts, close to that of the water/oxygen couple, which is +818 millivolts (154). Thus, ferric iron is an excellent electron acceptor, comparable to oxygen. What allows ferrous iron to serve as an electron donor in biosynthetic reductions is the energy provided by UV light, just as water can act in the same capacity in higher photosynthetic organisms, thanks to the energy supplied by visible light. If, as I have proposed, prebiotic oxidations were coupled with an energy-retrieval system (thioester formation, phosphorylation), as are contemporary biological oxidations, the $Fe^{2+}/Fe^{3+}$ cycle would have served as a transducer of light energy into a biologically useful form, in the same way as the $H_2O/O_2$ cycle does today. But with an important difference. Ferrous ions act as the catalysts of their own photooxidation and can do so in an unstructured aqueous medium. The photooxidation of water requires a highly complex catalytic system assembled within the framework of a proton-tight membrane. Thus, the iron cycle appears as the more advantageous of the two, and one may well ask why, if it ever operated, it came to be replaced by the oxygen cycle. As will be seen, molecular oxygen was probably the culprit.

The banded-iron formations mentioned in Chapter 6 offer suggestive evidence of the occurrence of an iron cycle, perhaps operating in a wavelike manner such that iron

**7–5** *The iron cycle.*
(1) Thanks to the UV-supported photooxidation of $Fe^{2+}$ ions, $CO_2$ and other inorganic precursors are reduced to prebiotic building blocks, with the consumption of protons (see Figure 6–1, page 127). (2) Oxidation of the synthesized materials takes place with $Fe^{3+}$ ions as electron acceptors and is coupled to thioester-dependent substrate-level phosphorylations, capable, in turn, of supporting work. Thanks to this cycle, UV-light energy is made to support vital work.

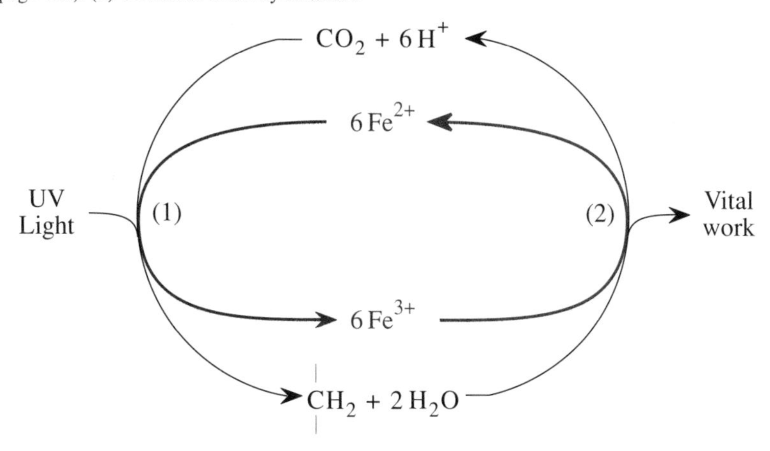


oxidation and reduction (corresponding to autotrophy and heterotrophy) alternatively predominated. The widespread distribution of these formations over the whole surface of our planet and the fact that their deposition continued steadily over more than 2 billion years attest to what must have been a major biogeological phenomenon. The decline of this process, beginning about 2 billion years ago and continuing until its extinction about 1.5 billion years ago, could have been due to quenching by molecular oxygen, produced in increasing quantities by cyanobacteria-like phototrophic organisms equipped with photosystem II. As a consequence, the UV-powered $Fe^{2+}/Fe^{3+}$ cycle would have been replaced progressively by the $H_2O/O_2$ cycle supported by visible light, which dominates the biosphere today.

Should this interpretation be correct, photosystem II could have appeared much later in the evolution of phototrophy than has been inferred by some authors from the existence of stromatolites that date as far back as 3.5 billion years ago (Chapter 5). The kind of bacterial mats from which these stratified rocks are believed to have originated are covered by cyanobacteria in the contemporary world. This does not necessarily mean, however, that their ancient counterparts were covered by similar organisms and not, for example, by more primitive phototrophs. On the other hand, it is also possible that conditions may long have remained such as to allow a balance to be maintained between the production of oxygen by cyanobacteria-like organisms and its removal by the generation of enough reducing equivalents. The $Fe^{2+}/Fe^{3+}$ and $H_2O/O_2$ cycles could have overlapped and, perhaps, interacted for a very long time, before the latter eventually took over.

Interestingly, some banded-iron formations go back to the earliest geological record of 3.8 billion years ago, quite possibly, therefore, to prebiotic times. This indicates that ferric iron may well have been the first outside electron acceptor used by emerging life. It could have supported the oxidative synthesis of thioesters at an early stage.

Also attesting to the importance of iron—and of sulfur—in the prebiotic world is the early importance of iron-sulfur proteins, already mentioned above. This fact raises the possibility that iron came to acquire its central role in biological electron transport, now exercised to a considerable extent within the framework of the heme group, by way of iron-sulfur proteins. Indeed, evidence of a possible evolutionary kinship between some hemoproteins and some iron-sulfur proteins has been reported (396). All in all, the thioester world is perhaps better named the thioester-iron world.

## PUTTING PROTONS TO WORK

Contrary to some of the opinions quoted at the beginning of this chapter, I consider encapsulation by a membrane a late event in the origin of life (Chapter 9). According to my model, early protometabolism developed within what may be described as some sort of extended protocytosol, possibly in contact with catalytic surfaces or aggregates; and the first energy-retrieval systems were soluble substrate-level phosphorylation systems. I imagine that considerable metabolic diversification took place at this stage, possibly including the appearance of complex carbohydrates and lipids and of special molecules such as flavins, porphyrins, pterines, pyridoxal, thiamine, quinones, and carotenoids and other isoprene derivatives.

In line with the above developments, energy metabolism would have also diversified. An interesting acquisition would have been the coupling of substrate-level phosphorylation processes to the oxidation of mineral electron donors. One such system exists in certain sulfur bacteria, coupling ATP assembly to the oxidation of sulfite to sulfate. Some such process could have supported a fully autotrophic metabolism, as occurs in some contemporary chemolithotrophs.

Eventually, carrier-level phosphorylation would have had to emerge and, in due time, some means of harvesting and utilizing visible-light energy. From all that is known of present-day organisms, these innovations had to be preceded by the construction of an ion-tight membrane and required the development of mechanisms for generating and exploiting protonmotive force.

The events that led to the encapsulation of prebiotic systems inside membranous envelopes will be discussed in Chapter 9. Granting the existence of such envelopes, the question arises as to how two such apparently unrelated reactions as electron transfer and ATP assembly could have come to be coupled by the unlikely link of protonmotive force. I suggest that the link came first, in the form of the proton potential that would build up automatically across the membrane in the acidic medium I postulate, if metabolic reactions taking place inside the protocell tended to raise the internal pH (see Figure 7–6). Harnessing such a gradient for uphill electron transport or, alternatively, for ATP assembly could have happened independently. With either one of these two reactions and the appropriate ancillary support, the system could derive all its energy needs from the dissipation of the proton gradient maintained by the environmental acidity and metabolic proton consumption.

7–6 *The possible origin of protonmotive force.* The figure shows schematically how the internal metabolic alkalinization of a protocell occupying an acidic environment would create a proton gradient across the membrane. Harnessing of this gradient, whether for reverse electron transfer (from a reduced flavoprotein, $FpH_2$, to $NAD^+$) or for phosphorylation, would have conferred a considerable evolutionary advantage on the beneficiary of this modification. The subsequent emigration to neutral waters of an organism possessing both systems would have allowed coupling of the two systems by means of the protonmotive force generated by one of them running in reverse.

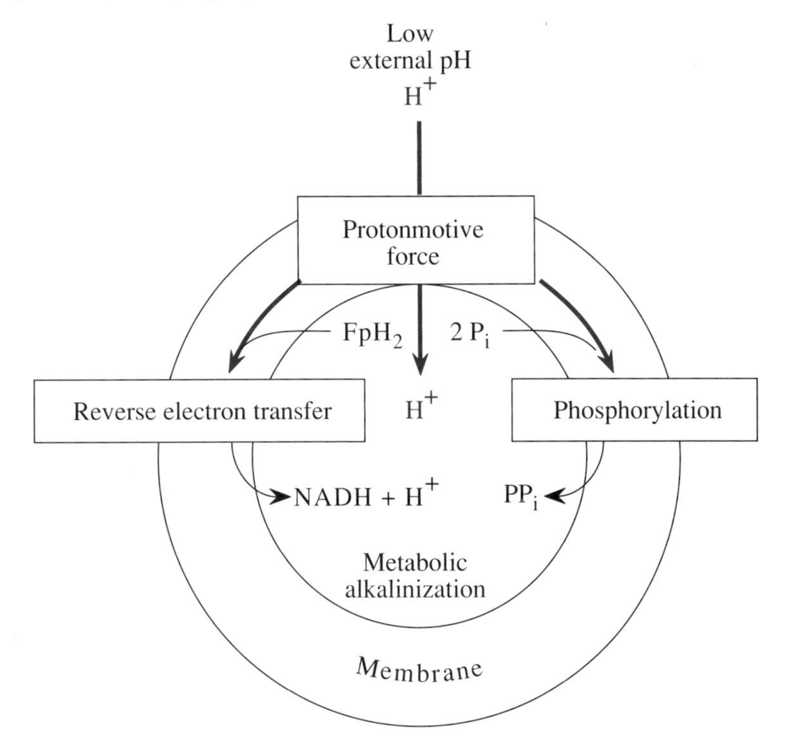


In a less acidic milieu, the two reactions would lose much of their value, unless they occurred side by side and became coupled by protonmotive force, one of them being reversed. ATP produced elsewhere within the system could support reverse electron transfer. Or downhill electron transfer could power ATP synthesis, as is more common. With electron acceptors of sufficiently high oxidation-reduction potential available in the environment, a respiratory chain with more than one phosphorylation site could become functional. Such coupled systems are now found almost universally.

As an alternative possibility, one could imagine that proton-extruding systems first emerged as mechanisms, presumably of great selective value, for raising the internal pH against a high external acidity. In one or the other event, the widespread use of protonmotive force for energy transduction throughout the living world today is explained as a legacy of a highly acidic prebiotic environment and may be viewed as a clue to the existence of such an environment. This is one more reason for believing that life originated in an acidic medium.

## THE HARNESSING OF VISIBLE LIGHT

The ultimate step toward bioenergetic autonomy was the acquisition of photochemical transducers capable of supporting a proton gradient with the help of visible light. The simplest such system is bacteriorhodopsin, the photochemical pigment of the archaebacterium *Halobacterium halobium* (Chapter 2). It has a retinoid (a relative of vitamin A) as a light-sensitive component and uses light energy directly to displace protons, apparently without the participation of electrons. Despite its simplicity, this system has had little evolutionary success, except possibly in vision, which similarly depends on a rhodopsin.

Phototrophy relies almost universally on the action of chlorophylls, which use light quanta to raise electrons from a low to a high level of energy, from which they are fed into phosphorylating electron-transport chains operating by protonmotive force (or directed towards biosynthetic reductions). Chlorophylls are magnesium-containing porphyrins, close relatives of the iron-containing porphyrins, or hemes. These are the active groups of cytochromes, which are key components of phosphorylating electron-transport chains throughout nature. It is tempting to believe that hemes developed first, as part of the iron-centered biogenetic process, and that chlorophylls came later. It would have been of little use for emerging life to develop a system for energizing electrons photochemically by means of chlorophylls if it had not had available a phosphorylating system (containing cytochromes) for exploiting the energized electrons.

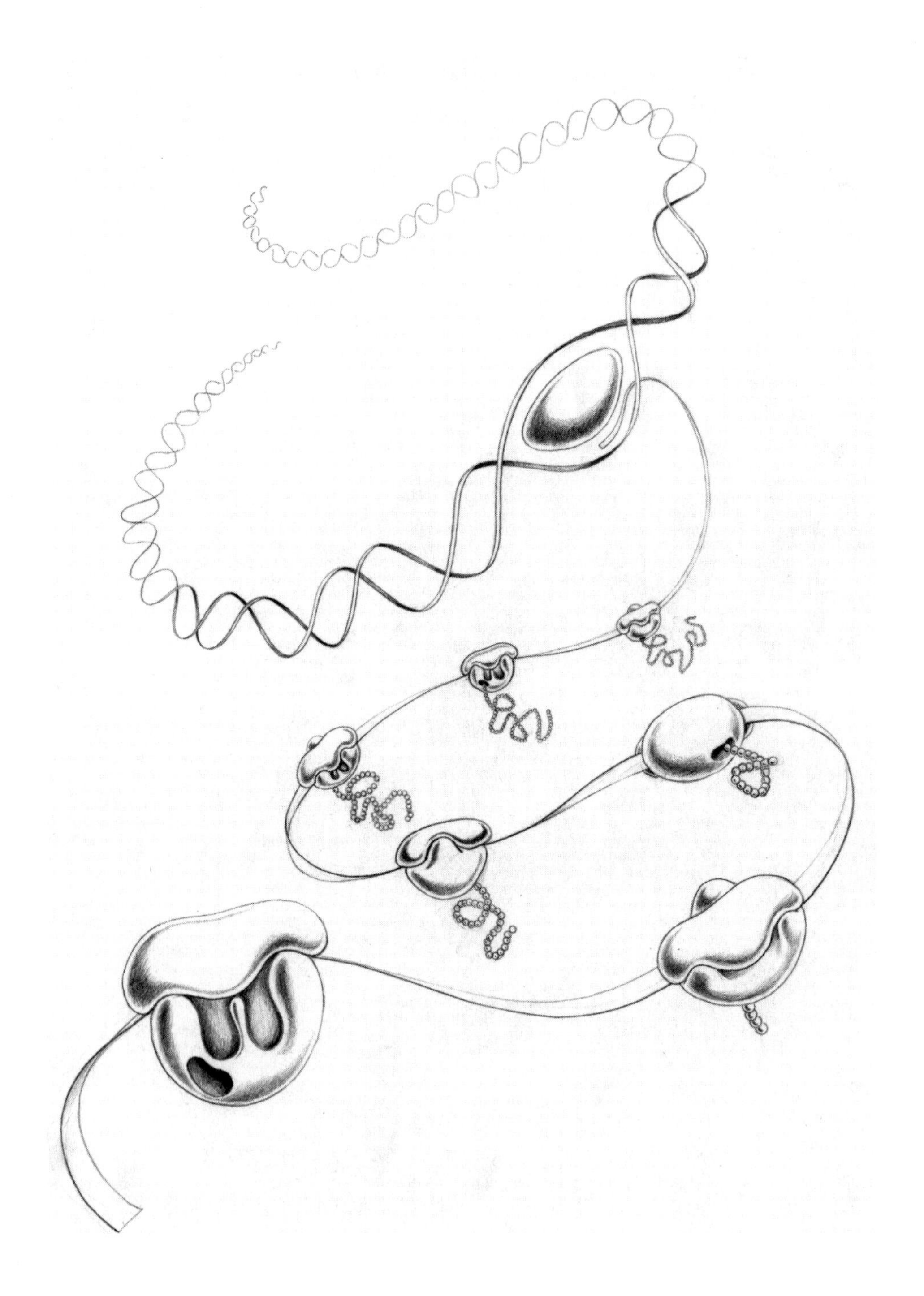

CHAPTER EIGHT

# Mastering Information

## PRELIMINARY CONSIDERATIONS

### *The First Genes*

Until it reached the RNA world, incipient life did not, so most theories would claim, have any means of storing, replicating, testing, or selecting information. This statement does not apply to my model, which, at every step of its development, rests squarely on the principle of chemical complementarity. It is through the recognition between catalysts and substrates that metabolic reactions emerge and new metabolites appear. And it is through the same process that some useful catalysts are selected for longer survival and possible improvement. As in all cases of complementarity, the relevant information is written twice: in the structure of the metabolites and in that of the catalysts. This is an important point. I shall come back to it.

With the advent of nucleotides, a new form of information is about to spread through the system. Also based on the principle of complementarity, it involves the cardinal pairing of G with C and of A with U, destined to become the very fabric of life's language. We are now entering what could be called an RNA world, although it differs greatly from what is generally understood by this term. It was reached by time-proven pathways that have left lasting imprints in the metabolism of all living organisms; and it is equipped with a varied set of multimeric catalysts, in addition to eventual "ribozymes." It does, however, share a major problem with the popular variety: how did RNA come to control protein synthesis?

Before we can address this question, a few points must be made concerning the RNA world. First, it is, indeed, an RNA world,[1] not a DNA world. In present-day life, DNA acts entirely by way of RNA, whereas RNA can act on its own. Viruses are a

[1] Because of the mechanistic difficulties raised by the abiotic assembly of the ribose-phosphate backbone of RNA, or even by the mere synthesis of D-ribose (390), several authors have proposed the existence of simpler, replicatable RNA precursors, including hydroxyapatite-bonded mononucleotides (267) and pseudo-polynucleotides with a glycerol-phosphate or some other flexible, acyclic backbone (222, 273, 380), or with a triose-phosphate hemiacetal backbone (274), or even with a pyrophosphate backbone (363). An all-purine RNA precursor, with xanthine and isoguanine replacing uracil and cytosine, respectively, has been proposed by Wächtershäuser (274, 276) to account for the allegedly easier prebiotic formation of purines as compared to pyrimidines. With protoenzymes to guide the formation of correct mono- and oligonucleotides from the start, my model has no need for such precursors.

case in point. Therefore, RNA must have preceded DNA in the history of life on earth. This is generally agreed on.

Next, we may take it that replication started almost as soon as the first oligonucleotides made their appearance. Experiments by Orgel (112, 224) have shown that such a process can take place with very crude catalysts, zinc ions, for example. Furthermore, the possibility of an RNA-catalyzed replication of RNA may also be envisaged (320, 417). The likelihood of replication is even greater in the system considered here, which is assumed by definition to contain all necessary catalysts. The term "replication" refers, of course, to the familiar two-step process that is based each time on the synthesis of a complementary copy directed by base pairing. This process is believed to have been rather imprecise in its prebiotic form and to have yielded what Woese (66) has called "statistical molecules," and Eigen (220, 250) "quasi-species," that is, populations of closely similar, but not identical, molecules (which could be called molecular variants or mutants).

An important consequence of the appearance of replicatable oligo- or polynucleotides is to render possible a process of Darwinian evolution at the molecular level. First demonstrated by Sol Spiegelman (408), this remarkable phenomenon was subsequently investigated in detail by the groups of Orgel (224, 342) and Eigen (220). The facts are as follows: RNA replicating in a test tube under the influence of the replicase enzyme from the phage Q$\beta$ rapidly evolves, as a result of replication errors and of competition between the "mutants," toward a stable quasi-species of molecules exhibiting maximal fitness under the conditions chosen. For molecules, Eigen and Peter Schuster write, "selective value is defined as an optimal combination of structural stability and efficiency of faithful replication" (252, page 357). If conditions are changed—by the addition of an inhibitor, for example—molecular evolution produces a different kind of RNA, one adapted to the new conditions. Thus, a new selection factor, namely the ability to be replicated more efficiently, has been added to molecular survival, already mentioned in Chapter 6. This point will be of importance in the discussions that follow. Recent work has shown that more subtle selection factors, such as the display of some catalytic activity, may also be made to intervene (403a, 422a).

Most likely, the first RNA "gene" coded only for itself and was, therefore, shaped by molecular selection. Eigen and Ruthild Winkler-Oswatitsch (378) have tried to make an educated guess as to what such a molecule would be like. Their recipe for the "Ur-Gen" goes like this: 1) The molecule is between 50 and 100 nucleotides long. A shorter one would be too unstable; a longer one would succumb to replication errors.[2] 2) The molecule is made mostly of GC (GC pairs are stronger than AU pairs) to insure greater stability and, especially, stronger bonding during the slow process of replication; but it should be "doped" with AU (about one for every 10 GC) for easier

[2] Eigen and Schuster (250) have derived a rule that allows one to calculate the maximum length a replicatable polymer may attain without losing its information upon repeated replication cycles as a function of replication fidelity and as a function of a parameter measuring the reproductive performance of the polymer. In rough approximation, this length is equal to the inverse of the average error rate per monomer inserted. The maximum length of 100 nucleotides corresponds to an average error rate of 1% for the primitive replication of a GC-rich RNA.

melting during replication. 3) Both halves of the molecule are complementary in antiparallel fashion, so as to fold into a stable, double-helical hairpin structure, which is identical, except at the level of its hinge, in the + and – replicative strands. According to the authors, vestiges of such a structure can be found in today's tRNAs, all of which they believe to be descended from a single ancestral molecule or, more precisely, molecular quasi-species, answering their description of the "Ur-Gen" (see Figure 8–1, page 177).

This partly deductive, partly inductive, reconstruction of the first gene is consistent with the widely accepted view that some sort of proto-tRNA initiated the involvement of RNA in protein synthesis. I shall examine this point below. A corollary of the hypothesis is that the first proto-mRNAs must have been siblings of proto-tRNAs, which implies that we should find traces of a primitive code in the sequence of the "Ur-Gen."

## *The Origin of the Genetic Code*

Few topics hold more fascination for biologists of every kind, and even for non-biologists, than the process whereby 20 amino acids became selected for protein synthesis and matched with informational nucleotide triplets. In spite of an enormous amount of both experimental and theoretical work,[3] the answer to this central problem remains elusive. But a few basic points have emerged.

1. The code developed gradually. As pointed out by Woese (66), early translation must have been very inaccurate, and it is likely that the first codons coded for groups of similar amino acids and not for single ones. Orgel (266) has even suggested that the first code had only two codons, one for hydrophilic amino acids (philines) and the other for hydrophobic amino acids (phobines).
2. Early proteins were made from a small number of amino acid species, perhaps as few as 4. The present number of 20 was reached progressively by incorporation of new amino acids into the system. Attempts at reconstructing the order in which amino acids were adopted will be considered below.
3. In spite of the small number of initial amino acids, the first code used triplet codons and not the doublets that might have sufficed. This statement has been defended by Crick on the basis of the "principle of continuity." As he writes: "A change in codon size necessarily makes nonsense of *all* previous messages and would almost certainly be lethal" (242, page 372). Which does not mean, he hastens to add, that the primitive code was a triplet code. The third base of each codon could have carried no information, for example, as has been suggested by Thomas Jukes (130). As I shall mention later, there are good steric reasons for the magic number three. Closer spacing would have made it difficult for two charged tRNAs to sit

[3] It is not possible within the limited framework of this book to review in any detail all the investigations and speculations devoted to the origin of the genetic code. Readers may get an idea of the diversity of approaches and viewpoints in relation to this problem by consulting References 66, 73, 130, 242, 243, 249, 252, 258–260, 268, 271, 280–282, 291, 315.

**8–1** *Models of primitive tRNAs, as reconstructed by Eigen and Winkler-Oswatitsch (378).* The original publication should be consulted for understanding the details of the proposed structures. R = purine; Y = pyrimidine; N = any base. **(a)** Statistical "master sequence" of 3′ half of the tRNA molecule, computed from 144 tRNA sequences known at the time. Colored circles indicate deviations with respect to a coding sequence utilizing the two-dimensional RNY code depicted in Figure 8–2b, page 179. The adjacent sequence illustrates the structure of the corresponding peptide. Question marks refer to what the authors call "jokers." These are nucleotides occupying constant positions, which presumably reflect structural constraints specific to the tRNA molecules rather than primitive coding. **(b)** "Master sequence" (with triplet 61–63 omitted) folds into a hairpin structure that brings the acceptor and anticodon ends in close juxtaposition (see also Reference 261). Out of 15 joining base pairs, 7 follow Watson-Crick rules (solid bars), and 3 more are of R-Y type (broken bars). **(c)** "Grandfather sequence" of a hypothetical, more primitive, ancestral tRNA molecule obeying the one-dimensional GNC code shown in Figure 8–2a, page 179. Color illustrates the replacements required to derive this sequence from that of the "master sequence." Extent of correct base pairing is notable. **(d)** "Grandfather sequence" folds readily into a cloverleaf structure, again with extensive correct base pairing.

side by side along a messenger strand. Wider spacing would have complicated peptidyl transfer.

4. The first code was comma-less, by which is meant that messages were readable only in one frame. Jamming of the system by improperly positioned charged tRNAs was thereby prevented. According to Crick and coworkers (243), the first codons (in agreement with this requirement) had the structure PuPuPy. The Eigen group, for reasons that should be read in their original publications, prefer the structure PuNPy, more precisely GNC, which happens to code for the four most abundant amino acids obtained in prebiotic simulation experiments (see Tables 6–1, page 124, and 6–2, page 125): glycine (GGC) and alanine (GCC), both 20 times more abundant than the other amino acids; aspartic acid (GAC) and valine (GUC), which are 5 times more abundant. On the basis of this hypothesis, the authors have proposed an ingenious model of development from what they call a one-dimensional code (only the middle base is significant, framed by G and C) to the present, three-dimensional code (378). This model is illustrated in Figure 8–2, page 179. Its plausibility is supported by the sequences represented in Figure 8–1. As shown in Figure 8–2a, the reconstructed "master sequence" is suggestive of an erstwhile PuNPy code, which was followed at a time when members of the tRNA quasi-species also functioned as mRNAs. Figure 8–2c shows how a GNC structure can be derived from the "master sequence" by only 27 base substitutions (out of a total of 72), on condition that triplet 61–63 be excised.
5. An important, and still controversial, point concerns the structure of the code. As first noted by Sonneborn (136), this structure seems to be such as to minimize translational errors due to misreading and the deleterious effects of point mutations. In many cases, the replacement of one base by another either does not change the nature of the inserted amino acid or substitutes for it an amino acid with similar physical characteristics so that the functional properties of the altered protein are not significantly modified. For this reason, it has frequently been proposed that the genetic code was optimized by natural selection. Woese (66) was the first to point to a logical inconsistency in this proposal: any mutation that changed the meaning

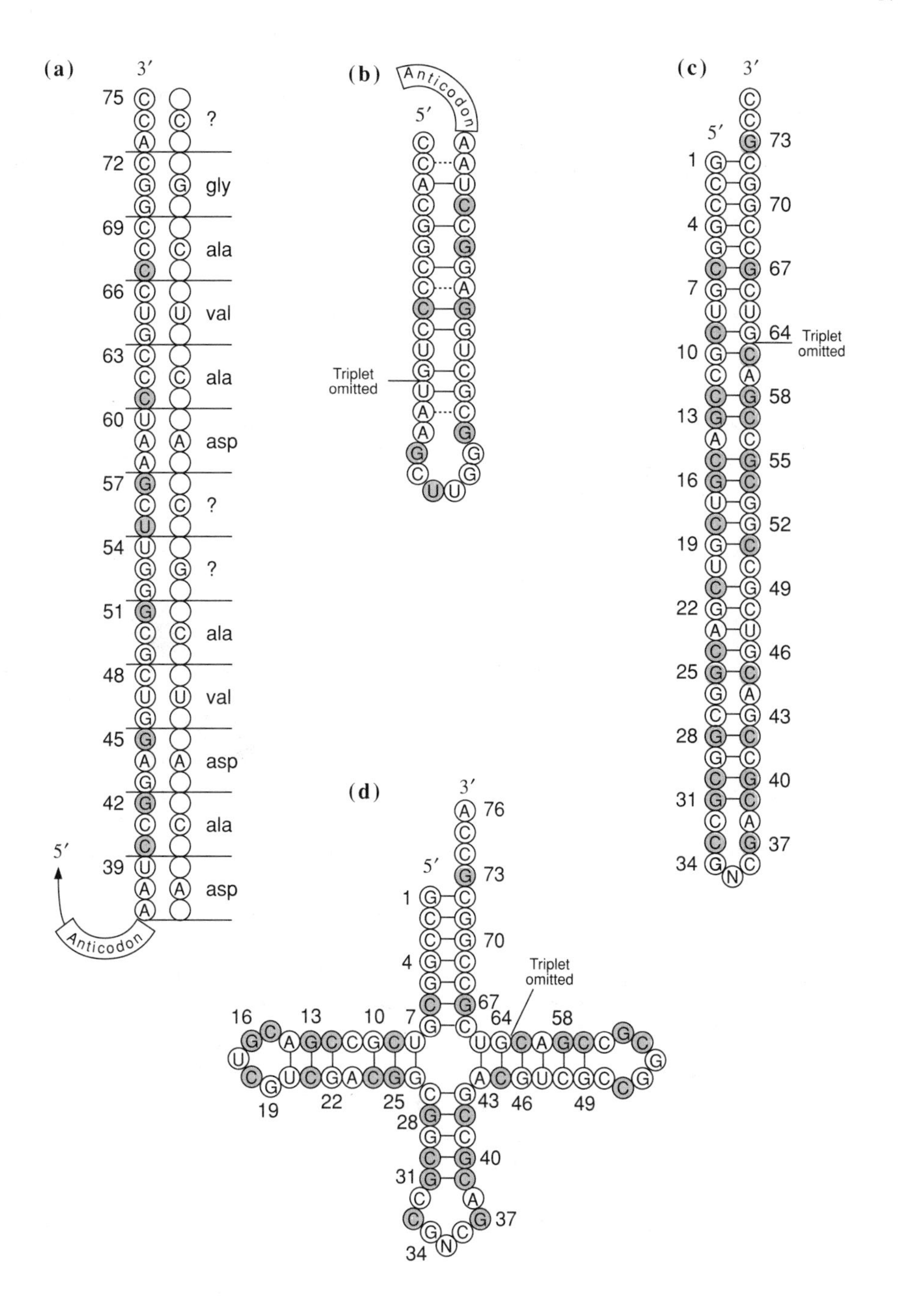
(a)
3′
5′
Anticodon
gly
ala
val
ala
asp
?
ala
val
asp
ala
asp
(b)
Anticodon
5′
Triplet omitted
(c)
3′
5′
Triplet omitted
(d)
3′
5′
Triplet omitted

**8–2** *Evolution of the genetic code, according to Eigen and Winkler-Oswatitsch (378).*
The frames show the first letter (left), second letter (top), and third letter (right) of the codons. R = purine; Y = pyrimidine; N = any base. (**a**) Ancestral one-dimensional GNC code (GNY if pyrimidine wobbling is allowed in the third position). (**b**) Two-dimensional RNY code arising from (a) by the addition of adenine in the first position. (**c**) Intermediate three-dimensional RNN code derived from (b) by the addition of purines in the third position (with wobbling). (**d**) Fully degenerate three-dimensional NNN code.

of a codon would most likely be lethal because it would cause most proteins in the organism to be altered at the same time. The mutant would have no chance to survive, and natural selection no opportunity to select.

Another serious objection to the selection theory is that its assumed basis could be incorrect. If synonyms are discounted, there is no statistically valid evidence that replacement of one base by another in a codon will cause the corresponding amino acid to be replaced by one with similar physical properties more frequently than by one with different properties (281, 404). What, instead, seems to be favored by probability is that the amino acid will be replaced by one that is metabolically related to it, either as a precursor or as a product. This finding has led J. Tze-Fei Wong to propose an interesting coevolutionary theory linking the development of the genetic code to the progressive appearance of new amino acids (280–283). Like many other authors, he assumes that translation started with a small number of amino acids (those most readily made by abiotic mechanisms) that divided most or all of the available codons among themselves and that new amino acids entered the system by appropriating codons from their erstwhile owners. What is original to his theory is the mechanism Wong proposes for this appropriation. New amino acids, he suggests, stole codons from their metabolic precursor amino acids, either competitively, by vying with the precursors for a specific tRNA, or biosynthetically, by actually arising from tRNA-linked precursors, as still happens today in the formation of formyl-methionine (301), glutamine in a number of instances (350, 367), and selenocysteine (333). The codons of the new amino acids would thus be related to those of their metabolic precursors because they originally belonged to the precursors. Wong's model is illustrated in Figure 8–3, page 180, which may profitably be compared with Figure 8–2.

6. Another important point, related to the preceding one, concerns the objective basis, if there is one, of codon assignments. Ever since the genetic code was deciphered—and even before (see Reference 255)—investigators have searched for possible stereochemical relationships between amino acids and their codons or, preferably, their anticodons. Except for bringing up a certain correlation in hydrophobicity (223), this search has not been very fruitful so far. (See, however, Shimizu's model below.) Consequently, many workers have subscribed to the theory, proposed by Crick (242), that codon assignments were essentially accidental and that the code's structure eventually became "frozen" when changes would have been too disruptive to be compatible with life. The fact that changes did sometimes occur—with suppression mutations, for example (see Chapter 3, Footnote 24, page 93), and, especially, in mitochondrial systems (see Table 3–3,

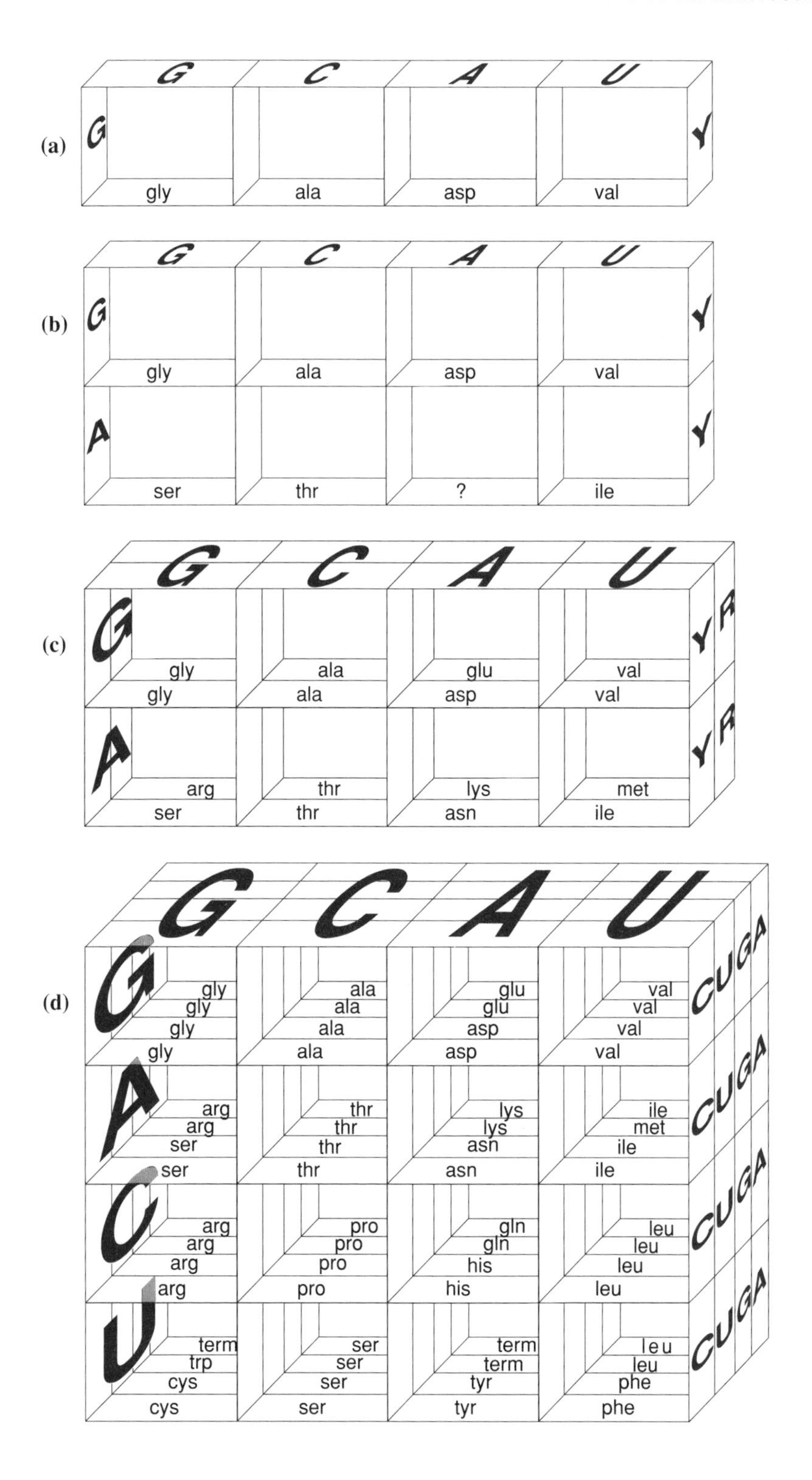
(a)
G C A U
G Y
gly ala asp val
(b)
G C A U
G Y
gly ala asp val
A Y
ser thr ? ile
(c)
G C A U
G YR
gly ala glu val
gly ala asp val
A YR
arg thr lys met
ser thr asn ile
(d)
G C A U
G CUGA
gly ala glu val
gly ala glu val
gly ala asp val
gly ala asp val
A CUGA
arg thr lys ile
arg thr lys met
ser thr asn ile
ser thr asn ile
C CUGA
arg pro gln leu
arg pro gln leu
arg pro his leu
arg pro his leu
U CUGA
term ser term leu
trp ser term leu
cys ser tyr phe
cys ser tyr phe

**8–3** *Model proposed by Wong (280) for the coevolutionary development of the genetic code and of amino acids.*

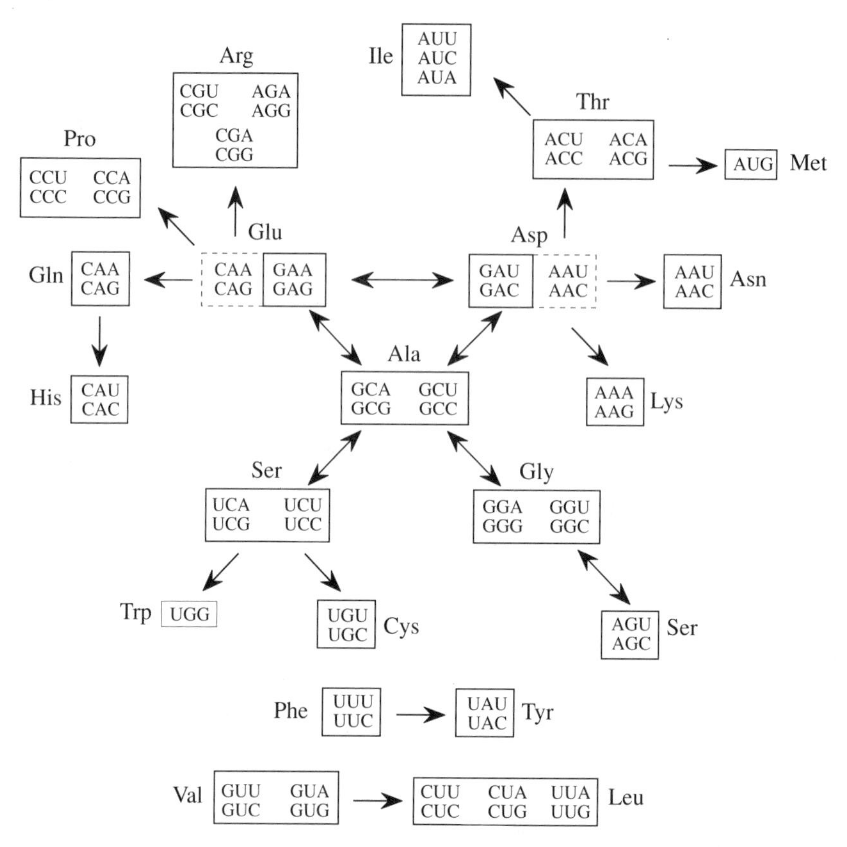


page 93)—clearly demonstrates the nonimmutability of codon assignments (130). This view calls for some comments.

First, if there is any substance to developmental models of the kind illustrated in Figures 8–2 and 8–3, codon assignment was far from accidental but, rather, occurred on a "first-come-first-served" basis. Instead of a "frozen accident," the code structure would be better described as a "historically necessary accident."

Second, the stereochemical theory may not be dead after all. Prompted by early work that had suggested that the fourth nucleotide from the 3′ end of the tRNA molecule may have a special "discriminator" function (319),[4] Mikio Shimizu has made the intriguing observation that if the molecule is folded so as to bring this nucleotide in contact with the anticodon, a "pocket" is formed that specifically accommodates the cognate amino acid (271). If proto-tRNAs had the structure shown in Figure 8–1b, page 177, at the time codon assignment took place, such a

[4] Paul Schimmel has pointed out to me that more recent results do not entirely bear out this conclusion.

pocket could have guided selection, as has, indeed, been proposed by J. J. Hopfield (261). This, however, would mean that all 20 proteinogenic amino acids were selected by proto-tRNA molecules having this kind of primitive structure.

Related to the problem of codon assignment is the hotly debated question of "tRNA identity," with which I have associated the notion of a "second genetic code" (418).

## *The Second Genetic Code*

As explained in Chapter 1, the real act of translation is accomplished by the aminoacyl-tRNA synthetases, the enzymes that attach the amino acids (protein language) to their cognate tRNAs (nucleic acid language). The reaction catalyzed by these enzymes takes place in two steps (see Figure 1–12, page 25). First, the amino acid is activated to an enzyme-bound aminoacyl adenylate (perhaps in equilibrium with a covalent aminoacyl-enzyme combination linked by a thioester bond, see Reference 303). Next, the aminoacyl group is transferred to the tRNA. In terms of information, the first step involves recognition of the amino acid, the second one, that of the nucleic acid. These two recognition sites, imprinted in the structure of the enzyme, thus govern the primary act of translation. There are 20 such enzymes, one for each amino acid, irrespective of the number of distinct tRNA molecules (up to six) that carry the same specificity. Consequently, there are 20 pairs of recognition sites (assuming that homologous tRNAs are recognized in the same way, which seems probable). These 20 pairs define the primary dictionary of genetic translation, the primary code.

Considerable interest has been devoted in recent years to what is called tRNA identity, that is, the structural features in the tRNAs that are recognized by the aminoacyl-tRNA synthetases (for reviews, see References 181, 194, 202–204, 423, 426). In a number of cases, these features are represented mainly by the anticodon or include the anticodon as a significant component (see, in particular, References 181, 339, 351, 352). To this extent, the primary code coincides with the classical genetic code, as any lover of logical simplicity would have predicted a priori. However, there are about as many cases in which the anticodon has little or nothing to do with tRNA identity. The simple fact that a number of aminoacyl-tRNA synthetases recognize two or more distinct tRNA molecules bearing different anticodons (see Figure 1–13, page 26) already points to the existence of other recognition features more invariant than the anticodons. As pointed out by Paul Schimmel, this fact "implies that, for at least some of the enzymes, the anticodon is not the primary determinant for recognition" (203, page 2,747). The alanine-specific tRNA is particularly impressive in this respect. Its identity seems to depend exclusively on the presence of a G3-U70 wobble base pair in the acceptor stem of the molecule (179, 328, 336). Even a small minihelix lacking the anticodon but containing the characteristic feature is efficiently charged by the enzyme (321). However, it is not entirely clear whether the bases are recognized as such or by some structural irregularity that they produce (335). For several other tRNAs, the determinants are more scattered but remain concentrated on the acceptor stem and in its neighborhood (i.e., near the carried aminoacyl group).

It follows from what has just been seen that the primary code overlaps only partly with the classical genetic code. Because of this fact, which may be of crucial importance with respect to the origin of life, I have put forward the concept of a "second genetic code imprinted into the structure of aminoacyl-tRNA synthetases," and I have proposed the name "paracodon" for whatever structural features of the tRNA molecule are recognized by the enzyme (418). According to this definition, paracodons and anticodons are distinct but with a zone of overlap, like the codes they define. These proposals have not been received kindly by many experts. The "second genetic code" is "an unfortunate term," according to LaDonne Schulman and John Abelson, because it "implies that a common set of rules governs tRNA recognition by aminoacyl-tRNA synthetases" (423, page 1,591). Michael Yarus voices a similar objection. "I do not believe," he writes, "that this is a useful metaphor for the tRNA-aminoacyl-tRNA synthetase interaction. The idea of coding, of uniform transliteration between two sets of uniform symbols, brilliantly captures the essence of message action. But tRNA identity needs a more flexible notion that recognizes the geometrical and biochemical complexity of the identity set" (426, page 741).

This dispute is over semantics. Whether it is called a second code or otherwise, the underlying concept remains of central importance. What strikes me as particularly significant is the nondegenerate character of the second code. It includes no synonyms: 20 synthetases for 20 amino acids, whatever the number of tRNAs. This indicates that paracodons, even when they include the anticodons, could be subject to a very strict structural determinism. Also impressive is the fact that the second code seems to be highly conserved in evolution. Such is the case for the G3-U70 determinant of the alanine-specific tRNA (329) and, perhaps, for several others. For complementarities between nucleic and peptidic structures, such properties are remarkable. As will be seen, the second code could be the older one. It could go back to the very dawn of the biogenic process.

## THE ORIGIN OF TRANSLATION

### *The Primary Interaction*

Let us recall the setting. Our system, presumably as yet unconfined (Chapter 9), depends on a primitive, thioester-energized, autotrophic protometabolism, catalyzed by rudimentary, multimeric protoenzymes assembled from thioesters. It has reached a state where oligonucleotides are being formed, replicated, and selected on the basis of their stability and replicatability (molecular evolution). Logically, the first step toward translation in a system of this sort must have been the occurrence of interactions between certain oligonucleotides, destined to become tRNAs (proto-tRNAs), and amino acids. Advocated in one form or another by Woese (66), Dillon (53), Folsome (56), Hopfield (261), and Orgel (268), among others, this theory is widely accepted, whatever "world" is postulated as the setting. It is consistent with Eigen's reconstruction of the earliest RNAs as being ancestral to tRNAs.

A primary interaction between proto-tRNAs and amino acids raises four questions. What was the source of the energy of the aminoacyl-RNA bond, which is of high-energy type? How was the reaction catalyzed? How and when did information enter the system? What kind of selective forces drove the evolutionary development of the process?

In the framework of the present model, taken to be abundantly supplied with activated amino acids, the energetic aspect of the interaction poses no problem. We need only assume an attack on an activated aminoacyl group by the 3′-terminal hydroxyl group of the proto-tRNA (represented as RNA — OH).[5] The objects of the attack could have been thioesters:

$$\mathrm{R'{-}S{-}\overset{\overset{\displaystyle O}{\|}}{C}{-}R + RNA{-}OH \rightleftharpoons R'{-}SH + RNA{-}O{-}\overset{\overset{\displaystyle O}{\|}}{C}{-}R} \qquad \textbf{(41)}$$

Or, more likely, they would have been acyl adenylates formed either from thioesters by Reaction 35, page 155, or, as they are in present-day systems, by transadenylylation from ATP (the reverse of Reaction 36, page 155):

$$\mathrm{AMP{-}O{-}\overset{\overset{\displaystyle O}{\|}}{C}{-}R + RNA{-}OH \rightleftharpoons RNA{-}O{-}\overset{\overset{\displaystyle O}{\|}}{C}{-}R + AMP{-}OH} \qquad \textbf{(42)}$$

The catalytic aspect of the reaction, which is sometimes attributed by advocates of a protein-less RNA world to the activity of some "ribozyme" or to that of the proto-tRNA itself, should not worry us either inasmuch as appropriate multimeric catalysts are part of the proposed model. Of greater interest are the questions related to information and to selection.

Let us consider the last question first. Translation still being a long way off, only molecular properties, mainly increased stability and more efficient replicatability, could have operated in the selection of the proto-tRNAs. We must, therefore, assume that those oligonucleotides that came to bear aminoacyl groups on their 3′-terminal hydroxyl turned out to be more stable than uncharged oligonucleotides, or that they were better substrates for replication, or both. These possibilities are not implausible, considering that replication starts at the 3′ end of the template. It is thus possible that aminoacylated oligonucleotides shared a 3′-terminal conformation that favored the onset of replication or, as suggested by Orgel (268), prevented the initiation of replication at internal positions of the template. It is also possible that this conformation rendered the molecules more resistant to breakdown.

For such a selection to occur, some oligonucleotides must have been more likely than others to capture aminoacyl groups. Had the process been purely random, there would have been no basis for selection. Perhaps capture was facilitated by some structural feature, such as a CCA 3′-terminus, that promoted acceptance of any sort of aminoacyl group. Molecules with additional binding sites that were specific for a given side chain or for a group of side chains would have enjoyed a further advantage if the kind of folded conformation induced by such a binding enhanced their selective

[5] Alan Weiner and Nancy Maizels (365) have made an alternative proposal adapted to the concept of a primitive RNA world. In their model, the energy necessary for the activation of aminoacyl groups is assumed to come from RNA phosphodiester bonds taken to be attacked by free amino acids.

value (compare structure (b) to structure (a) in Figure 8–5, page 189). I find this possibility particularly attractive because it provides a simple answer to the information problem: matching of amino acids with tRNAs started as a result of direct interactions, endowed with a measure of specificity, between the two partners. The indirect, enzyme-mediated form of matching came later, after the translation machinery had been set into place. That a specific interaction can indeed occur between an amino acid and an RNA has been shown for arginine (368).

Here is where the notions of paracodons and of a second genetic code become relevant. If the prime mover of specific associations between amino acids and certain proto-tRNAs was a direct molecular interaction, one might hope to find traces of it in the mechanism catalyzed today by aminoacyl-tRNA synthetases. Such could well be the case. There are kinetic indications that the bound aminoacyl group influences tRNA binding (for reviews, see References 202, 204). An old experiment has even shown that a tRNA can play a role in the process whereby mismatched aminoacyl groups are edited off before their transfer (294). This observation prompted Woese to write more than 20 years ago: "The fact that there is any distinction made between two amino acids at this second step . . . is intriguing, because it suggests the possibility that the tRNA may indeed be looking at the amino acid it carries" (66, page 124). This perceptive comment has lost none of its timeliness.

There is, of course, no proof that the tRNA "looks" at the amino acid directly, nor that this interaction has anything to do with a primeval interaction that allegedly drove a proto-tRNA and an amino acid to a common destiny leading to protein synthesis. But the hypothesis is attractive. It would explain the nondegenerate, and apparently deterministic, character of the second genetic code. Indeed, one is inclined to suspect that a nucleotide housing capable of specifically harboring a given amino acid can have but few degrees of freedom. It would also be understandable, in the framework of this hypothesis, that tRNA-identity features are frequently situated in the vicinity of the acceptor end of the tRNA. Anticodons would seem too far away to "look" at the amino acid.

According to my proposal, therefore, what I have called the second genetic code could, in fact, be first in time and based on stereochemical relationships—or, at least, it could be an evolutionary offshoot from such a primitive code. In contrast, the classical code could, indeed, be Crick's "frozen accident," which came to be superimposed on the older code. This implies that anticodons came on the scene later—which, we shall see, is a very likely possibility—and that they substituted for, or added to, the original paracodons in some instances.[6]

[6] The detailed structure of *E. coli* glutaminyl-tRNA synthetase complexed with its tRNA and with ATP has recently been elucidated at 2.8 Å resolution by the group of Thomas Steitz (349). According to the authors, "There is no direct interaction observed between the amino acid and the tRNA in the glutaminyl-tRNA synthetase complex. . . . Thus, the suggestion of a recognition code that arose through specific tRNA interactions with specific amino acids appears unlikely" (page 1,141). I would rather conclude that, if a recognition code arose in this way, it left little trace in the structure of this particular enzyme. It should be noted also that the interaction I postulate would not be with a free amino acid but with an aminoacyl group, either thiol-linked (Reaction 41) or AMP-linked (Reaction 42).

Whatever the relationship between the features that define tRNA identity today and those that may have determined the original matching between proto-tRNAs and amino acids (or rather aminoacyl groups), the hypothesis that this matching rested on direct interactions between the two partners appears conceptually simpler and more plausible than the alternative hypothesis that postulates the intervention of some coupling agents prefiguring the aminoacyl-tRNA synthetases. This opinion is shared by a number of authors but not by all. Orgel, for example, writes: "The first step in the evolution of protein synthesis almost certainly was the appearance of RNA adaptors that captured amino acids from their environment" (268, page 469). Hartman, on the other hand, prefers to assume "that the existence of an activating enzyme was a necessity at the beginning . . . (and that) there was no specific physical-chemical interaction between the amino acid and the tRNA" (258, page 423). In line with Cairns-Smith's theory (50), he suggests that clays may have served as mediators of the union between the two partners. A major advantage of the hypothesis of a direct interaction is that it readily explains how RNA came to be involved in protein synthesis in the first place. It could even account for the choice of building blocks, including their chirality. There are no answers to these questions, but they deserve to be raised. Why are proteins made of only 20 amino acids and not the 32 (or rather 31, to account for stop codons) allowed by wobbling? Why the L and not the D isomers? Why the 20 that we know and not others? Why valine, for example, and not isovaline? Why not $\alpha$-aminobutyric acid, in spite of its probably having been one of the most abundant prebiotic building blocks (see Tables 6–1, page 124, and 6–2, page 125)? Perhaps the proto-tRNAs helped to select the winners. Shimizu has made a similar point, attributing the sifting role to his anticodon-discriminator "pockets" (271).

## *The Assembly of Peptides*

Granting the formation of aminoacyl-proto-tRNA complexes, the next logical step would be the participation of such complexes in acyl-transfer reactions leading to the assembly of peptides. For such a reaction to take place, the two partners need to be aligned next to each other in a configuration conducive to transfer and in contact with a catalytic center capable of mediating this transfer. This is essentially what happens on the surface of ribosomes.

It is tempting to assume, in view of the inherent ability of nucleotide sequences to join by base pairing, that RNA molecules were instrumental in bringing together the substrates of the primeval acyl-transfer reaction. It is possible that RNA molecules also assumed the catalytic role, as adherents of the RNA-first hypothesis like to believe. Instead or in addition, some multimeric catalyst, as foreseen in my model, could have been involved. So far, present-day ribosomes have given little information one way or the other. It is not yet known with certainty whether their peptidyl-transferase activity is due to protein components or to RNA components or to both.[7]

[7] Participation of rRNAs in the peptidyl-transferase activity of ribosomes is suspected by a number of workers (100, 101, 193).

If RNA molecules were involved in early peptide synthesis, what relationship did they bear to present-day RNAs? Because of the importance of bringing the aminoacylated 3′ ends of the proto-tRNAs together in a catalytic environment, I like to think of the first helper RNAs as being precursors of ribosomal RNAs, proto-rRNAs. On the other hand, an early involvement of proto-mRNAs has to be predicated if we accept Eigen's model of an ancestral RNA quasi-species that included both proto-tRNAs and proto-mRNAs. Perhaps both kinds were needed for chain elongation to take place effectively. An important factor in this connection is strength of bonding. Primitive peptide assembly must have been clumsy and slow. Therefore, attachment of the reaction partners to their support must have been a durable one, so as to keep them in place long enough for a full cycle of peptidyl acceptance and donation to occur. This is one more reason why Eigen likes "stickier," GC-rich molecules. To ensure sufficient bonding strength with proto-mRNAs alone as stabilizers, Crick and coworkers (243) have postulated a flip-flop mechanism that allows a total of five bases to participate in proto-tRNA/proto-mRNA binding. This model has been adopted by the Eigen group (252). An interesting suggestion has been made by Hans Kuhn and Jürg Waser (262) on the assumption that the proto-tRNAs had the hairpin structure illustrated schematically in Figure 8–1c, page 177. They propose that such structures adopted a left-handed, double-helical configuration and formed stable stacks along a "collector strand," in the manner depicted in Figure 8–4 (see also Reference 263a). The authors see amino acids as attaching later to such stacks, but this is probably not essential to the validity of their model. Presumably, the kind of lateral joining that they postulate could occur as well with aminoacyl-bearing proto-tRNA molecules. What is less clear are the conditions that would be required to cause such structures to join as proposed, in spite of their high negative charge: high ionic strength, low temperature, calcium ions, other factors? The authors remain vague on

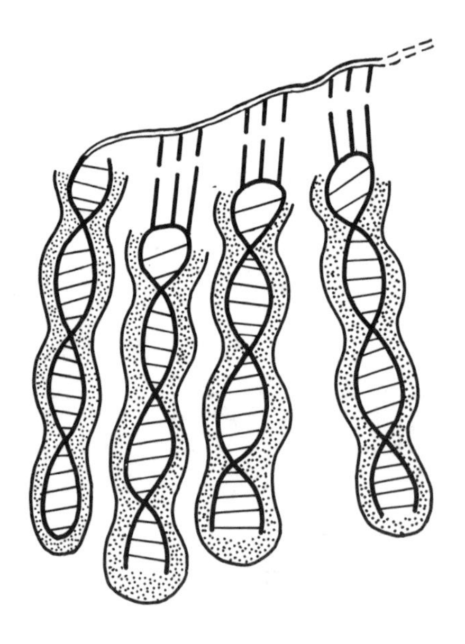

**8–4** *Scheme illustrating possible stacking of hairpin-structured proto-tRNAs along a collector strand, according to Kuhn and Waser (262).*

this topic. Note that all these various models apply to a protein-less RNA world. In the setting that I propose, multimers may be expected to participate structurally in consolidating the proto-RNA complexes and catalytically in helping peptidyl transfer.

Whichever way peptide assembly from aminoacyl-proto-tRNAs came to be initiated, the question arises as to what selective advantage it could have conferred upon the RNA species involved. Perhaps there was none. Or, rather, kinship with selected varieties enjoying positive selection (proto-tRNAs) sufficed. It is also conceivable that participation in the reaction complexes improved the stability of the molecules involved or, less evidently, their replicatability. Perhaps, as suggested by Orgel (268), it was advantageous for proto-tRNAs to bear a peptidyl, instead of an aminoacyl, group.

Eventually, however, selection would have to be influenced by the products of peptide synthesis. This is a crucial evolutionary turning point, in that a reflective loop is entering the system for the first time. RNA molecules are no longer selected on the basis of their intrinsic capacity to survive and multiply but on that of their ability to facilitate the production of something that helps them survive and multiply. Most models run into a difficulty at this stage. To close the loop, they need information (i.e., translation), which implies that they have to presuppose what is to be explained in the first place. Eigen and Schuster, for example, preface an analysis of their "realistic hypercycle" by stating: "Let us for the time being assume that a crude replication and translation machinery, functioning with adequate precision and adapted to a sufficiently rich alphabet of molecular symbols, has come into existence by some process not further specified" (252, page 341). But this clearly needs to be specified if we are to understand the subsequent steps.

My model avoids this need to put the cart before the horse. Even if the first RNA-mediated peptide assemblies were totally random or, to be more precise, noninformed and restricted only by steric and kinetic factors,[8] the mere ability to make peptides in this way could in itself have had some survival value. Presumably, the products of this activity would have included rudimentary catalysts, functionally similar to those found among the primary multimers postulated by my model, as well as other molecules of value, because of structural reasons, for example. Therefore, systems able to make peptides by the RNA-mediated pathway would have acquired a second means of meeting their needs for catalytic and structural multimers. The peptides made by the new pathway may not have been better initially than the thioester-generated multimers, but they had the enormous advantage of perfectibility by natural selection because of their dependence on replicatable molecules. However, we are not there yet, as translation is yet to come.

What is perfectible at this stage is not yet the quality of the peptides, but only the ability to make them. It is conceivable that this ability was enough of an asset to drive selection. The asset, however, must have been supramolecular. It could no longer have consisted simply in better stability and replicatability of the RNA molecules

[8] The term "random," in the sense of "with equal chances for each item" (OED), does not apply to the products of undirected chemical reactions. It must be understood as "giving each item a chance equal to its probability." This point was discussed in Chapter 6.

involved. It must have affected the survival value of a more complex entity. Therefore, at this stage, if not earlier, prebiotic systems must have become partitioned into a multiplicity of discrete units able to compete in true Darwinian fashion. I shall consider this question of encapsulation in the next chapter.

In the meantime, let us take stock of the new situation we have reached. RNA subspecies are emerging, not simply because of their intrinsic capacity for survival and multiplication, but because of functional properties that favor the (random) formation of substances (oligopeptides) useful for the survival and multiplication of the entity to which they belong. Darwinian selection can act, though in a way that is not as yet dependent on translation. Its screening criterion is simply efficiency in oligopeptide assembly. Selection favors proto-rRNAs and/or proto-mRNAs best able to help charged proto-tRNAs transfer aminoacyl and peptidyl groups from one to the other and, in a coevolutionary process, proto-tRNAs best able to interact with the helper RNAs.

## *Translation*

If the protein-synthesizing machinery first developed without an informational element, how did this element enter the system? In addressing this question, we must remember that, in living organisms today, mRNAs do not only serve to dictate the sequence of amino acid assembly. They also play an essential role in the strategic positioning of aminoacyl-tRNA and peptidyl-tRNA complexes on the surface of ribosomes. It is thus conceivable that the conformational function preceded the informational one and that it was, in fact, instrumental in bringing about the latter. Proto-mRNAs could have entered the system as structural adjuncts of the peptide-synthesizing machinery.

The model I propose is illustrated schematically in Figure 8–5. The catalytic part of the "protoribosome" is made up of a proto-rRNA, perhaps assisted by a multimeric proto-peptidyl transferase, and consolidation is ensured by a proto-mRNA joined to the two aminoacylated proto-tRNAs by triplets of complementary bases. As already mentioned, the association by triplets is imposed by topological factors. It ensures an optimal spacing of the partners for efficient aminoacyl or peptidyl exchange. As concerns the complementarity—at least partial (wobbling allowed)—of the joining triplets, it is mandated for obvious reasons. Without it, there would be no joining. Thus, the key arrangement of translation is put into place by exclusively conformational factors.

In such a system, things would be simple if amino acids and protoanticodons were specifically matched to start with, as foreseen, for example, in Shimizu's model (271). Translation would take place from the very beginning, and authentic Darwinian evolution would start right away.

Even if precise matching between amino acids and proto-anticodons did not exist initially, its progressive appearance would, predictably, be favored by natural selection. Consider that we are dealing with a random peptide synthesizer of which there are many copies situated in distinct, competing entities. Mutations of the RNAs involved create the diversity on which selection acts, but within stringent constraints.

**8–5** *Hypothetical steps in the development of RNA-mediated peptide synthesis.*
**(a)** Unspecific proto-tRNAs capture aminoacyl groups. **(b)** Specific proto-tRNAs capture aminoacyl groups. **(c)** Proto-rRNA aligns charged proto-tRNAs, and peptidyl transferase (part of proto-rRNA or separate catalyst) catalyzes aminoacyl (peptidyl) transfer. This step could be dispensed with, (b) leading directly to (d). **(d)** Proto-mRNA consolidates synthetic machinery. **(e)** (Not shown.) Proto-tRNAs and proto-mRNAs coevolve toward codon-anticodon specificity.

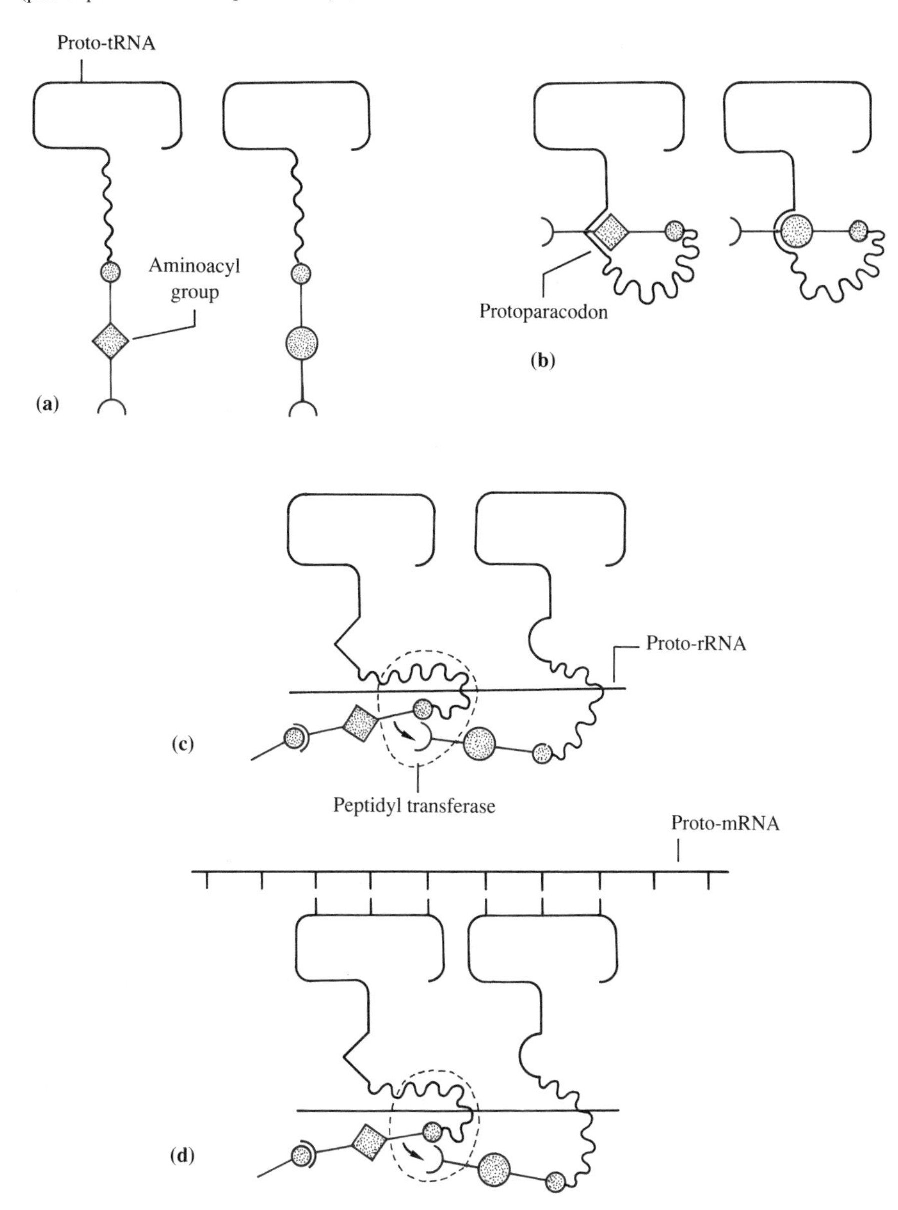

First, only mutations that respect the topological factors just defined will be tolerated, as others disrupt the machinery. Furthermore, the critical mutations will be those that affect the proto-tRNAs rather than those that affect proto-mRNAs, as it is not yet the quality of the messages that counts, but that of the parts of the synthetic machinery. In such a context, units possessing unambiguous proto-tRNAs will have a manifest advantage over those that have proto-tRNAs associating the same proto-anticodon with different amino acids, the advantage being the possibility of faithful reproduction of a message. In other words, from the moment translation became mechanistically possible, its emergence was obligatory.

## RETRIEVAL OF THE PRIMEVAL INFORMATION

When, according to my model, RNA-made oligopeptides replaced the earlier, thioester-derived multimers as catalysts, what happened to the original protometabolic network? Was it lost and did a new one have to be developed? Or could the original network be salvaged? As mentioned at the beginning of this chapter, the information relative to this network was written both in the structures of the protoenzymes and in those of the protometabolites. There are thus two ways in which this information could have been retrieved.

### *Retrieval from the Protoenzymes*

The most straightforward way of retrieving the protometabolic network would have been by copying the catalysts and transferring the relevant information to the RNA machinery. Mechanisms for doing this do not exist in the present-day living world. They are, in fact, ruled out by what Crick has called the Central Dogma.[9] However, today's heresy could well have been yesterday's orthodoxy, especially with a four-billion-year interval. The possibility of retrieval from the catalysts thus deserves to be considered.

Indeed, as I mentioned in Chapter 6, several authors unwilling to subscribe to the RNA-first theory have postulated a protein-replicating mechanism. Among them are Dillon, who, at the end of a searching and well-documented analysis of the problem of the origin of life, writes in unambiguous terms: "The first protobiont consisted of a primitive polyamino acid system which was capable of replicating itself" (53, page 477). Shapiro also advocates the possibility of early protein copying and imagines the participation of "a set of two-handed interpreter molecules" that would "recognize an

[9] The Central Dogma—more correctly termed the Central Postulate (17, 21)—was defined by Crick in 1957 as follows: "This states that once 'information' has passed into protein *it cannot get out again*. In more detail the transfer of information from nucleic acid to nucleic acid, or from nucleic acid to protein may be possible, but transfer from protein to protein, or from protein to nucleic acid is impossible" (96, page 153). This affirmation is almost universally accepted. It corresponds to a fundamental tenet of modern molecular biology and implies the rejection of the generally discredited Lamarckian theory of the heredity of acquired characters.

amino acid bound within a protein chain and the same amino acid in the free state" (64, page 284). Shapiro does not speculate further on the nature of these interpreter molecules. It is of some interest that the function he attributes to them could conceivably have been carried out by the proto-tRNAs of my model.

All we need to assume is that the structural features that, in the oligonucleotide molecules, were responsible for the specific recognition of aminoacyl groups—the "paracodons" (418) or their precursors (protoparacodons)—could similarly have recognized the same amino acids as residues in oligopeptides. Imagine some system that would allow two aminoacylated proto-tRNAs to sit side by side on a peptide template[10] in such a way as to allow acyl transfer, and you have the makings of a possible peptide-replicating machinery (see Figure 8–6a, page 192).

For the proposed mechanism to work, binding to the template should be restricted to aminoacylated proto-tRNAs. Otherwise, uncharged molecules would jam the system. A priori, one would rather expect the opposite to occur, as charged proto-tRNAs are likely to have their protoparacodon pocket blocked by the attached aminoacyl group and thereby prevented from interacting with a residue of the template. However, this argument can be turned around. Aminoacyl groups bound to the protoparacodon pocket could have been unavailable as acceptors for aminoacyl or peptidyl groups until they had been dislodged by the amino acid residue in the template. Such a condition could have imposed a kinetic impediment, but it could also have been beneficial by restricting oligopeptide assembly to the template-directed form. Another requirement of the system would have been a sufficiently unfolded template, but this would not have been a problem with the kind of short oligopeptides involved.

We cannot know whether the proposed mechanism operated in early prebiotic times, nor even whether it could have worked. But it deserves consideration because it would have provided for the direct retrieval of the structural information written into the sequences of early oligopeptide catalysts. It would have given organisms an efficient and economical way of making their own catalysts, and it would have allowed them unlimited time for developing a translation machinery without the risk of running into some metabolic logjam. Not only that. Peptide copying would have allowed true Darwinian selection to operate directly on a peptide genome long before the translation machinery was completed; it could even have played an important role in the development of this machinery.

Assuming such a mechanism, we would still have to explain how the information stored in the peptide genome was transferred to an RNA genome. We would still be left with the problem of the origin of translation, with perhaps the additional difficulty

[10] Admittedly, only regular peptides could have been copied in this manner, leaving out all the more heterogeneous multimers. This limitation could have been less severe than it seems. As I pointed out in Chapter 6, it is not inconceivable that the catalytic multimers that emerged as long-time survivors may have included a large proportion of chirally homogeneous peptides, possibly, even preferentially, of the L variety. In such event, peptide copying would have been a means of selectively retrieving the most interesting members of the multimer population.

that the evolutionary advantages likely to have driven the development of such a process might have been weaker in systems endowed with a peptide genome.

There is a way out, namely reverse translation (see Figure 8–6b). The topology of protein synthesis in contemporary organisms tells us that aminoacylated tRNAs can align their anticodons in register with an mRNA while they exchange a peptidyl group. It is thus plausible that the evolutionary precursors of the tRNAs could have adopted a similar configuration, held together, not by an mRNA, but by a peptide template, perhaps with the additional help of a proto-rRNA. And it becomes at least conceivable that, in doing so, they offered properly aligned nucleotide triplets to a replicating catalyst that would copy them into a continuous complementary strand.

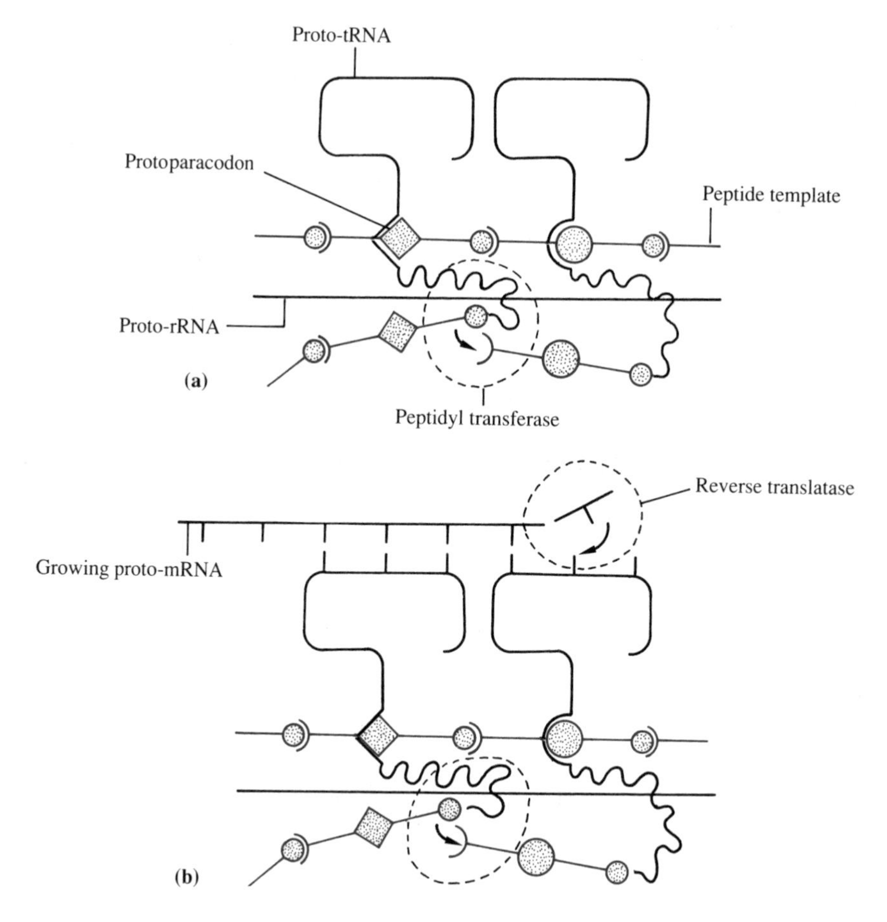


**8–6** *Possible mechanisms of peptide copying and reverse translation.*
**(a)** Aminoacylated proto-tRNAs are selectively aligned on a peptide template by their amino-acid recognition sites (protoparacodons). Proto-rRNA and peptidyl transferase catalyze aminoacyl (peptidyl) transfer. **(b)** Proto-mRNA is assembled on protoanticodons displayed by proto-tRNAs on the peptide-copying machinery.

The first proto-mRNAs would thereby have arisen by a process of authentic reverse translation. At first, the resulting messages would have been mostly nonsense because of the probable haphazard nature of the protoanticodons displayed for replication. Should, however, a system have developed for translating the messages back into peptides, selection would have favored a set of proto-tRNA molecules fitted with anticodons such that the products of reverse translation could provide direct translation with faithful, unambiguous, and comma-less messages.

The advantage of this mechanism is that it would have ensured a direct continuity between the earlier and the later catalysts. The primeval catalysts would, indeed, have been the true protoenzymes, the actual molecular ancestors of present-day enzymes. The proposal is, however, likely to be viewed by most as too heretical to merit serious consideration. It is therefore fortunate that a more orthodox hypothesis may also be entertained.

### *Retrieval from the Protometabolites*

If we assume that similar elements were involved in the selection of the oligopeptides assembled with the help of RNAs and in that of the multimers made from thioesters (Chapter 6), we may expect to find among the later set the main functional properties of the earlier one. At this stage, therefore, for each protometabolic reaction, there must have been a catalytic multimer made from thioesters and an oligopeptide synthesized by the RNA machinery. We have no way of guessing the extent to which the members of each pair resembled each other chemically. No matter. As long as the later set included all the catalysts needed for the continuing operation of protometabolism, the protometabolic network would have provided an effective screen for the selection of increasingly useful RNA molecules. At first, while peptide synthesis was still random, selection would have been driven by the advantages of an increasingly efficient synthetic machinery (better functional RNAs). After translation was developed, mRNAs coding for catalytic peptides would have been singled out and subjected to selection on the basis of the qualities of their products, in standard Darwinian fashion. Therefore, even though the information inscribed in the sequences of the primary multimeric catalysts could not be retrieved directly, the functional essence of this information could nevertheless be salvaged by way of the protometabolites. In other words, in spite of the participation of different catalysts in protometabolism and in metabolism, a direct filiation would, nevertheless, obtain between the two.

## THE RNA GENOME

The first RNA genes must have been short if, as seems likely, the accuracy of their replication was poor. As shown by the Eigen school (220, 250, 376), if a message is too long with respect to the error rate of replication, it cannot survive through successive replication cycles and degenerates irreversibly. As mentioned above, the maximum length estimated for a primitive GC-rich RNA gene would be of the order of 100 nucleotides (see Footnote 2, page 174). With increasing AU participation, it would become correspondingly shorter.

These estimates are consistent with the assumed roles of the early genes. First to come, according to my proposed model, were the proto-tRNAs, which have remained short up to the present day. Ribosomal RNAs have grown much longer, but a short oligonucleotide stretch could well have sufficed in their early precursors for the main job of correctly aligning two charged proto-tRNAs. As to the first proto-mRNAs, the length of their coding parts could very well have lain in the neighborhood of 30 to 60 nucleotides, corresponding to oligopeptides of 10 to 20 residues. This, as discussed in detail in Chapter 6, is about the maximum range compatible with the presence of active catalysts in a mixture of multimers originating by an initially random assembly process. Thus, whether one starts from the peptide or from the RNA, the estimates agree.

Dynamically, the early RNA genome must have been a very haphazard and disorderly hodgepodge. Genes were, presumably, present in multiple copies of their two complementary forms, which tangled in ways that must have variously affected their stability, their replicatability, and their ability to function effectively. We may take it that the main task of selection in those early days was to achieve the best possible compromise between these three to some extent mutually exclusive properties, all critically dependent on base pairing. An interesting question relates to the possible effect of a high temperature (Chapter 7) on these phenomena. The answer to this question depends on which—of the benefits of flexibility or the disadvantages of disruption—would have outweighed the others.

It is out of this jumble that coding slowly emerged, no doubt, as a result of innumerable trial fittings of the three types of RNAs. During all that time, the protoenzymes had to remain short. Only after their selective synthesis by translation had become possible could their length be allowed to increase beyond the limit set by random assembly and molecular selection. This increase would have taken place at the mRNA level, and this raises two problems.

The first one arises from Eigen's rule. Genes could be lengthened only to the extent that the accuracy of their replication increased. Thus, the premium mutations were those that resulted in a gain in replication accuracy. These must have set the pace of evolutionary progress. Thanks to these mutations, RNA genes could increase in length without succumbing to replication errors, up to the maximum of 5,000 to 10,000 nucleotides allowed by the accuracy of the best RNA replicases. Memories of this outcome may have been conserved in the RNA viruses (excluding retroviruses, which are replicated via DNA), whose genes are, indeed, no more than 10,000 nucleotides long.

The second problem posed by gene lengthening concerns its mechanism. In light of previous discussions on "exon shuffling" (Chapter 3), it is tempting to assume that a major role in the process was played by *trans* (intermolecular) splicing of existing short genes. Such a process would have allowed extensive combinatorial trials and would have led in a relatively short time to the development of polypeptides of increasing length (not by progressive elongation, but by modular increments) and greater sophistication. The mechanism involved could have resembled RNA splicing as we know it today. But other mechanisms, dependent, for example, on insertion or on the replication of RNA segments held end to end, are conceivable.

Needless to add, other factors of genetic variability, such as replication errors and other kinds of mutations, would also have played a role in evolution at this stage. But the tremendous possibilities offered by combining genes in various ways must have made this process a particularly important source of variation.

Linking minigenes by *trans* splicing would have yielded good modular associations only if the spliced products were translatable in phase. This may not have been so in a large number of cases, and coding sequences could frequently have been separated by meaningless, phase-breaking, intervening sequences. It is possible that *cis* (intramolecular) splicing was developed as a means of rectifying this situation.

There is thus a good likelihood that many of the first full-length genes were split genes in need of processing for conversion to effective mRNAs. This fact is highly relevant to the origin of RNA splicing. As was already pointed out in Chapter 3, it is likely that splicing mechanisms arose originally in the form of simple RNA-catalyzed processes. We now see that they may have played an important role in the early evolution of the RNA genome, clearly sufficient to account for their selection. It seems likely that RNA splicing in contemporary organisms dates back to this early phase in the history of life. But this does not necessarily mean that today's exons are direct descendants of primeval RNA minigenes, nor that the introns that separate the exons have come to us from the first intervening sequences that were formed when the game of gene shuffling started. I shall address this question at the end of this chapter.

Given the two conditions just defined—increased replication accuracy and the existence of splicing mechanisms—genes of increasing length could be assembled according to a wealth of different combinations. Thanks to translation, they could be tested by their products and screened by natural selection. Acquisition of increasingly better enzymes, of new enzymes, of regulatory mechanisms of various kinds, of stronger feedback links controlling RNA replication, splicing, and translation, are among the more obvious gains that would have driven Darwinian selection. Readers with a taste for theoretical modelling may be interested in the "hypercycle" concept developed in elaborate detail by the Eigen school (220, 250–252). Those who are frightened by complicated equations may find solace in Ninio's comments on their utility (73).

## THE COMING OF DNA

### *From RNA to DNA*

At some stage in the development of the first living cells, enzymes appeared that could remove the 3′-oxygen from ribose and add a 5-methyl group to uracil, so that dATP, dGTP, dCTP, and dTTP entered the scene. Subsequently, some RNA replicase evolved into a mutant variety with greater affinity for the deoxynucleoside triphosphates than for the substrates of RNA synthesis. Reverse transcriptase was born and, with it, the ability to store genetic information into DNA. Further mutants provided the means for replicating the DNA and for transcribing it back into RNA.

We do not know that things happened that way, of course. But the proposed script seems by far the simplest and most likely one for moving from the RNA world, which all agree came first, to today's DNA-RNA world. Retroviruses, so a popular theory asserts, have preserved unto this day traces of those fateful events.

Once again, we have to ask the same questions: What drove the process? What benefits did the cells derive from adding DNA to their genetic equipment? It is sometimes pointed out in this connection that DNA is stabler than RNA, especially in neutral or alkaline medium. The hydroxyl group in position 2′ of the pentose represents a dangerous nucleophilic neighbor for the phosphodiester bond; it is involved in RNA hydrolysis and splicing and in RNA catalysis. This property, no doubt, explains the greater versatility of RNA but does not necessarily disqualify it as a depository of genetic information. After all, RNA presumably played this role for a long time. It still does in many viruses. And it remains a central and reliable mediator in all forms of expression of genetic information.

Therefore, not so much chemical stability as lack of biochemical versatility may be one of the advantages of DNA. If our picture of the RNA genome is correct, one of its main characteristics was that it lent itself to widespread combinatorial experimentation. This was all to the good as long as full-length genes were still in the making. The more elaborate and perfected the genome became, however, the more of a liability rather than an asset versatility would have become. The existence of a genetic storage form informationally compatible with RNA but unable to enter the genetic "melting pot" would then have become increasingly advantageous. In this connection, the emergence of the enzyme that removes the 2′-oxygen from ribose (ribonucleoside diphosphate reductase) may not have been as critical to the advent of DNA as is sometimes believed. This enzyme was indispensable but, perhaps, not decisive. It could have been around for a long time and could have remained unexploited for the construction of DNA until a second storage form of genetic information became clearly advantageous. In other words, life may have played with DNA for a long time in what might be called a half-hearted way and have adopted it definitely only when the need for it made itself felt through selective benefits.

Another major advantage of DNA is that it would have allowed a clearcut separation between the replication of genetic information and its expression. An RNA genome, simultaneously offering multiple copies of genes and of their complementary forms for replication and for translation, is intrinsically "messy" and difficult to control. With the advent of DNA, genetic information could be stored in single copies or in the minimum number of copies needed for adequate expression (rDNAs, for instance). It could be replicated in coordinate fashion and at the appropriate time. Gene expression and amplification could be ensured independently of replication and controlled individually by the newly developed process of transcription.

A third advantage of DNA was that it rendered possible the grouping of all the genes of an organism into a single, continuous genome, controlled by a single replication origin. This, however, would have required improving the fidelity of replication. As we know, the much greater length of DNA chromosomes, as compared to the longest RNA genes, is rendered possible by the existence of proofreading mechanisms (Chapter 1). The fact that such mechanisms apparently exist only for DNA replication or, at least, are much more efficient in the case of DNA replication, makes one wonder whether there is something special about DNA with respect to the possibility of proofreading. If so, the pressure in favor of DNA would have been considerable.

Remarkably, the enzyme that allegedly engineered the fateful transfer of genetic information from RNA to DNA has largely disappeared from the present-day living world, where it subsists mainly in retroviruses, plasmids, and transposable elements. Long believed to be present only in eukaryotes, it has now been detected also in prokaryotes (425). This is important with respect to the possible ancient origin of the enzyme. So far, however, the function of reverse transcriptase, wherever it is found, seems to be related mainly to the reproduction of the stretch of DNA that includes its own gene. It apparently belongs to an essentially "selfish" gene (see Chapter 6, Footnote 2, page 130). It could, however, have some interesting evolutionary descendants, for example, telomerase. This ribonucleoprotein enzyme, which builds the telomeric, single-stranded, G-rich, 3′-ends of eukaryotic chromosomes, has recently been found to be a specialized reverse transcriptase that uses a segment of its own RNA as template (353, 369). There is also the possible involvement of a reverse transcriptase-related protein in intron transfer, which has been mentioned earlier (see Chapter 3, Footnote 14, page 81).

RNA replicases, which, presumably, played a key role in the replication of RNA genes, have likewise largely disappeared, except in viruses. Such exclusions must be meaningful and indicative of strong selective pressures against such enzymes, especially in view of the continuing persistence of the analogous DNA-informed RNA and DNA polymerases. Once DNA was firmly in place, it obviously had to be protected against intrusions—oncogenes tell us one reason why—and it had to have the monopoly over replication and amplification. It also became the main site of genetic rearrangements. Presumably, adequate genetic control depended on such centralization.

Assuming that events took place more or less as outlined, at what stage of the development of the RNA genome did DNA take over? In other words, what was the size of the RNA stretches that were transcribed into DNA? This is a very interesting question, directly related to another central problem: under what form—RNA or DNA—did the assembly of full-length genes mostly take place? In the preceding account, I have taken it for granted that much of the process of gene assembly involved RNA "minigenes." But there is no compelling reason for this. If, as suggested, DNA was waiting in the wings, so to speak, fidelity of replication could have been an important factor. Whichever nucleic acid could be replicated more accurately, and therefore could provide the longer genes, would have been advantaged. We must remember also that present-day examples of gene combination—in the immune system, for example—occur at the DNA level. RNA mechanisms are restricted mostly to alternative splicing. On the other hand, one cannot help associating the advent of DNA with a degree of metabolic sophistication that could have been achieved only with enzymes, and therefore genes, of respectable length.

## *The Origin of Introns*

We have reached the following conclusions, largely by logical deduction: 1) the first genes must have been very short; 2) full-length genes most likely arose from these early genes by a combinatorial process; and 3) this process must have been such that its building blocks were often separated by intervening sequences that had to be spliced out for the product of assembly to be meaningful. Logic thus supports the "antiquity of introns" (432a).

Logic, however, also supports their early loss. Especially if gene assembly took place at the RNA level, there was every opportunity for introns to have been lost by the preferential replication and, later, reverse transcription of the spliced RNAs, as these were probably more abundant, and certainly more accurately replicatable, than their unspliced precursors. Could the advantages of continuing "exon shuffling" have been sufficient to prevent this loss? The question is at least debatable.

Furthermore, a direct continuity between present-day exons and primeval minigenes is by no means certain. Not all of today's exons code for protein domains or modules. Some introns show clear evidence of recent insertion, and the possibility that some may have arisen by reverse splicing followed by reverse transcription is no longer fanciful (see Footnote 16, page 82). Furthermore, the average size of exons—135 nucleotides (98)—is distinctly larger than the maximal estimated length of the primeval minigenes—30 to 60 nucleotides at most. These and other inconsistencies can easily be explained away by saying "much can happen in four billion years." However, this argument works both ways. The continuity that is invoked could concern the ability to split genes rather than the genes themselves. It is conceivable that the ancestral split genes went through a "streamlining" eclipse at the common-ancestor stage and resuscitated progressively in a different form in eukaryotes. This could have happened thanks to the persistence or acquisition of some splicing mechanism, helped, perhaps, by hidden "scars" from the primeval process that facilitated the insertion of introns into certain sites. We cannot even exclude the possibility that the ability to split genes was discovered entirely de novo in the protoeukaryote line and developed progressively (Chapter 3) thanks to the selective advantages of exon shuffling.

CHAPTER NINE

# Confining Life

## ENCAPSULATION

### *Early or Late?*

Because the cell is the basic unit of life, many biologists believe that life must have started with the formation of some sort of primitive "protocell." This theory was first proposed by Oparin (61). In the 1920s, when little was known of the origin of our planet, Oparin and, independently, Haldane proposed the hypothesis that conditions on the prebiotic earth favored the synthesis of a variety of organic chemicals, which, as Haldane put it, "must have accumulated until the primitive oceans reached the consistency of hot dilute soup" (28, page 107). The concept of a primeval soup filled with edibles of all kinds is generally attributed to this quotation, although Darwin had already, half a century earlier, conjured up "some warm little pond, with all sorts of ammonia and phosphoric salts, light, heat, electricity, etc. present" (18, page 202). According to Oparin, as the primeval soup thickened, some of its constituents condensed into "coacervates." This word was coined by the Dutch physical chemist Henrik G. Bungenberg de Jong (142), who had found that certain aqueous mixtures of polymers—gum arabic and gelatin, for example—would, under given conditions, break spontaneously into suspensions of particles or vesicles of microscopic size. This phenomenon was viewed as relevant to the so-called "sol-gel" transformation in the then-popular colloidal theory of protoplasm. For Oparin, it held the key to the origin of life: the primeval coacervates were the seeds from which the first living cells developed by accretion and acquisition of new properties. He defended this theory until the end of his life, in 1980, and enriched it with numerous experimental observations.

Coacervation is a rather trivial property of large molecules that possess both hydrophilic and hydrophobic parts. The phenomenon is explained by physical interactions between nonpolar groups on one hand and between polar groups, water, and ions on the other. It is often reversible and critically dependent on such physical factors as temperature, ionic strength, pH, presence of certain ionic species, etc. It has attracted the attention of a number of workers who, like Oparin, have credited it with a seeding role in the origin of life. The most assertive defender of this theory is Fox, who has devoted a lifetime to investigating the "proteinoid microspheres" that he produces by stirring heat-polymerized amino acids with water (57, 58, 107, 254). These structures are said to display "lifelike" properties, including growth, budding, division, fusion, catalysis, even "behavioral" traits, which Fox describes as motility, excitability, communication, and "sociality." On the significance of his results, he writes: "in view of the ease of spontaneous formation of a protoorganism, no other conceivable pathway of evolution would have been able to compete" (107, page 41). He does, however, concede: "The possibility that the essential processes of peptide and internucleotide bond synthesis could have developed outside a cell is perhaps not rigorously ruled out; the protocell conceivably might have arisen later" (107, page 43).

In recent years, the primeval-protocell theory has fallen under increasing criticism on the grounds that the prebiotic earth was not likely to have offered conditions suitable for coacervation or for the formation and microspherulation of proteinoids. Many authors, often with different axes to grind, have underscored the unlikelihood of organic building blocks attaining even moderate concentrations in the prebiotic soup, mostly because of the many destructive influences to which they would have been subjected. In addition, it is pointed out that polymerization by dehydration could not have taken place in a watery milieu. For Woese (279), the "Oparin ocean" is an impossibility. In the words of Thaxton, Bradley, and Olsen: "It is becoming clear that however life began on earth, the usually conceived notion that life emerged from an oceanic soup of organic chemicals is a most implausible hypothesis. We may therefore with fairness call this scenario 'the myth of the prebiotic soup' " (65, page 66), an expression that was later adopted by Shapiro (64). Wächtershäuser's categorical condemnation of the prebiotic soup theory has already been cited (Chapter 6, page 131). We have also seen that such experts of prebiotic syntheses as Miller and Orgel (60), who, unlike Thaxton and colleagues, cannot be suspected of creationist leanings, consider the fragility of prebiotic products a serious problem.

Several authors have postulated an original protocell for energetic, rather than morphological, reasons. Some, as was mentioned in Chapter 7, can be quite forceful in expressing their beliefs. Incipient life, they claim, needed a membrane to accommodate the photochemical transducer necessary to capture indispensable energy from the sun. Folsome, one of the advocates of this theory, has found a way around the concentration problem. Large amounts of polymeric material, he points out, invariably are formed, alongside the well-advertised small organic molecules, in Miller-type prebiotic simulation experiments. This material "appears as a thin oily scum on the water's surface. When disturbed by motion, this scum separates from the surface to form spherules ranging in size from 1 to 20 micrometers in diameter, as well as more complex structures" (56, page 88). Such spherules, Folsome speculates, could, with the help of some hydrophobic photoreceptor, such as a porphyrin, give rise to "organic automata" that would use ultraviolet light to power amino acid polymerization and other reactions. These structures would thereby grow, divide, diversify, compete for available resources, and eventually evolve into the first protobionts. Woese also believes in primordial photochemical protocells. But he does not visualize these as membrane-bounded entities arising out of the Oparin ocean. He sees them instead as minuscule water droplets suspended in the atmosphere around an extremely warm planet that did not yet possess any surface water (279).

All such proposals are highly conjectural. Our ignorance of the conditions on the prebiotic earth is still enormous, as shown by the many settings that have been, and continue to be, proposed for the origin of life. As regards concentration, in particular, even accurate knowledge of the global geochemical conditions cannot suffice for plausible evaluations. We still need to know a number of local factors, such as the nature of the particular environment in which the early prebiotic syntheses took place, the rates of synthesis of the various compounds, the degree to which they might have been sheltered or otherwise protected against damaging processes, to name only a few. Unless we subscribe to special creation, we must assume that, where life started, local conditions were adequate to overcome the concentration problem. In my opinion, postulating a protocell that used light energy to raise its internal concentration of essential chemicals begs the question. It amounts to predicating an effect to account for its cause; for how would active transport systems have come together in the first place? For these reasons, I have chosen to meet the problem head-on. At least in my model, only a few basic building blocks needed to be present at relatively high concentration initially. This, as already pointed out, is an advantage of the model over others.

This model does not require early cellularization. The prebiotic metabolic network that I envision could well have developed without being separated from the environment by a membrane. Such a separation could even have created serious permeation problems. We must remember that these problems are solved in contemporary cells by a unique combination of integral membrane proteins and lipid bilayers. It is unlikely that they could have been solved by either proteinoids or oily polymers alone. This is also the opinion of Eigen and coworkers, who write: "Organization into cells was surely postponed as long as possible. Anything that interposed spatial limits in a homogeneous system would have introduced difficult problems for prebiotic chemistry. Constructing boundaries, transporting things across them and modifying them when necessary are tasks accomplished today by the most refined cellular processes. Achieving analogous results in a prebiotic soup must have required fundamental innovations" (220, page 107).

If the protocell was preceded by some kind of extended "protocytosol," at what stage in the development of such a system would compartmentation have become useful, or even necessary? I have mentioned two. One is associated with the emergence of energy-retrieval mechanisms dependent on protonmotive force (Chapter 7), the other, with the onset of authentic Darwinian competition and selection (Chapter 8). Of the two, only the second represents an inescapable deadline. Prebiotic metabolism could have been supported indefinitely by soluble substrate-level phosphorylation systems if conditions were right; many contemporary organisms can survive on such processes exclusively. In contrast, once information had to be tested and screened at a higher than strictly molecular level, competing systems had to be individualized. As expressed by Eigen and coworkers: "Cellular organization was needed because it was the only way to solve the one problem of information processing in evolution that self-replicative competition and hypercyclic cooperation were unable to address: the evaluation of the information in genetic messages" (220, page 107).

If encapsulation was postponed to the point where it became a condition of improvement by natural selection, then it occurred at the time replicating systems had acquired the ability to sustain growth at a supramolecular level. This means, in the framework of the developmental model outlined in Chapter 8, as soon as the first RNA-dependent machinery for making oligopeptides had emerged. As pointed out in that chapter, the machinery need not yet have developed true translation for selection to start operating.

### *Mechanisms*

Whatever its timing, the creation of membranes must have been an event of considerable complexity. It has no counterpart in the living world. Biological membranes always grow out of preexisting membranes; they do not arise de novo. Bilayers can, however. When vigorously agitated, phospholipid micelles in aqueous suspension automatically become organized into closed vesicles limited by one or more molecular bilayers (liposomes). This would seem to solve the problem, since causes of turbulence must have been plentiful in the prebiotic world. However, a liposome origin of biological membranes poses two difficulties.

First, there is the problem of synthesis. Phospholipids, whether of the ether or ester variety, are complex molecules that are not expected to arise readily by random chemistry. This is a serious difficulty for all membrane-first theories, but it does not affect my model, which, by definition, is equipped with appropriate catalysts. There could, perhaps, have been a need for some hydrophobic site. In living organisms, several steps of phospholipid synthesis take place in or on lipid bilayers, which provide the hydrophobic milieu needed by water-insoluble intermediates. Obviously, the first phospholipids could not have arisen on lipid bilayers. However, the prebiotic world should not have lacked hydrophobic materials. The tarlike material assumed to arise abiotically—Folsome's "scum"—has already been mentioned. In addition, my postulated multimerization process must itself be expected to produce a mass, perhaps even a preponderance, of hydrophobic and amphiphilic molecules. A glance at the alleged building blocks of the process (see Tables 6–1, page 124, and 6–2, page 125) makes this clear. Such multimers, kept together by mutual attractions, would have covered the seas with oily slicks, coated rocks with greasy films, filled the waters with aggregates and micelles of all kinds, providing plenty of likely sites for the assembly of the first phospholipids.

The second problem is more serious. Liposomes are essentially impermeable, not only to macromolecules, but even to most hydrophilic ions and molecules. The formation of a liposome is an all-or-none phenomenon, so systems undergoing this kind of encapsulation would move without transition from total openness to a state of hermetic confinement. They would suddenly become entirely dependent on their own ability to manufacture the multimeric catalysts they needed to continue operating. Even if they could face this emergency adequately, as assumed in my model of late encapsulation, they would still be exposed to immediate starvation and early demise unless polypeptides with hydrophobic segments happened to be immediately available for the creation of appropriate channels or carriers.

Gunter Blobel (310) has proposed an ingenious solution to this problem. He assumes that the first vesicles were empty (see Figure 9–1). Left to float around in what I have called the protocytosol, these "inside-out" cells slowly grew into structures of increasing complexity, which included sites of various metabolic reactions as well as anchoring points for chromosomes, ribosomes, and other complexes. Eventually, the vesicles folded around their attached organelles and progressively closed into double-membranous envelopes. Cavalier-Smith (121) has adopted this model and used it in support of his contention that the first cells were of the gram-negative kind (Chapter 4).

This hypothesis is readily fitted into my model provided the time of closing of the vesicles is moved back to an earlier developmental stage. For reasons made clear in Chapter 8, structures as complex as chromosomes and ribosomes could not have

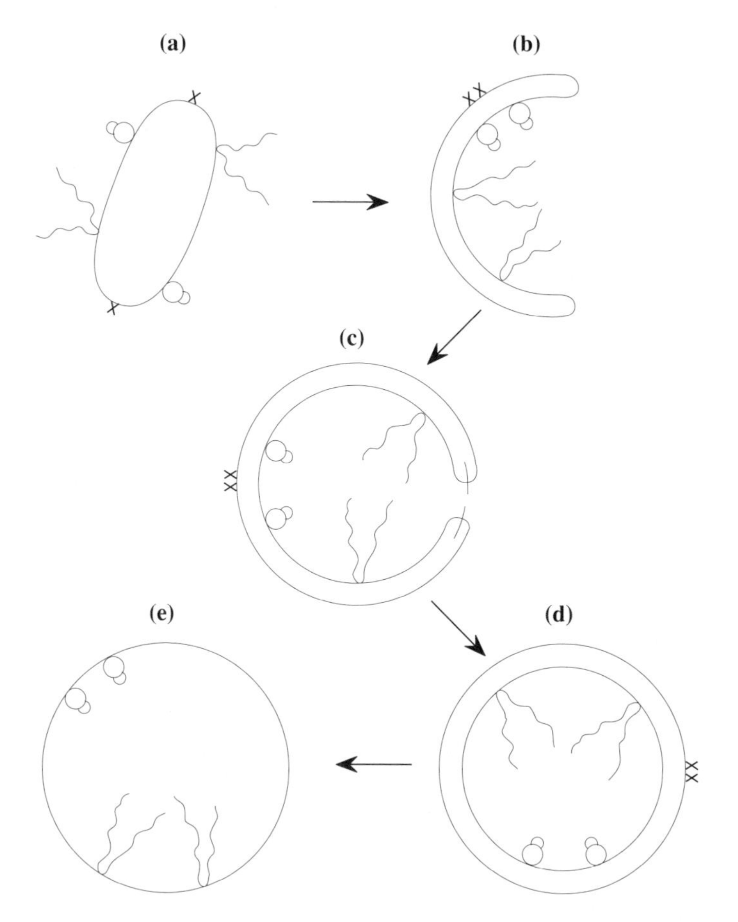


**9–1** *Formation of protocells from inside-out vesicles.*
**(a)** Primitive vesicle binds macromolecules (X) and macromolecular complexes, including ribosomes and chromatin, on its outer surface. **(b)** Bound components adopt a nonrandom distribution on the vesicle's surface, and vesicle starts folding. **(c)** Vesicle closes into a "gastruloid" with an orifice that can open and close. **(d)** Fusion at the orifice results in a primordial cell delimited by two membranes, possibly ancestral to gram-negative bacteria. **(e)** Loss of outer membrane gives rise to a cell delimited by a single membrane, possibly ancestral to gram-positive bacteria.

According to Blobel (310).

arisen in an unpartitioned system. There does, however, remain the problem of how the transmembrane peptides that were required for the exchanges of matter with the environment arose and came to be inserted into the lipid bilayer. They would have been useful only after closure of the protocellular envelope. Yet, they must, by necessity, have been present before this event for the fenced-off protocells to be viable.

In view of this difficulty, one may wonder whether encapsulation was not initiated by hydrophobic peptides rather than by phospholipids. Such peptides could have created loose meshworks sufficiently leaky to permit free exchanges with the environment and sufficiently hydrophobic to shelter lipid metabolism. Membranes would then have arisen through the gradual insertion of lipids between what were to become the transmembrane segments of integral membrane proteins, not the other way round. It is perhaps revealing in this respect that the plasma membrane of *Halobacterium* can be delipidated without loss of structural integrity, thanks to the cohesiveness of the closely packed bacteriorhodopsin molecules, which possess no fewer than seven hydrophobic transmembrane segments and make up 75% of the total mass of the membrane (304). As just mentioned, materials needed for making such "protomembranes" would have been present in abundance right from the start among the products of multimerization.

## FROM PROTOCELLS TO CELLS

### *The First Protocells*

At the time of its birth, even the most advanced protocell would still have been a long distance away from the simplest bacterial cell. We can get a measure of this distance by considering the maximum degree of development a prebiotic system could have attained by the time confinement became a necessary condition of further progress.

If my model is to be believed, the metabolism of the protocells resembled present-day metabolism in its main lines, due allowance being made for the acidic medium in which it may have developed. It followed roughly the same pathways, to make, at the very least, oligopeptides, oligonucleotides, phospholipids, and all the necessary precursors and intermediates. These reactions would have been sluggish and imprecise, with the rudimentary nature of the catalysts involved being compensated for to some extent by high concentrations of intermediary metabolites (though hardly more than centimolar, on average, if hundreds of intermediates were to be accommodated).

The protocells' dependence on the environment could have ranged anywhere between strict heterotrophy and complete autotrophy. It is, however, tempting to assume that a measure of self-sufficiency had been attained by the time encapsulation took place and that the protocellular stage of life no longer depended on the abiotic syntheses that had first set the biogenic process on its course. This autonomy is not an obligatory condition and could have been achieved later.

For their energy supply, the protocells probably relied exclusively on substrate-level phosphorylations coupled to electron-transfer reactions. As we have seen in

Chapter 7, ferric ions could have served as exogenous electron acceptors in these processes. But other electron-consuming reactions may be envisaged, including the reduction of $H^+$ ions and $CO_2$ to methane, which is known to support certain forms of autotrophy. Most likely, ATP had already become the main phosphoryl carrier, but inorganic pyrophosphate could still have played a significant role in this respect.

A minimum of several hundred protoenzymes must have been needed to catalyze these various reactions. According to my model, they were oligopeptides and other multimers for which two assembly mechanisms—both noninformed—were available: the primeval mechanism dependent on thioesterification and the newly developed, RNA-mediated one. It seems reasonable to assume that the two mechanisms coexisted in the protocells, as they do in many bacteria today. For reasons that have been explained in Chapter 6, the types of building blocks used in the synthesis of the protoenzymes must have been few in number (hardly more than eight), and the average chain length must have been short (10–20 residues at most). Help from various coenzymes and from catalytic oligonucleotides could have become available at this stage.

As to the genome of the protocells, it would have consisted of a motley collection of replicatable oligonucleotides that included proto-tRNAs, proto-rRNAs, and proto-mRNAs, so far selected only for their intrinsic molecular qualities. Because of the poor fidelity of replication, the selected molecules would have had to be short, probably between 50 and 100 nucleotides, as we have seen in Chapter 8.

The membranes surrounding the protocells must also have been of a rudimentary nature. We may visualize them as simple porous structures, capable of retaining large molecules but allowing small molecules to pass through with, as yet, little specificity control of the traffic.

One last property required by the first protocells would have been the ability to grow and produce offspring by division and, thereby, to participate in the great Darwinian game.

## *The Rise in Complexity*

As pointed out, my reconstructed protocell is about the most complex such entity that can be imagined. Greater complexity could probably not have been achieved without prior cellularization. Yet, compared with even the most rudimentary of existing forms of life, my protocell is still a highly primitive system. I suspect that it may correspond in a certain measure to what Woese had in mind when he formulated his concept of the progenote (238). How did it evolve into the common ancestral cell that I attempted to sketch in Chapter 4?

Needless to say, we have no precise answer to that question. But we can make some general statements with a fair amount of confidence. 1) The process took a great many steps. 2) Each step was signalled by some mutation that conferred a selective advantage and resulted in the mutated protocells producing more offspring than the others. 3) Essential early mutations were those that increased the fidelity of replication and, after it was developed, of translation. As we saw in Chapter 8, increased fidelity of these processes is the sine qua non of complexification by way of more sophisticated polypeptides. 4) Some obvious milestones included the development of translation and of the genetic code, the entrance of DNA, the conversion of the protocell boundary into an authentic membrane fitted with appropriate transport systems, the development of membrane-bound electron-transport chains and of carrier-level phosphorylation, and, perhaps, that of a photosynthetic machinery.

So much for generalities. What about pathways? Were multiple avenues open at each critical step, and did chance decide on one among many possible end points? Or were the constraints—both intrinsic and environmental—such that only a single course could be pursued to a successful conclusion, all others necessarily leading into dead ends? This question is so important as to deserve a separate chapter.

ΔΗΜΟΚΡΙΤΟΣ

CHAPTER TEN

# CHANCE OR NECESSITY

## THE PROBABILITY OF LIFE

In 1970, the celebrated French biologist Jacques Monod published under the title *Le Hasard et la Nécessité* an "Essay on the Natural Philosophy of Modern Biology" that soon became a worldwide best-seller, sparking passionate discussions that have not yet died down (31). The title of the book was borrowed from the Greek philosopher Democritus, who wrote: "Everything existing in the universe is the fruit of chance and of necessity." Rarely was a quotation more aptly chosen. It condenses in remarkably pithy fashion the essence of the problem the existence of life poses to the philosopher: To what extent is the life phenomenon the product of blind chance? To what extent is it the obligatory outcome of deterministic[1] forces?

Very different answers have been given to these questions. Biochemists generally agree with George Wald who, in 1964, alluded to "the dawning realization . . . that life in fact is probably a universal phenomenon, bound to occur wherever in the universe conditions permit and sufficient time has elapsed" (113, page 120). Eigen expresses the same view even more forcefully. "We may furthermore conclude," he wrote in 1971, "that the evolution of life, if it is based on a derivable physical principle, must be considered an *inevitable* process despite its indeterminate course . . . it is not only inevitable in principle but also sufficiently probable in a realistic span of time. It requires appropriate environmental conditions (which are not fulfilled everywhere) and their maintenance. These conditions have existed on earth and must still exist on many planets in the universe" (249, page 519). In his classic textbook, *Biochemistry,* Albert Lehninger also saw the development of life in a deterministic fashion, as "the result of a long chain of single events, so that each stage in their evolution developed from the preceding one by only a very small change." And he adds, "each single step in the evolution of the first cells must have had a reasonably high probability of happening in terms of the laws of physics and chemistry" (2, 1st Edition, page 771). I made the same point in Chapter 5.

[1] Here and elsewhere in this book, I use the term "deterministic" in a phenomenological, not in a doctrinal, sense. I apply it to the phenomena that led to the appearance of life, to mean that these phenomena were determined entirely by the physical-chemical conditions that prevailed at the time they took place and, therefore, that they should be reproducible under the same conditions. I leave to the physicists and philosophers the more general question whether the entirety of the universe obeys, or not, such a determinism (see, notably, Reference 34).

In contrast, less chemically oriented biologists, physicists, and philosophers often tend to be impressed with the unlikelihood of anything as complex as a living cell's arising spontaneously and are inclined to invoke a fantastic, and no doubt unique, stroke of luck (if not the intervention of some supernatural agency). Even Monod, though a staunch defender of mechanistic materialism, gives chance the star role. "The universe," he writes, "was not pregnant with life" (31, page 145). Together with his colleague Jacob, he is struck by the universality of the genetic code and concludes that life may have arisen only once. "Which would mean," he adds, "that its a priori probability was virtually zero" (31, page 145). Jacob, more cautious, points out that "there is no measurable probability for an event that took place only once" (20, page 306).[2] The philosopher Popper has adopted Monod's point of view and used it as an argument against philosophical determinism (see Footnote 1). He writes: "Thus we may be faced with the possibility that the origin of life (like the origin of the universe) becomes an impenetrable barrier to science, and a residue to all attempts to reduce biology to chemistry and physics" (34, page 148).

Some authors have sought an escape from the probability quandary by assuming that life arose hierarchically in some sort of stepwise fashion. This theory has been vividly illustrated by Herbert Simon (272) in his allegory of two watchmakers, Hora and Tempus. Hora assembled his watches element by element and had to start all over again every time he was interrupted by the telephone. Tempus, on the other hand, made simple stable subsets, which he then combined into modules of greater complexity, repeating this process stage by stage. He was able to conduct his work to a successful end in spite of repeated telephone calls. A somewhat similar model has been proposed by Ilya Prigogine and Grégoire Nicolis (269), who emphasized the ability of certain systems to land by random fluctuations into an ordered structure that is stabilized and maintained far from equilibrium by a continual generation of entropy. Life, they suggest, could have climbed up a tall ladder of improbability by jumping successively from one such "dissipative structure" to another of higher complexity. The philosopher-theologian Arthur Peacocke hails this proposal as a major breakthrough: "Because of the discovery of these dissipative systems," he writes, "and of the possibility of 'order-through-fluctuations,' it is now possible to regard as highly probable the emergence of ordered, self-reproducing molecular structures—that is, of living systems" (33, page 63).

Then, there are the many—among them some scientists (65)—who reason that the emergence of life is such an improbable event that it could not have happened without some special act of creation.

The great diversity of these opinions reflects their largely subjective nature. Individual viewpoints often reveal ideological, philosophical, or religious biases more than they express objective appraisals, for the simple reason that not enough elements are available for objective analysis. My model is similarly burdened with uncertainties and biases—mostly those of a biochemist, as it happens. But, within its hypotheti-

[2] I have supplied my own translations of the quotations from Monod and Jacob. They conform more closely to the original than their published translations.

cal framework, it is sufficiently detailed and precise to allow a clearcut conceptual characterization: it is emphatically and unambiguously *deterministic*. If the model is correct, then anywhere in the universe where the conditions that prevailed in the early days of our planet should obtain, life would develop in the same way, to the point of using the same amino acids and other structural building blocks, the same metabolic pathways, the same sorts of enzymes and coenzymes, the same kinds of proteins and nucleic acids, the same blueprint—if not in every small detail, at least in all essential features. Only after all the main lines of the blueprint had been fulfilled would complexity reach a degree sufficient for chance to begin playing a role and bring in diversity and unpredictability. This statement is intrinsic to the model. It has nothing to do with the actual probability of appropriate biogenic conditions being realized elsewhere in the universe, a topic on which experts disagree as much as they do on the probability of life.

Consider first the primary building blocks produced by abiotic syntheses. Clearly, their formation owes nothing to chance. Given a certain set of conditions, they will arise reproducibly, as they do in the laboratory. In this case, we happen to know that they can also arise elsewhere in the universe, but this point is not germane to the main argument, which, as just mentioned, is not concerned with the probability of the conditions, but only with the probability of their outcome.

The next step in the model—and one of its central props—is the formation of catalytic multimers by random (undirected) assembly of simple building blocks. It is a characteristic of the model that all the catalysts needed for protometabolism to develop and be maintained must be made continually by this random mechanism until some replication process becomes operative. Only by exploiting randomness fully and leaving nothing to chance can this result be achieved. An analogy with the immune system can be detected here. The immune system covers an immense number of contingencies by means of a vast repertoire of antigen-recognition sites but makes use of only a few of these in reality. It is the same with the multimers in my model. Most of them are useless, but they have to be made in order for the useful ones to be present. In the immune system, the useful entities are selected by antigens, that is, ligands that bind to them. In my model, substrate binding may play a similar role in the selection of useful multimers. But the analogy stops here. In the immune system, selection occurs by enhanced replication and leads to considerable amplification. In my model, selection relies on enhanced survival, and the resulting amplification is limited and generally modest. As we saw in Chapter 6, the most important selective forces probably act through the preferential synthesis and enhanced stability of the multimers.

The third key element of the model is energy, in the form of thioesters arising spontaneously thanks to favorable environmental conditions. Again no miracle, not even some moderately chancy accident, need be invoked. The postulated phenomena are simple and straightforward and should be perfectly reproducible.

Given a defined set of substrates, a complete assortment of rudimentary catalysts, and a reproducible supply of energy-rich thioesters, metabolism must perforce develop along lines that, though invisible, are determined by the properties of the catalysts and by the nature of their substrates. This kind of determinism could extend very far. Should, for example, my proposal that proto-tRNAs "fished out" amino acids for protein synthesis be correct, even the nature of the 20 proteinogenic amino acids could be a standard feature of life, wherever it arose.

Perfect reproducibility of conditions being a myth, we must allow for statistical fluctuations and must therefore ask at what stage such fluctuations could have significantly affected the direction of the biogenic process, to the point of allowing a plurality of outcomes under the same conditions. As far as I can see, the strong internal coherence of the system I postulate would preclude such an occurrence until the number of possible random events started to exceed the limits imposed by the constraints of time and space. Take the length of catalytic peptides, for example. If they are decapeptides constructed from eight species of amino acids, several hundred of each possible variety of decapeptide could be fitted inside a single cell (see Table 6–3, page 138, and Figure 6–4, page 139). An exhaustive exploration of chance is possible. Make the peptides 50 residues long, using all 20 amino acids, and all the water in the oceans would not suffice to accommodate one molecule each of all possible varieties. To run them all for selective sifting through a population of cells would take immeasurably more time than the universe has existed. At this level of complexity, therefore, random mutations can explore only an infinitesimal fraction of the polypeptide landscape. Chance must perforce decide which of the innumerable potential choices are offered to natural selection in reality. Or so it appears.

This reasoning would be valid if the game were indeed played at the individual amino acid (or nucleotide) level. But this is almost certainly not the way things happened. Emerging life worked like Tempus in Simon's allegory (272), not like Hora (see page 212). Complexity was achieved by the modular assembly of subsets, by the combination of minigenes and, later, of exons (Chapter 8). This completely changes the rules of the game, at least if selection sufficiently restricts the number of available modular building blocks. Should, for example, little more than 30,000 decapeptides made with eight amino acid species survive out of the billion that are possible, the number of 20-mers that could arise from them by modular assembly would be of the order of one billion, or one-billionth of all possible 20-mers. This means that, if selection acts in a sufficiently restrictive fashion at each level of complexity, the system could continue for a long time to explore all the options left available.

If, as seems to me likely, this is how things happened, the road taken by nascent life could have been entirely deterministic up to the progenote, perhaps even up to the common ancestral cell, leaving to chance only such rare, perhaps nonexistent, events when two or more equally advantageous choices were offered and selection decided on the toss of a coin. Some codon choices may fall in this category, as was mentioned in Chapter 8, but even that is far from certain.

Does this mean that there is only one basic blueprint for life? I am tempted to answer "yes" to this question, except for the unknown influence of the environment. I believe that the pathway followed by the biogenic process up to the ancestral cell was almost entirely preordained by the intrinsic properties of the materials involved, given a certain kind of environment or succession of environmental conditions. Could it have followed a different course in a different scene? Or is the environment itself part of the deterministic picture because only a single kind of environment can foster a biogenic process? The question remains open. We know that, at some stage of the process, the road did divide into two separate directions, presumably bent on invading two distinct niches. We obviously cannot exclude the earlier occurrence of viable bifurcations of which one line died out because the appropriate environment was not available or for some other accidental reason. Even so, I strongly suspect that if the progeny of this line had survived, we would have no difficulty in recognizing in it the common blueprint of life as I have attempted to delineate it in Chapter 1.

Once the first historically documented fork split the pathway of emerging life into two branches, authentic diversification started. Remarkable in this process is the extremely long time required for the transition from the prokaryotic to the eukaryotic way of life: some two billion years, or about 1,000 times the time it took for an ape to evolve into a human. Are we to read in this enormous time span a greater intervention of chance? Did life have to wait for something close to a miracle in order to move in the eukaryotic direction? Not necessarily.

We must remember, first, that the transition from prokaryote to eukaryote must have taken a very large number of successive steps. At the risk of offending our human vanity, we must admit that, in terms of complexity, the span that exists between a bacterium, such as *Escherichia coli,* and the primitive eukaryote *Giardia lamblia* is immensely wider than the distance that separates us from our simian ancestors. As we saw in Chapter 3, a large number of new proteins had to be developed, tooled, and honed to an extraordinary degree of precision for a typical protoeukaryotic cell to emerge out of a prokaryote. As far as we know, our prohominid forebears may not have had to invent a single protein to become human.

It is also possible that some of the intermediates in the transition—naked, bulky, sluggish, largely defenseless except for their digestive advantage—were particularly fragile and environmentally demanding. The contribution of chance may have been made more at the level of the environment than at that of the evolutionary process itself. Perhaps the process started many times, only to abort repeatedly due to unfavorable conditions. It is also possible that the Earth had first to be covered by bacterial life and profoundly transformed by it before being able to provide the right milieu for primitive eukaryotes to arise and thrive. Could it be significant, for example, that the emergence of eukaryotes coincided with the rise in the oxygen content of the atmosphere?

For all these reasons, it is possible that even the appearance of eukaryotic cells may constitute an obligatory and reproducible outcome of the development of life, provided certain environmental conditions be realized according to a given temporal succession. In formulating this hypothesis, I am no longer on the relatively secure terrain offered by my model of the primary biogenic process, whose intrinsically deterministic character I have emphasized above. I invoke, rather, the probabilistic argument, reiterated many times in this book and that I believe to be inescapable: the biogenic process, including the development of the first eukaryotic cells, required an immense number of steps, most of which were almost bound to occur under the prevailing conditions. Their cumulative unfolding is explainable only if it was practically enforced, written into the very tissue of matter, waiting only for favorable conditions in order to take place. As to the degree of probability or improbability of these conditions, opinions are greatly divided at the present time (see, for example, References 35, 36, 40). There are many, however, who believe with Wald (113), Eigen (249), and others that there is nothing exceptional about these conditions and that they could very well be realized in many other parts of the universe.

What now of the further evolution of the eukaryotes? How much does it owe to chance? How much to necessity? In its ultimate form, this question leads directly to the age-old search by humankind for the meaning, if any, of its own existence. This book is devoted mainly to the nature and origin of cellular life, not to its later evolution. I do not feel prepared at this stage to tackle the evolutionary problem, which is currently the object of an enormous amount of experimental and conceptual research and of many impassioned debates among experts. Let me just confess to a personal bias, easily understandable in the context of the theory defended in this book. I suspect that chance may have played a lesser role, and necessity a more important one, than Monod and many present-day molecular biologists and evolutionists believe. This suspicion does not rest on any sort of solidly structured argument. Perhaps it is little more than wishful thinking, inspired by the fact that I do not, like Monod, relish the thought of being "alone in the indifferent vastness of the Universe out of which [I] emerged only by chance" (31, page 180). More rationally, it is based on the feeling that the principle of modular assembly, with its consequent restrictions, has continued to play a major role in evolution, using modules of increasing size and complexity. The "constraints of chance" (246) could be much more compelling than is generally assumed.

## THE PACE OF BIOGENESIS

An interesting corollary of my deterministic model of the early biogenic process is that it could have been quite fast. Consider, for example, a primitive catalyst with a turnover number of 10 molecules of substrate per molecule of catalyst per minute—as opposed to values of the order of 1,000 to 100,000 for present-day enzymes. Assume further that it is present at 0.1 nanomolar concentration and is kept fully saturated with substrate. It will need little more than one century to raise the concentration of its product (assumed to be thermodynamically favored and to suffer no destruction) to 10 millimolar. This calculation is, of course, subject to many corrective factors. But it suffices to make clear that the time span needed for the development of protometabolism as I envisage it, up to the formation of the first oligonucleotides, could be measured in thousands of years, at most in tens of thousands of years—as opposed to the hundreds of millions of years that seem to have been available according to geochemical data. This means that the launching of life may not have required an inordinately long stability of environmental conditions and also that it could have been attempted many times in different places and at different times.

As complexity increased, however, the process could have slowed down progressively. The time from scratch to the first protocell could have been relatively short, that from protocells to the first prokaryote much longer, and, as we have just seen, that from prokaryotes to the first eukaryote very much longer—the main reason being that so many more steps were required for each successive transition. Then, after the eukaryotes appeared, the pace of evolutionary complexification picked up progressively again, in almost autocatalytic fashion. One would like to know what lies behind this kinetic peculiarity. But that is another story.

## EPILOGUE

I have quoted Monod's declaration "The Universe was not pregnant with life," to which he added "nor the biosphere with man" (31, page 145). I have made it clear that I disagree with his first statement. Life belongs to the very fabric of the universe. Were it not an obligatory manifestation of the combinatorial properties of matter, it could not possibly have arisen naturally. By ascribing to chance an event of such unimaginable complexity and improbability—remember Hoyle's allegory of the Boeing 747 emerging from a junkyard (Chapter 5)—Monod does, in fact, invoke a miracle. Much as he would have refused this description, he sides with the creationists.

As to the second half of Monod's affirmation, I also tend to disagree with it, as I have already mentioned. But it would need another book and much more than biochemistry and molecular biology to substantiate my belief. When I listen to music, wander through a gallery, feast my eyes on the pure lines of a Gothic cathedral, read a poem or a scientific article, watch my grandchildren at play, or simply reflect on the fact that I can do all those things, including reflecting on my ability to do them, I find it impossible to conceive of the universe of which I am part as not bound by its very nature to give birth somewhere, sometime, perhaps in many places and at many times, to beings capable of enjoying beauty, experiencing love, seeking truth, and apprehending mystery. This, no doubt, classifies me as a romantic. So be it.

# BIBLIOGRAPHY

The list that follows includes all the references cited by numbers in the text and a few others of general interest. It has been classified and annotated so as also to serve as a guide to further reading. As in all classifications, there are borderline cases that could fall under one heading or under another. In particular, the distinction between cell biology and molecular biology has become so blurred in recent years that it was deemed better to group the two topics under a single heading.

## GENERAL GUIDE

1 **de Duve, C.** *A guided tour of the living cell.* New York: Scientific American Books, Inc.; 1984.

Complementary to the present book, this illustrated tour of the cell blends cell biology, biochemistry, and molecular biology under the general aegis of bioenergetics (principles summarized in an Appendix). It can serve as the recommended source of additional information and illustrative details that were omitted for the sake of brevity.

## TEXTBOOKS

### *Biochemistry*

2 **Lehninger, A. L.** *Biochemistry.* 3rd ed. New York: Worth Publishers, Inc.; 1980.

A major work when it first appeared in 1970. It remains a classic.

3 **Lehninger, A. L.** *Principles of biochemistry.* New York: Worth Publishers, Inc.; 1982.

A somewhat condensed version of the former, described by the author, who died in 1986, as replacing a new edition that inclusion of recent advances would have made too big for an undergraduate textbook.

4 **Rawn, J. D.** *Biochemistry.* Burlington, NC: Neil Patterson Publishers; 1989.

A newcomer on the scene, richly presented and illustrated.

5 **Stryer, L.** *Biochemistry.* 3rd ed. New York: W. H. Freeman and Company; 1988.

It initiated a new style in textbooks when it first appeared in 1975. It is still a favorite.

*Cellular and Molecular Biology*

6 **Alberts, B.; Bray, D.; Lewis, J.; Raff, M.; Roberts, K.; Watson, J. D.** *Molecular biology of the cell.* 2nd ed. New York: Garland Publishing, Inc.; 1989.

When it came out in 1983, this textbook represented the first attempt at comprehensively covering eukaryotic molecular biology, with excursions into immunology, developmental biology, and neurobiology.

7 **Darnell, J.; Lodish, H.; Baltimore, D.** *Molecular cell biology.* New York: Scientific American Books, Inc.; 1986.

This book is comparable to the preceding one in the scope, level, and quality of its coverage, differing from it in the relative emphasis given to different topics. It is often profitable to consult both.

8 **De Robertis, E. D. P.; De Robertis, E. M. F., Jr.** *Cell and molecular biology.* 7th ed. Philadelphia: W. B. Saunders Company; 1980.

First published in 1946 by the senior author, in collaboration with F. A. Saez and W. N. Nowinski, this elementary textbook has adequately incorporated in successive editions the many advances of biochemistry, cytology, and molecular biology.

9 **De Robertis, E. D. P.; De Robertis, E. M. F., Jr.** *Essentials of cell and molecular biology.* Philadelphia: Saunders College Publishing; 1981.

A summary of the preceding, for the benefit of advanced high-school and junior-college students.

10 **Jensen, W. A.; Park, R. B.** *Cell ultrastructure.* Belmont, CA: Wadsworth Publishing Company, Inc.; 1967.

A "mini-atlas" of 60 pages illustrating all major cell structures as viewed by electron microscopy.

11 **Kessel, R. G.; Kardon, R. H.** *Tissues and organs.* San Francisco: W. H. Freeman and Company; 1979.

A compendium of the structures of all major mammalian tissues and organs, illustrated mostly by scanning electron micrographs.

12 **Stent, G. S.; Calendar, R.** *Molecular genetics: an introductory narrative.* 2nd ed. San Francisco: W. H. Freeman and Company; 1978.

Although outdated, this text remains useful for its emphasis on the historical development of the concepts of molecular genetics.

13 **Watson, J. D.; Hopkins, N. H.; Roberts, J. W.; Steitz, J. A.; Weiner, A. M.** *Molecular biology of the gene.* Vol. 1 and 2, 4th ed. Menlo Park, CA: The Benjamin/Cummings Publishing Company, Inc.; 1987.

This classic, which was first published by the senior author in 1965, has been enlarged to cover the molecular biology of both prokaryotes and eukaryotes. RNA biochemistry and molecular biology are treated in particular detail.

*Microbiology*

14 **Atlas, R. M.** *Microbiology, fundamentals and applications.* New York: Macmillan Publishing Company; 1984.

An introduction to modern bacteriology.

15 **Margulis, L.; Corliss, J. O.; Melkonian, M.; Chapman, D. J.** *Handbook of protoctista.* Boston: Jones and Bartlett Publishers; 1990.

This comprehensive survey provides a morphological and physiological description of single-celled eukaryotes, organized according to a new taxonomy resolutely based on endocytobiological principles.

16 **Sleigh, H.** *Protozoa and other protists.* 2nd ed. London: Edward Arnold (A division of Hodder and Stoughton); 1989.

An elementary textbook, mainly focused on the organisms formerly known as protozoa.

## ESSAYS AND MONOGRAPHS

### *Historical and Biographical*

17 **Crick, F.** *What mad pursuit.* New York: Basic Books, Inc., Publishers; 1988.

A pleasantly written story of his own life by one of the two discoverers of the "double helix."

18 **Darwin, F.** *The life and letters of Charles Darwin.* Vol. 2. New York: D. Appleton; 1887.
Remembered here for a famous "warm little pond."

19 **Farley, J.** *The spontaneous generation controversy from Descartes to Oparin.* Baltimore: The Johns Hopkins University Press; 1977.

A historical introduction to the problem of the origin of life, it includes some interesting details on Alexander Oparin, the father of the subject.

20 **Jacob, F.** *La logique du vivant: une histoire de l'hérédité.* Paris: Editions Gallimard; 1970. Translated as *The logic of life.* New York: Pantheon Books; 1973.

A broad fresco of genetics by one of the founders of molecular biology.

21 **Judson, H. F.** *The eighth day of creation: the makers of the revolution in biology.* New York: Simon and Schuster; 1979.

A history of molecular biology, based largely on extended personal interviews with some of the protagonists and written in a vivid journalistic style.

22 **Lipmann, F. A.** *Wanderings of a biochemist.* New York: Wiley-Interscience; 1971.

This autobiography of the creator of the "squiggle" is illustrated by key reprints selected by the author (see also References 82, 149b).

23 **McCarty, M.** *The transforming principle.* New York: W. W. Norton and Company; 1985.

A sober account of the work that established DNA as the genetic material, by the last surviving member of the Avery-MacLeod-McCarty team that made this historical discovery.

24 **Sapp, J.** *Beyond the gene: cytoplasmic inheritance and the struggle for authority in genetics.* New York: Oxford University Press; 1987.

A well-documented history of the political and ideological undercurrents that pitted Mendelian and non-Mendelian geneticists against each other between the 1920s and the early 1980s.

25 **Watson, J. D.** *The double helix: a personal account of the discovery of the structure of DNA.* Stent, G. S., ed. New York: W. W. Norton and Company; 1980.

A re-edition of the notorious tale that shook the scientific world by its uninhibited candor when it first appeared in 1968. It is accompanied here by a number of commentaries and reviews.

*Philosophical and General Essays*

26 **Bradbury, R.** *Zen in the art of writing: essays on creativity.* Santa Barbara, CA: Joshua Odell Editions, Capra Press; 1989.

A collection of enjoyable essays that provided me with a quotation for my preface.

27 **Crick, F.** *Life itself.* New York: Simon and Schuster; 1981.

One of the most famous scientists of our time exposes his view of life—presented as an import from some distant planet—in an entertaining style directed at the general public.

28 **Haldane, J. B. S.** *On being the right size, and other essays.* Smith, J. M., ed. Oxford: Oxford University Press; 1985.

A collection of penetrating and witty essays, including one on the origin of life, that were contributed to the *Daily Worker* and other left-wing publications by the celebrated British geneticist.

29 **Jacob, F.** *Le jeu des possibles: essai sur la diversité du vivant.* Paris: Librairie Arthème Fayard; 1981. Translated as *The possible and the actual.* Seattle: University of Washingon Press; 1982.

A humanist's view of evolution.

30 **Medawar, P. B.** *The uniqueness of the individual.* London: Methuen and Company, Ltd.; 1957.

Eight elegantly written essays by a great immunologist and thinker.

31 **Monod, J.** *Le hasard et la nécessité: essai sur la philosophie naturelle de la biologie moderne.* Paris: Editions du Seuil; 1970. Translated as *Chance and necessity.* New York: Knopf; 1971; London: Collins; 1972.

In this book, which has stirred considerable controversy, a master of modern biology defends a stoically (and romantically) despairing, existentialist view of human life.

32 **Morowitz, H. J.** *Cosmic joy and local pain: musings of a mystic scientist.* New York: Charles Scribner's Sons; 1987.

The author, who has devoted much thought to the origin of life, ventures beyond the strict area of science, to arrive at a mystical pantheistic philosophy.

33 **Peacocke, A. E.** *God and the new biology.* New York: Harper and Row; 1986.

A valiant and erudite attempt at reconciling theology and science.

34 **Popper, K.** *The open universe: an argument for indeterminism.* Totowa, NJ: Rowman and Littlefield; 1982.

A distinguished philosopher of science explains in detail why he does not subscribe to determinism.

Also References 59, 65, 68, 74.

*Cosmology*

35 **Breuer, R.** *Contact with the stars.* San Francisco: W. H. Freeman and Company; 1982.

Written at a time when the possibility of communicating with extraterrestrial civilizations was evoking a considerable amount of interest, this sober appraisal leads the author to conclude that "the chances of coming upon traces of or signals from extraterrestrial civilizations are very slim."

36 **Feinberg, G.; Shapiro, R.** *Life beyond earth.* New York: William Morrow and Company; 1980.

This very readable book addresses in a popular style the probability of there being life elsewhere in the universe. It ends with the conclusion that "the universe is full of life."

37 **Hoyle, F.; Wickramasinghe, N. C.** *Lifecloud.* New York: Harper and Row; 1978.

In this book and the following two, the authors propose the theory that life originated in outer space and that it continues to do so up to the present day, showering the Earth with as many as $10^{20}$ "germs" a year.

38 **Hoyle, F.; Wickramasinghe, N. C.** *Diseases from space.* New York: Harper and Row; 1979.

39 **Hoyle, F.; Wickramasinghe, N. C.** *Evolution from space.* New York: Simon and Schuster; 1981.

40 **Rood, R. T.; Trefil, J. S.** *Are we alone?* New York: Charles Scribner's Sons; 1981.

After filling in the most plausible values for all the variables in the so-called Green Bank equation and taking into account a number of other factors, one author answers: "Yes, we are alone, because we are special;" the other concludes: "Perhaps we are not alone, but, most likely, the others don't care."

41 **Urey, H.** *The planets: their origin and development.* New Haven: Yale University Press; 1952.

This book does not deal directly with the origin of life, but it has had a considerable influence on the subject by its depiction of likely prebiotic conditions, which were then used by Stanley Miller in his first laboratory simulation experiments.

*Biochemistry and Bioenergetics*

42 **Broda, E.** *The evolution of the bioenergetic processes.* Oxford: Pergamon Press; 1975.

An elementary text, of interest because of its emphasis on evolutionary aspects.

43 **Harold, F. M.** *The vital force: a study of bioenergetics.* New York: W. H. Freeman and Company; 1986.

An introduction to bioenergetics, highlighting the role of protonmotive force.

44 **Haurowitz, F.** *Chemistry and biology of proteins.* New York: Academic Press; 1950.

An advanced survey in its day, remembered here for its model of protein replication.

45 **Racker, E.** *Mechanisms in bioenergetics.* New York: Academic Press; 1965.

Straight out of the laboratory, this monograph by one of the founders of bioenergetics remains enlightening by its step-by-step approach to substrate-level phosphorylation.

46 **Racker, E.** *A new look at mechanisms in bioenergetics.* New York: Academic Press; 1976.

A reappraisal written after protonmotive force came on the scene.

Also Reference 22.

*Cell Biology*

46a **Preston, T. M.; King, C. A.; Hyams, J. S.** *The cytoskeleton and cell motility.* Glasgow: Blackie & Son Ltd; 1990.

This book covers the main aspects of cellular and intracellular motility considered in molecular and functional terms.

47 **Pugsley, A. P.** *Protein targeting.* New York: Academic Press; 1989.

A recent update on a rapidly expanding topic.

Also References 20, 21, 23, 24, 25, 27, 29, 31, 53, 66, 73, 74.

*Origin of Life*

48 **Bernal, J. D.** *The origin of life.* London: Weidenfeld and Nicholson; 1967.

One of the first books on the topic, written by an eminent British physical chemist and crystallographer.

49 **Brooks, J.; Shaw, G.** *Origin and development of living systems.* New York: Academic Press; 1973.

Written from the point of view of organic chemists, this book covers, in relatively simple terms, a wide field ranging from the beginning of the Universe to the origin of life. It is outdated but remains useful as a guide to the early literature.

50 **Cairns-Smith, A. G.** *Genetic takeover and the mineral origins of life.* Cambridge: Cambridge University Press; 1982.

Starting with a severe—some would say damning—indictment of current models of the abiotic synthesis of informational macromolecules, the author offers his own theory of a first phase of life based on clay genes. In addition to an ingenious thesis, the book offers useful discussions, comments, and suggestions.

51 **Calvin, M.** *Chemical evolution.* New York: Oxford University Press; 1969.

Known for unravelling the main pathway of biological carbon dioxide assimilation, the author approaches the problem of the origin of life from the point of view of the expert chemist and biochemist. His book remains actual.

52 **Day, W.** *Genesis on planet earth.* 2nd ed. New Haven: Yale University Press; 1984.

This very readable and balanced account of the problem of the origin of life envisages a sequence of events similar in several respects to that proposed in the present book. It includes a useful reading list.

53 **Dillon, L. S.** *The genetic mechanism and the origin of life.* New York: Plenum Press; 1978.

This detailed and strongly documented analysis is clearly written, though without concession to the reader. Its bibliography includes more than 3,000 references.

54 **Dyson, F.** *Origins of life.* Cambridge: Cambridge University Press; 1985.

This booklet summarizes in less than 80 pages a well-known theoretical physicist's venture into biology influenced by the ideas of Motoo Kimura (69).

55 **Florkin, M.** *L'Evolution biochimique*. Paris: Masson et Cie; 1944. Translated as *Biochemical evolution*. New York: Academic Press; 1949.

One of the pioneers of comparative biochemistry views the origin of life from the point of view of the biochemical knowledge of his time. A Gallic fondness for conceptual analysis and classification is apparent.

56 **Folsome, C. E.** *The origin of life: a warm little pond*. San Francisco: W. H. Freeman and Company; 1979.

This concise little book includes a number of original ideas and ingenious models.

57 **Fox, S. W.; Dose, K.** *Molecular evolution and the origin of life*. New York: Marcel Dekker; 1977.

A good introduction to the "proteinoid microspheres," which provide the main basis of the theory that Fox has defended during more than 25 years.

58 **Fox, S. W.** *The emergence of life*. New York: Basic Books, Inc., Publishers; 1988.

Written primarily for "those who have had a high school course in biology," this book provides the latest summary of the author's highly personal views.

59 **Kenyon, D. H.; Steinman, G.** *Biological predestination*. New York: McGraw-Hill; 1969.

Biochemical considerations lead the authors to views not too different from those defended in the last chapter of the present book.

60 **Miller, S. L.; Orgel, L. E.** *The origins of life*. Englewood Cliffs, NJ: Prentice Hall; 1973.

An introduction to the chemical problems raised by the origin of life, written by two experts of abiotic chemistry.

61 **Oparin, A. I.** *The origin of life on the earth*. 3rd ed. New York: Academic Press; 1957.

First published in Moscow (Moskovskii Rabochii) in 1924, this book approaches the problem of the origin of life in the context of dialectical materialism. It went through several revisions and had a considerable impact. It is generally considered as having opened the origin of life to scientific inquiry.

62 **Orgel, L. E.** *The origins of life: molecules and natural selection*. New York: John Wiley and Sons; 1973.

An elementary account for the benefit of readers "who have a limited background in chemistry or biology."

63 **Quastler, H.** *The emergence of biological organization*. New Haven: Yale University Press; 1964.

An interesting discussion of the problem, written from the point of view of probability and information theories.

64 **Shapiro, R.** *Origins, a skeptic's guide to the creation of life on earth*. New York: Summit Books; 1986.

This entertaining and critical overview of the various theories that are presently vying to explain the origin of life is recommended as an introduction to the subject. It includes vignettes of the main workers in the field, together with accounts of a number of personal interviews.

65 **Thaxton, C. B.; Bradley, W. L.; Olsen, R. L.** *The mystery of life's origin: reassessing current theories.* New York: Philosophical Library; 1984.

Written by a chemist, a material scientist, and a geochemist, this book contains a critical and well-documented discussion of the main theories of the origin of life. Unable to accept any of these theories as plausible, the authors adopt a creationist conclusion.

66 **Woese, C.** *The genetic code: the molecular basis for genetic expression.* New York: Harper and Row; 1967.

Not specifically directed at the problem of the origin of life, this book includes a number of pertinent discussions on the origin of translation and of the genetic code.

Also References 19, 27, 28, 31, 32, 36, 37, 38, 39.

*Evolution*

67 **Dawkins, R.** *The selfish gene.* New York: Oxford University Press; 1976. New edition 1989.

A highly readable transposition in modern words of the old theme: a chicken is an egg's means of making another egg. In the new edition, the part devoted to human behavior is expanded and the author answers a number of criticisms.

68 **Dawkins, R.** *The blind watchmaker.* New York: W. W. Norton and Company; 1986.

An uncompromising exposition of Darwinian evolutionary theory, accompanied by some strongly worded statements of philosophical nature.

69 **Kimura, M.** *The neutral theory of molecular evolution.* Cambridge: Cambridge University Press; 1983.

By applying a statistical treatment of molecular evolution, the author arrives at the conclusion that genetic drift is more important than natural selection as an evolutionary driving force. His mathematical treatment has been adopted by Freeman Dyson to build a model of the origin of life that circumvents the need for an early replicating system (54).

70 **Margulis, L.** *Origin of eukaryotic cells.* New Haven: Yale University Press; 1970.

This book played an important role in reviving the endosymbiotic theory of the development of eukaryotic cells.

71 **Margulis, L.** *Symbiosis in cell evolution.* San Francisco: W. H. Freeman and Company; 1981.

This updated version of the former includes some considerations on the origin of life.

72 **Margulis, L.; Sagan, D.** *Micro-Cosmos.* New York: Summit Books; 1986.

A colorful popularization of Margulis's theory, which, in combination with James Lovelock's Gaia hypothesis, presents the biosphere as a giant superorganismic colony of bacteria.

73 **Ninio, J.** *Approches moléculaire de l'évolution.* Paris: Masson; 1979. Translated as *Molecular approaches to evolution.* Princeton: Princeton University Press; 1983.

An imaginative and critical essay, written in a lively and sometimes irreverent style.

74 **Ruffié, J.** *Traité du vivant.* Paris: Arthème Fayard; 1982.

This vast compendium covers the whole field of evolution from population biology to sociology by way of molecular biology. Current controversies on evolutionary mechanisms are well presented.

Also References 18, 20, 29.

### *Microbiology*

75 **Doflein, F.; Reichenow, E.** *Lehrbuch der Protozoenkunde.* 6th ed. Jena: Fischer Verlag; 1949.

This classic textbook is cited here in a historical context.

76 **Koshland, D. E., Jr.** *Bacterial chemotaxis as a model behavioral system.* New York: Raven Press; 1980.

An account of how bacteria "swim" in response to outside stimuli, with a surprising final extrapolation to human behavior.

77 **Martinez-Palomo, A.** *The biology of* Entamoeba histolytica. Chichester, UK: Research Studies Press; 1982.

A survey of the morphological and functional properties of an intriguing phagocyte seemingly devoid of an internal cytomembrane system.

## MULTI-AUTHOR BOOKS AND SYMPOSIUM PROCEEDINGS

### *General Essays*

78 **Handler, P., ed.** *The scientific endeavor, history of the universe, nature of matter: the determinants and evolution of life.* New York: The Rockefeller University Press; 1963.

A panorama of the state of knowledge at the time of publication, by a galaxy of experts brought together on the occasion of the hundredth anniversary of the National Academy of Sciences of the United States.

### *Biochemistry, Cellular and Molecular Biology*

79 **Baltscheffsky, H.; Jörnvall, H.; Rigler, R., eds.** *Molecular evolution of life. Chem. Scr.* 26B; 1986.

Proceedings of a conference covering a wide spectrum of subjects.

79a **Chater, K. F.; Brewin, N. J.; Casey, R.; Roberts, K.; Wilson, T. M. A.; Flavell, R. B.** Protein targeting. *J. Cell Science.* Supplement 11; 1989.

A good survey of the topic covering both prokaryotic and eukaryotic mechanisms.

80 **Cold Spring Harbor Laboratory.** *Evolution of catalytic function. Cold Spring Harbor Symp. Quant. Biol.* 52; 1987.

Report of a conference of special interest to the readers of this book. Provides a considerable amount of important facts, while illustrating at the same time the bold speculative tendencies of contemporary molecular biology.

81 **Fahimi, H. D.; Sies, H., eds.** *Peroxisomes in biology and medicine.* Berlin and New York: Springer-Verlag; 1987.

Proceedings of a symposium that gathered most of the experts in the field.

82 **Kleinkauf, H.; von Döhren, H.; Jaenicke, L., eds.** *The roots of modern biochemistry: Fritz Lipmann's squiggle and its consequences.* Berlin: Walter de Gruyter; 1988.

This compendium reports the proceedings of a meeting in which many leading biochemists paid homage to the memory of the founder of modern bioenergetics.

83 **Rechsteiner, M., ed.** *Ubiquitin.* New York: Plenum Publishing Corporation; 1988.
A complete overview of the properties of a fascinating molecule.

84 **Slonimski, P.; Borst, P.; Attardi, G., eds.** *Mitochondrial genes.* Cold Spring Harbor, NY: *Cold Spring Harbor Monograph Ser.* 12; 1982.
Proceedings of a conference on a subject that remains actual.

85 **Terzaghi, E. A.; Wilkins, A. S.; Penny, D., eds.** *Molecular evolution: an annotated reader.* Boston: Jones and Bartlett Publishers, Inc.; 1984.
A collection of reprints of seminal papers.

86 **Weissmann, G.; Claiborne, R., eds.** *Cell membranes, biochemistry, cell biology and pathology.* New York: Hospital Practice Publishing Company, Inc.; 1975.
Remains a good introduction to the subject, in spite of its age.

*Origin of Life*

87 **Dose, K.; Fox, S. W.; Deborin, G. A.; Pavlovskaya, T. E., eds.** *The origin of life and evolutionary biochemistry.* New York: Plenum Press; 1974.
Proceedings of a conference with emphasis on so-called abiotic chemistry.

88 **Fox, S. W., ed.** *The origins of prebiological systems and of their molecular matrices.* New York: Academic Press; 1965.
An early conference on the origin of life.

88a **Oparin, A. I.; Pasynskii, A. G.; Braunshtein, A. E.; Pavlovskaya, T. E., eds.** *The origin of life on the earth.* London: Pergamon Press; 1959.
This first international symposium on the origin of life, attended by many of the great biochemists of the day, launched the topic as a respectable subject of inquiry.

89 **Schopf, J. W., ed.** *Earth's earliest biosphere: its origin and evolution.* Princeton: Princeton University Press; 1983.
A compendium of detailed data on the cradle of life.

*Evolution*

90 **Lee, J. J.; Fredrick, J. F., eds.** *Endocytobiology III. Ann. N. Y. Acad. Sci.* 503; 1987.
A mine of interesting information on the origin of eukaryotic cells.

91 **Nardon, P.; Gianinazzi, V.; Grenier-Pearson, A. M.; Margulis, L.; Smith, D. C., eds.** *Endocytobiology IV.* Paris: INRA (Institut National de la Recherche Agronomique); 1990.
A valuable follow-up on the preceding.

*Microbiology*

92 **Hayflick, L., ed.** *The Mycoplasmatales and the L-phase of bacteria.* New York: Appleton-Century-Crofts; 1969.
An overview of these smallest of all living organisms, of special interest in relation to the origin of life.

93 **Lewin, R. A.; Cheng, L., eds.** *Prochloron: a microbial enigma.* New York: Routledge, Chapman and Hall; 1989.
The complete information on a possible contemporary relative of a chloroplast ancestor.

94 **Schleifer, K. H.; Stackebrandt, E., eds.** *Evolution of prokaryotes.* New York: Academic Press; 1985.

A useful guide to a rapidly changing subject.

## CONTRIBUTIONS TO SYMPOSIA

### *Biochemistry, Cellular and Molecular Biology*

95 **Baltscheffsky, H.; Lundin, M.; Luxemburg, C.; Nyren, P.; Baltscheffsky, M.** Inorganic pyrophosphate and the molecular evolution of biological energy coupling. *Reference 79:*259–262; 1986.

96 **Crick, F. H. C.** On protein synthesis. *Symp. Soc. Exp. Biol.* 12:138–163; 1958.

97 **de Duve, C.** Peroxisomes and related particles in historical perspective. *Ann. N. Y. Acad. Sci.* 386:1–4; 1982.

98 **Gilbert, W.** The exon theory of genes. *Reference 80:*901–905; 1987.

99 **Gō, M., Nosaka, M.** Protein architecture and the origin of introns. *Reference 80:*915–924; 1987.

100 **Moore, P. B.** On the modus operandi of the ribosome. *Reference 80:*721–728; 1987.

101 **Nomura, M.** The role of RNA and protein in ribosome function: a review of early reconstitution studies and prospects of future studies. *Reference 80:*653–663; 1987.

102 **To, L. P.** Are centrioles semiautonomous? *Reference 90:*83–91; 1987.

103 **Weiner, A. M.** Summary. *Reference 80:*933–941; 1987.

104 **Wittmann, H. G.; Yonath, A.** Architecture of ribosomal particles as investigated by image reconstruction and X-ray crystallographic studies. *Reference 82;* 1988:481–492.

### *Origin of Life*

105 **de Duve, C.** Prebiotic syntheses and the mechanism of early chemical evolution. *Reference 82;* 1988:881–894.

106 **Ferris, J. P.** Prebiotic synthesis: problems and challenges. *Reference 80:*29–35; 1987.

107 **Fox, S. W.** A model for protocellular coordination of nucleic acid and protein syntheses. In: Kageyama, M.; Nakamura, K.; Oshima, T.; Ushida, T., eds. *Science and scientists.* Tokyo: Japan Scientific Societies Press; 1981:39–45.

108 **Lederberg, J.** Preface. In: Wood, J. A.; Chang, S., eds. *The cosmic history of the biogenic elements and compounds.* Washington, DC: NASA, Ames Research Center; 1985:vii–viii.

109 **Lipmann, F.** Projecting backward from the present stage of evolution of biosynthesis. *Reference 88;* 1965:259–280. Also reprinted in *Reference 22;* 1971:212–226.

110 **Miller, S. L.** Current status of the prebiotic synthesis of small molecules. *Reference 79:*5–11; 1986.

111 **Miller, S. L.** Which organic compounds could have occurred on the prebiotic earth? *Reference 80:*17–27; 1987.

112 **Orgel, L. E.** Evolution of the genetic apparatus: a review. *Reference 80:* 9–16; 1987.

112a **Pirie, N. W.** Chemical diversity and the origins of life. *Reference 88a:* 76–83; 1959.

113 **Wald, G.** The origin of life. *Reference 78;* 1964:113–134.

114 **Wieland, T.** Sulfur in biomimetic peptide syntheses. *Reference 82;* 1988:213–221.

*Evolution*

115 **Benner, S. A.** Reconstructing the evolution of proteins. In: Benner, S. A., ed. *Redesigning the molecules of life.* Berlin: Springer Verlag; 1988:115–175.

116 **Bereiter-Hahn, J.** Compartmentation of calcium and energy metabolic pathways: implications for eukaryote evolution and control of cell proliferation. *Reference 90:*372–379; 1987.

117 **Bermudes, D.; Fracek, S. P., Jr.; Laursen, R. A.; Margulis, L.; Obar, R.; Tzertzinis, G.** Tubulinlike protein from *Spirochaeta bajacaliforniensis. Reference 90:*515–527; 1987.

118 **Bermudes, D.; Margulis, L.; Tzertzinis, G.** Prokaryotic origin of undulipodia: application of the panda principle to the centriole enigma. *Reference 90:*187–197; 1987.

119 **Cavalier-Smith, T.** The origin of cells: a symbiosis between genes, catalysts, and membranes. *Reference 80:*805–824; 1987.

120 **Cavalier-Smith, T.** The origin of eukaryote and archaebacterial cells. *Reference 90:*17–54; 1987.

121 **Cavalier-Smith, T.** The simultaneous symbiotic origin of mitochondria, chloroplasts, and microbodies. *Reference 90:*55–71; 1987.

122 **de Duve, C.** Evolution of the peroxisome. *Ann. N. Y. Acad. Sci.* 168:369–381; 1969.

123 **Delihas, N.; Fox, G. E.** Origins of the plant chloroplasts and mitochondria based on comparisons of 5S ribosomal RNAs. *Reference 90:*92–102; 1987.

124 **Erdmann, V. A.; Wolters, J.; Pieler, T.; Digweed, M.; Specht, T.; Ulbrich, N.** Evolution of organisms and organelles as studied by comparative computer and biochemical analyses of ribosomal 5S RNA structure. *Reference 90:*103–124; 1987.

125 **Gellissen, G.; Michaelis, G.** Gene transfer: mitochondria to nucleus. *Reference 90:*391–401; 1987.

126 **Halmann, M.** Evolution and ecology of phosphorus metabolism. *Reference 87;* 1974:169–182.

127 **Jeon, K. W.** Change of cellular "pathogens" into required cell components. *Reference 90:*359–371; 1987.

128 **John, P.** *Paracoccus* as a free-living mitochondrion. *Reference 90:*140–150; 1987.

129 **Joyard, J.; Block, M. A.; Dorne, A.-J.; Douce, R.** Comparison of plastid envelope membranes and outer membranes from cyanobacteria: relevance to chloroplast evolution. *Reference 91:*527–536; 1990.

130 **Jukes, T. H.** Evolution of the amino acid code. In: Nei, M.; Koehn, R. K., eds. *Evolution of genes and proteins.* Sunderland, MA: Sinauer Associates Inc., Publishers; 1983:191–207.

131 **Lake, J. A.** Prokaryotes and archaebacteria are not monophyletic: rate invariant analysis of rRNA genes indicates that eukaryotes and eocytes form a monophyletic taxon. *Reference 80:*839–846; 1987.

132 **Margulis, L.; Hinkle, G.; Tzertzinis, G.** Symbiosis in the origin of eukaryotic motility: current status. *Reference 91:*523–525; 1990.

133 **Orgel, L. E.** Was RNA the first genetic polymer? In: Grunberg-Manago, M.; Clark, B. F. C.; Zachau, H. G., eds. *Evolutionary tinkering in gene expression.* New York: Plenum Press; 1989:215–224.

134 **Schenk, H. E. A.; Bayer, M. G.; Zook, D.** Cyanelles: from symbiont to organelle. *Reference 90:*151–167; 1987.

135 **Searcy, D. G.** Phylogenetic and phenotypic relationships between the eukaryotic nucleocytoplasm and thermophilic archaebacteria. *Reference 90:*168–179; 1987.

136 **Sonneborn, T. M.** Degeneracy of the genetic code: extent, nature, and genetic implications. In: Bryson, V.; Vogel, H. J., eds. *Evolving genes and proteins.* New York: Academic Press; 1965:377–397.

137 **Stanier, R. Y.** Some aspects of the biology of cells and their possible evolutionary significance. *Symp. Soc. Gen. Microbiol.* 20:1–38; 1970.

138 **Taylor, F. J. R.** An overview of the status of evolutionary cell symbiosis theories. *Reference 90:*1–16; 1987.

139 **Zillig, W.** Eukaryotic traits in archaebacteria: could the eukaryotic cytoplasm have arisen from archaebacterial origin? *Reference 90:*78–82; 1987.

## REVIEWS

### *Biochemistry*

140 **Borsook, H.** Peptide bond formation. *Adv. Protein Chem.* 8:127–174; 1953.

141 **Bowie, J. U.; Reidhaar-Olson, J. F.; Lim, W. A.; Sauer, R. T.** Deciphering the message in protein sequences: tolerance to amino acid substitutions. *Science* 247:1306–1310; 1990.

142 **Bungenberg de Jong, H. G.** Die Koazervation und ihre Bedeutung für die Biologie. *Protoplasma* 15:110–173; 1932.

143 **Cammack, R.** Evolution and diversity in the iron-sulfur proteins. *Chem. Scr.* 21:87–95; 1983.

144 **Dolin, M. I.** Cytochrome-independent electron transport. In: Gunsalus, I. C.; Stanier, R. Y., eds. *The Bacteria.* Vol. 2. New York: Academic Press; 1961:425–460.

145 **Granick, S.** The structural and functional relationships between heme and chlorophyll. *Harvey Lectures* 44:220–245; 1950.

146 **Jones, W. J.; Nagle, D. P., Jr.; Whitman, W. B.** Methanogens and the diversity of archaebacteria. *Microbiol. Rev.* 51:135–177; 1987.

147 **Kerscher, L.; Oesterhelt, D.** Pyruvate:ferredoxin oxidoreductase—new findings on an ancient enzyme. *Trends Biochem. Sci.* 7:371–374; 1982.

148 **Kleinkauf, H.; von Döhren, H.** Biosynthesis of peptide antibiotics. *Annu. Rev. Microbiol.* 41:259–289; 1987.

149 **Lechner, J.; Wieland, F.** Structure and biosynthesis of prokaryotic glycoproteins. *Annu. Rev. Biochem.* 58:173–194; 1989.

149a **Lipmann, F.** Attempts to map a process evolution of peptide biosynthesis. *Science* 173:875–884; 1971.

149b **Lipmann, F.** A long life in times of great upheaval. *Annu. Rev. Biochem.* 53:1–33, 1984.

150 **Müller, M.** Energy metabolism of protozoa without mitochondria. *Annu. Rev. Microbiol.* 42:465–488; 1988.

151 **Oesterhelt, D.; Tittor, J.** Two pumps, one principle: light-driven ion transport in halobacteria. *Trends Biochem. Sci.* 14:57–61; 1989.

152 **Reeves, R. E.** Metabolism of *Entamoeba histolytica* Schaudinn, 1903. *Adv. Parasitol.* 23:105–142; 1984.

153 **Schneider, D. L.** Proton pump ATPase of lysosomes and related organelles of the vacuolar apparatus. *Biochim. Biophys. Acta* 895:1–10; 1987.

154 **Thauer, R. K.; Jungermann, K.; Decker, K.** Energy conservation in chemotrophic anaerobic bacteria. *Bacteriol. Rev.* 41:100–180; 1977.

155 **Wood, H. G.** Some reactions in which inorganic pyrophosphate replaces ATP and serves as a source of energy. *Fed. Proc.* 36:2197–2205; 1977.

156 **Wood, H. G.** Inorganic pyrophosphate and polyphosphates as sources of energy. *Curr. Topics Cell Regul.* 26:355–369; 1985.

### *Cellular and Molecular Biology*

157 **Attardi, G.; Schatz, G.** Biogenesis of mitochondria. *Annu. Rev. Cell Biol.* 4:289–333; 1988.

158 **Bishop, W. R.; Bell, R. M.** Assembly of phospholipids into cellular membranes: biosynthesis, transmembrane movement and intracellular translocation. *Annu. Rev. Cell Biol.* 4:579–610; 1988.

159 **Borst, P.** How proteins get into microbodies (peroxisomes, glyoxysomes, glycosomes). *Biochim. Biophys. Acta* 866:179–203; 1986.

160 **Brodsky, F. M.** Living with clathrin: its role in intracellular membrane traffic. *Science* 242:1396–1402; 1988.

161 **Cech, T. R.** RNA as an enzyme. *Sci. Am.* 255(5):64–75; 1986.

162 **Cech, T. R.; Bass, B. L.** Biological catalysis by RNA. *Annu. Rev. Biochem.* 55:599–629; 1986.

163 **Clayton, D. A.** Transcription of the mammalian mitochondrial genome. *Annu. Rev. Biochem.* 53:573–594; 1984.

164 **Davis, B. D.; Tai, P.-C.** The mechanism of protein secretion across membranes. *Nature* 283:433–438; 1980.

165 **Dawidowicz, E. A.** Dynamics of membrane lipid metabolism and turnover. *Annu. Rev. Biochem.* 56:43–61; 1987.

166 **de Duve, C.** Microbodies in the living cell. *Sci. Am.* 248(5):74–84; 1983.

167 **de Duve, C.; Wattiaux, R.** Functions of lysosomes. *Annu. Rev. Physiol.* 28:435–492; 1966.

168 **Dingwall, C.; Laskey, R. A.** Protein import into the cell nucleus. *Annu. Rev. Cell Biol.* 2:367–390; 1986.

169 **Eilers, M.; Schatz, G.** Protein unfolding and the energetics of protein translocation across biological membranes. *Cell* 52:481–483; 1988.

170 **Emr, S. D.; Hall, M. N.; Silhavy, T. J.** A mechanism of protein localization: the signal hypothesis and bacteria. *J. Cell Biol.* 86:701–711; 1980.

171 **Farquhar, M. G.** Progress in unraveling pathways of Golgi traffic. *Annu. Rev. Cell Biol.* 1:447–488; 1985.

172 **Gerace, L.; Burke, B.** Functional organization of the nuclear envelope. *Annu. Rev. Cell Biol.* 4:335–374, 1988.

173 **Griffiths, G.; Simons, K.** The *trans* Golgi network: sorting at the exit site of the Golgi complex. *Science* 234:438–443; 1986.

174 **Grivell, L. A.** Mitochondrial DNA. *Sci. Am.* 248(3):78–89; 1983.

175 **Hartl, F.-U.; Neupert, W.** Protein sorting to mitochondria: evolutionary conservations of folding and assembly. *Science* 247:930–938; 1990.

176 **Heath, I. B.** Variant mitoses in lower eukaryotes: indicators of the evolution of mitosis? *Int. Rev. Cytol.* 64:1–80; 1980.

177 **Hirschberg, C. B.; Snider, M. D.** Topography of glycosylation in the rough endoplasmic reticulum and Golgi apparatus. *Annu. Rev. Biochem.* 56:63–87; 1987.

178 **Hortsch, M.; Meyer, D. I.** Transfer of secretory proteins through the membrane of the endoplasmic reticulum. *Int. Rev. Cytol.* 102:215–242; 1986.

179 **Hou, Y.-M.; Francklyn, C.; Schimmel, P.** Molecular dissection of a transfer RNA and the basis for its identity. *Trends Biochem. Sci.* 14:233–237; 1989.

180 **Jacob, F.; Monod, J.** Genetic regulatory mechanisms in the synthesis of proteins. *J. Mol. Biol.* 3:318–356; 1961.

181 **Kisselev, L. L.** The role of the anticodon in recognition of tRNA by aminoacyl-tRNA synthetases. *Prog. Nucleic Acid Res. Mol. Biol.* 32:237–266; 1985.

182 **Kornfeld, S.; Mellman, I.** The biogenesis of lysosomes. *Annu. Rev. Cell Biol.* 5:483–525; 1989.

183 **Kubai, D. F.** The evolution of the mitotic spindle. *Int. Rev. Cytol.* 43:167–227; 1975.

184 **Lazarow, P. B.; Fujiki, Y.** Biogenesis of peroxisomes. *Annu. Rev. Cell Biol.* 1:489–530; 1985.

185 **Leder, P.** The genetics of antibody diversity. *Sci. Am.* 246(5):102–115; 1982.

186 **Lee, C.; Beckwith, J.** Cotranslational and posttranslational protein translocation in prokaryotic systems. *Annu. Rev. Cell Biol.* 2:315–336; 1986.

187 **Levings, C. S., III; Brown, G. G.** Molecular biology of plant mitochondria. *Cell* 56:171–179; 1989.

188 **Mellman, I.; Fuchs, R.; Helenius, A.** Acidification of the endocytic and exocytic pathways. *Annu. Rev. Biochem.* 55:663–700; 1986.

189 **Nagley, P.; Devenish, R. J.** Leading organellar proteins along new pathways: the relocation of mitochondrial and chloroplast genes into the nucleus. *Trends Biochem. Sci.* 14:31–35; 1989.

190 **Nelson, W. G.; Pienta, K. J.; Barrack, E. R.; Coffey, D. S.** The role of the nuclear matrix in the organization and function of DNA. *Annu. Rev. Biophys. Chem.* 15:457–475; 1986.

191 **Newport, J. W.; Forbes, D. J.** The nucleus: structure, function, and dynamics. *Annu. Rev. Biochem.* 56:535–565; 1987.

192 **Niemann, H.; Mayer, T.; Tamura, T.** Signals for membrane-associated transport in eukaryotic cells. *Subcell. Biochem.* 15:307–365; 1989.

193 **Noller, H. F.** Structure of ribosomal RNA. *Annu. Rev. Biochem.* 53:119–162; 1984.

194 **Normanly, J.; Abelson, J.** tRNA identity. *Annu. Rev. Biochem.* 58:1029–1049; 1989.

195 **Ohkuma, S.** The lysosomal proton pump and its effect on protein breakdown. In: Glaumann, H.; Ballard, F. J., eds. *Lysosomes: their role in protein breakdown.* London: Academic Press; 1987:115–148.

196 **Perlman, P. S.; Butow, R. A.** Mobile introns and intron-encoded proteins. *Science* 246:1106–1109; 1989.

197 **Pfeffer, S. R.; Rothman, J. E.** Biosynthetic protein transport and sorting by the endoplasmic reticulum and Golgi. *Annu. Rev. Biochem.* 56:829–852; 1987.

198 **Prusiner, S. B.** Prions. *Sci. Am.* 251(4):50–59; 1984.

199 **Randall, L. L.; Hardy, S. J. S.; Thom, J. R.** Export of protein: a biochemical view. *Annu. Rev. Microbiol.* 41:507–541; 1987.

200 **Rose, J. K.; Doms, R. W.** Regulation of protein export from the endoplasmic reticulum. *Annu. Rev. Cell Biol.* 4:257–288; 1988.

201 **Sahagian, G. G.** The mannose 6-phosphate receptor: function, biosynthesis and translocation. *Biol. Cell.* 51:207–214; 1984.

202 **Schimmel, P.** Aminoacyl tRNA synthetases: general scheme of structure-function relationships in the polypeptides and recognition of transfer RNAs. *Annu. Rev. Biochem.* 56:125–158; 1987.

203 **Schimmel, P.** Parameters for the molecular recognition of transfer RNAs. *Biochemistry* 28:2747–2759; 1989.

204 **Schimmel, P.; Söll, D.** Aminoacyl-tRNA synthetases: general features and recognition of transfer RNAs. *Annu. Rev. Biochem.* 48:601–648; 1979.

205 **Schmidt, G. W.; Mishkind, M. L.** The transport of proteins into chloroplasts. *Annu. Rev. Biochem.* 55:879–912; 1986.

206 **Sharp, P. A.** Splicing of messenger RNA precursors. *Science* 235:766–771; 1987.

207 **Silhavy, T. J.; Benson, S. A.; Emr, S. D.** Mechanisms of protein localization. *Microbiol. Rev.* 47:313–344; 1983.

208 **Sprinzl, M.; Hartmann, T.; Meissner, F.; Moll, J.; Vorderwülbecke, T.** Compilation of tRNA sequences and sequences of tRNA genes. *Nucleic Acids Res.* 15(Suppl.):r53–r188; 1987.

209 **Tonegawa, S.** Somatic generation of immune diversity. *In Vitro Cell. Develop. Biol.* 24:253–265; 1988.

210 **Tzagoloff, A.; Myers, A. M.** Genetics of mitochondrial biogenesis. *Annu. Rev. Biochem.* 55:249–285; 1986.

211 **van Meer, G.** Lipid traffic in animal cells. *Annu. Rev. Cell Biol.* 5:247–275; 1989.

212 **Verner, K.; Schatz, G.** Protein translocation across membranes. *Science* 241:1307–1313; 1988.

213 **von Figura, K.; Hasilik, A.** Lysosomal enzymes and their receptors. *Annu. Rev. Biochem.* 55:167–193; 1986.

214 **Walter, P.; Gilmore, R.; Blobel, G.** Protein translocation across the endoplasmic reticulum. *Cell* 38:5–8; 1984.

215 **Walter, P.; Lingappa, V. R.** Mechanism of protein translocation across the endoplasmic reticulum membrane. *Annu. Rev. Cell Biol.* 2:499–516; 1986.

216 **Wickner, W.** Secretion and membrane assembly. *Trends Biochem. Sci.* 14:280–283; 1989.

217 **Wickner, W. T.; Lodish, H. F.** Multiple mechanisms of protein insertion into and across membranes. *Science* 230:400–407; 1985.

218 **Yonath, A.; Wittmann, H. G.** Challenging the three-dimensional structure of ribosomes. *Trends Biochem. Sci.* 14:329–335; 1989.

219 **Zurawski, G.; Clegg, M. T.** Evolution of higher-plant chloroplast DNA-encoded genes: implications for structure-function and phylogenetic studies. *Annu. Rev. Plant Physiol.* 38:391–418; 1987.

### *Origin of Life*

220 **Eigen, M.; Gardiner, W.; Schuster, P.; Winkler-Oswatitsch, R.** The origin of genetic information. *Sci. Am.* 244(4):88–118; 1981.

221 **Green, S.** Interstellar chemistry: exotic molecules in space. *Annu. Rev. Phys. Chem.* 32:103–138; 1981.

222 **Joyce, G. F.** RNA evolution and the origins of life. *Nature* 338:217–224; 1989.

223 **Lacey, J. C., Jr.; Mullins, D. W., Jr.** Experimental studies related to the origin of the genetic code and the process of protein synthesis: a review. *Orig. Life* 13:3–42; 1983.

224 **Orgel, L. E.** RNA catalysis and the origins of life. *J. Theor. Biol.* 123:127–149; 1986.

### *Evolution*

225 **Dodson, E. O.** Crossing the prokaryote-eukaryote border: endosymbiosis or continuous development? *Can. J. Microbiol.* 25:651–674; 1979.

226 **Ebringer, L.; Krajčovič, J.** Are chloroplasts and mitochondria the remnants of prokaryotic endosymbionts? *Folia Microbiol.* 31:228–254; 1986.

227 **Ebringer, L.; Krajčovič, J.** Prokaryotic character of chloroplasts and mitochondria—the present knowledge. *Folia Microbiol.* 32:244–280; 1987.

228 **George, D. G.; Hunt, L. T.; Yeh, L.-S. L.; Barker, W. C.** New perspectives on bacterial ferredoxin evolution. *J. Mol. Evol.* 22:20–31; 1985.

229 **Gray, M. W.** Organelle origins and ribosomal RNA. *Biochem. Cell Biol.* 66:325–348; 1988.

230 **Gray, M. W.** The evolutionary origins of organelles. *Trends Gen.* 5:294–299; 1989.

231 **Gray, M. W.; Doolittle, W. F.** Has the endosymbiont hypothesis been proven? *Microbiol. Rev.* 46:1–42; 1982.

232 **Gray, M. W.; Sankoff, D.; Cedergren, R. J.** On the evolutionary descent of organisms and organelles: a global phylogeny based on a highly conserved structural core in small subunit ribosomal RNA. *Nucleic Acids Res.* 12:5837–5852; 1984.

233 **Mahler, H. R.** The exon:intron structure of some mitochondrial genes and its relation to mitochondrial evolution. *Int. Rev. Cytol.* 82:1–98; 1983.

234 **Mahler, H. R.; Raff, R. A.** The evolutionary origin of the mitochondrion: a nonsymbiotic model. *Int. Rev. Cytol.* 43:1–124; 1975.

235 **Nelson, N.; Taiz, L.** The evolution of $H^+$-ATPases. *Trends Biochem. Sci.* 14:113–116; 1989.

236 **Sogin, M. L.** Evolution of eukaryotic microorganisms and their small subunit ribosomal RNAs. *Amer. Zool.* 29:487–499; 1989.

237 **White, H. B., III** Evolution of coenzymes and the origin of pyridine nucleotides. In: Everse, J.; Anderson, B.; You, K.-S., eds. *The pyridine nucleotide coenzymes.* New York: Academic Press; 1982:1–17.

238 **Woese, C. R.** Archaebacteria. *Sci. Am.* 244(6):98–122; 1981.

239 **Woese, C. R.** Bacterial evolution. *Microbiol. Rev.* 51:221–271; 1987.

## THEORETICAL ESSAYS, SPECULATIONS

### *Origin of Life*

240 **Black, S.** A theory on the origin of life. *Adv. Enzymol.* 38:193–234; 1973.

241 **Cedergren, R.; Grosjean, H.** On the primacy of primordial RNA. *BioSystems* 20:175–180; 1987.

242 **Crick, F. H. C.** The origin of the genetic code. *J. Mol. Biol.* 38:367–379; 1968.

243 **Crick, F. H. C.; Brenner, S.; Klug, A.; Pieczenik, G.** A speculation on the origin of protein synthesis. *Orig. Life* 7:389–397; 1976.

244 **Darnell, J. E., Jr.** Implications of RNA-RNA splicing in evolution of eukaryotic cells. *Science* 202:1257–1260; 1978.

245 **Darnell, J. E.; Doolittle, W. F.** Speculations on the early course of evolution. *Proc. Natl. Acad. Sci. USA* 83:1271–1275; 1986.

246 **de Duve, C.** Les contraintes du hasard. *Revue Générale* 1972(2):15–42; 1972.

247 **de Duve, C.** Selection by differential molecular survival: a possible mechanism of early chemical evolution. *Proc. Natl. Acad. Sci. USA* 84:8253–8256; 1987.

248 **Eakin, R. E.** An approach to the evolution of metabolism. *Proc. Natl. Acad. Sci. USA* 49:360–366; 1963.

249 **Eigen, M.** Selforganization of matter and the evolution of biological macromolecules. *Naturwissenschaften* 58:465–523; 1971.

250 **Eigen, M.; Schuster, P.** The hypercycle: a principle of self-organization. Part A: emergence of the hypercycle. *Naturwissenschaften* 64:541–565; 1977.

251 **Eigen, M.; Schuster, P.** The hypercycle: a principle of self-organization. Part B: the abstract hypercyle. *Naturwissenschaften* 65:7–41; 1978.

252 **Eigen, M.; Schuster, P.** The hypercycle: a principle of self-organization. Part C: the realistic hypercycle. *Naturwissenschaften* 65:341–369; 1978.

253 **Florkin, M.** Ideas and experiments in the field of prebiological chemical evolution. In: Florkin, M.; Stotz, E. H., eds. *Comprehensive biochemistry*. Vol. 29B. Amsterdam: Elsevier Scientific Publishing Company; 1975:231–260.

254 **Fox, S. W.** The evolutionary sequence: origin and emergences. *The American Biology Teacher* 48:140–149,169; 1986.

255 **Gamow, G.** Possible relation between deoxyribonucleic acid and protein structures. *Nature* 173:318; 1954.

256 **Gibson, T. J.; Lamond, A. I.** Metabolic complexity in the RNA world and implications for the origin of protein synthesis. *J. Mol. Evol.* 30:7–15; 1990.

257 **Hartman, H.** Speculations on the origin and evolution of metabolism. *J. Mol. Evol.* 4:359–370; 1975.

258 **Hartman, H.** Speculations on the evolution of the genetic code. *Orig. Life* 6:423–427; 1975.

259 **Hartman, H.** Speculations on the evolution of the genetic code: II. *Orig. Life* 9:133–136; 1978.

260 **Hartman, H.** Speculations on the evolution of the genetic code: III. The evolution of t-RNA. *Orig. Life* 14:643–648; 1984.

261 **Hopfield, J. J.** Origin of the genetic code: a testable hypothesis based on tRNA structure, sequence, and kinetic proofreading. *Proc. Natl. Acad. Sci. USA* 75:4334–4338; 1978.

262 **Kuhn, H.; Waser, J.** Molecular self-organization and the origin of life. *Angew. Chem.* 20:500–520; 1981.

263 **Lau, K. F.; Dill, K. A.** Theory for protein mutability and biogenesis. *Proc. Natl. Acad. Sci. USA* 87:638–642; 1990.

263a **Lehmann, U.; Kuhn, H.** Emergence of adaptable systems and evolution of a translation device. *Adv. Space Res.* 4(12):153–161; 1984.

264 **Morowitz, H. J.** Biological self-replicating systems. *Progr. Theor. Biol.* 1:35–58; 1967.

265 **Morowitz, H. J.; Heinz, B.; Deamer, D. W.** The chemical logic of a minimum protocell. *Orig. Life Evol. Biosp.* 18:281–287; 1988.

266 **Orgel, L. E.** A possible step in the origin of the genetic code. *Isr. J. Chem.* 10:287–292; 1972.

267 **Orgel, L. E.** Did template-directed nucleation precede molecular replication? *Orig. Life* 17:27–34; 1986.

268 **Orgel, L. E.** The origin of polynucleotide-directed protein synthesis. *J. Mol. Evol.* 29:465–474; 1989.

269 **Prigogine, I.; Nicolis, G.** Biological order, structure and instabilities. *Quart. Rev. Biophys.* 4:107–148; 1971.

270 **Sharp, P. A.** On the origin of RNA splicing and introns. *Cell* 42:397–400; 1985.

271 **Shimizu, M.** Molecular basis for the genetic code. *J. Mol. Evol.* 18:297–303; 1982.

272 **Simon, H. A.** The architecture of complexity. *Proc. Amer. Philos. Soc.* 106:467–482; 1962.

273 **Spach, G.** Chiral versus chemical evolution and the appearance of life. *Orig. Life* 14:433–437; 1984.

274 **Wächtershäuser, G.** Before enzymes and templates: theory of surface metabolism. *Microbiol. Rev.* 52:452–484; 1988.

275 **Wächtershäuser, G.** Pyrite formation, the first energy source for life: a hypothesis. *System. Appl. Microbiol.* 10:207–210; 1988.

276 **Wächtershäuser, G.** An all-purine precursor of nucleic acids. *Proc. Natl. Acad. Sci. USA* 85:1134–1135; 1988.

277 **Weber, A. L.** The triose model: glyceraldehyde as a source of energy and monomers for prebiotic condensation reactions. *Orig. Life Evol. Biosp.* 17:107–119; 1987.

278 **Westheimer, F. H.** Why nature chose phosphates. *Science* 235:1173–1178; 1987.

279 **Woese, C. R.** A proposal concerning the origin of life on the planet earth. *J. Mol. Evol.* 13:95–101; 1979.

280 **Wong, J. T.-F.** A co-evolution theory of the genetic code. *Proc. Natl. Acad. Sci. USA* 72:1909–1912; 1975.

281 **Wong, J. T.-F.** Role of minimization of chemical distances between amino acids in the evolution of the genetic code. *Proc. Natl. Acad. Sci. USA* 77:1083–1086; 1980.

282 **Wong, J. T.-F.** Evolution of the genetic code. *Microbiol. Sci.* 5:174–181; 1988.

283 **Wong, J. T.-F.; Bronskill, P. M.** Inadequacy of prebiotic synthesis as origin of proteinous amino acids. *J. Mol. Evol.* 13:115–125; 1979.

*Evolution*

284 **Cavalier-Smith, T.** The origin of nuclei and of eukaryotic cells. *Nature* 256:463–468; 1975.

285 **Cavalier-Smith, T.** The kingdoms of organisms. *Nature* 324:416–417; 1986.

286 **Cavalier-Smith, T.** Glaucophyceae and the origin of plants. *Evol. Trends Plants* 1(2):75–78; 1987.

287 **de Duve, C.** Origin of mitochondria. *Science* 182:85; 1973.

288 **Jacob, F.** Evolution and tinkering. *Science* 196:1161–1166; 1977.

289 **King, G. A. M.** Evolution of the coenzymes. *BioSystems* 13:23–45; 1980.

290 **Orgel, L. E.** The maintenance of the accuracy of protein synthesis and its relevance to ageing. *Proc. Natl. Acad. Sci. USA* 49:517–521; 1963.

291 **Osawa, S.; Jukes, T. H.** Evolution of the genetic code as affected by anticodon content. *Trends Genet.* 4:191–198; 1988.

292 **Sagan, L. (Margulis, L.)** On the origin of mitosing cells. *J. Theoret. Biol.* 14:225–274; 1967.

292a **Woese, C. R.; Kandler, O.; Wheelis, M. L.** Towards a natural system of organisms: proposal for the domains Archaea, Bacteria, and Eucarya. *Proc. Natl. Acad. Sci. USA* 87:4576–4579; 1990.

293 **Zillig, W.; Klenk, H.-P.; Palm, P.; Leffers, H.; Pühler, G. P.; Gropp, F.; Garrett, R. A.** Did eukaryotes originate by a fusion event? *Endocytobiosis & Cell Res.* 6:1–25; 1989.

## EXPERIMENTAL PAPERS

### *Biochemistry and Bioenergetics*

294 **Baldwin, A. N.; Berg, P.** Transfer ribonucleic acid-induced hydrolysis of valyladenylate bound to isoleucyl ribonucleic acid synthetase. *J. Biol. Chem.* 241:839–845; 1966.

295 **Clarke, D. J.; Fuller, F. M.; Morris, J. G.** The proton-translocating adenosine triphosphatase of the obligately anaerobic bacterium *Clostridium pasteurianum:* 1. ATP phosphohydrolase activity. *Eur. J. Biochem.* 98:597–612; 1979.

296 **Gevers, W.; Kleinkauf, H.; Lipmann, F.** Peptidyl transfers in gramicidin S biosynthesis from enzyme-bound thioester intermediates. *Proc. Natl. Acad. Sci. USA* 63:1335–1342; 1969.

297 **Jencks, W. P.; Gilchrist, M.** The free energies of hydrolysis of some esters and thioesters of acetic acid. *J. Amer. Chem. Soc.* 86:4651–4654; 1964.

298 **Kobayashi, H.; Suzuki, T.; Unemoto, T.** Streptococcal cytoplasmic pH is regulated by changes in amount and activity of a proton-translocating ATPase. *J. Biol. Chem.* 261:627–630; 1986.

299 **Koshland, D. E., Jr.** Application of a theory of enzyme specificity to protein synthesis. *Proc. Natl. Acad. Sci. USA* 44:98–104; 1958.

300 **Lynen, F.; Reichert, E.** Zur chemischen Struktur der "aktivierten Essigsaure." *Angew. Chem.* 63:47–48; 1951.

301 **Marcker, K.; Sanger, F.** N-Formyl-methionyl-sRNA. *J. Mol. Biol.* 8:835–840; 1964.

302 **Nasu, S.; Wicks, F. D.; Gholson, R. K.** L-Aspartate oxidase, a newly discovered enzyme of *Escherichia coli,* is the B protein of quinolinate synthetase. *J. Biol. Chem.* 257:626–632; 1982.

303 **Rapaport, E.; Remy, P.; Kleinkauf, H.; Vater, J.; Zamecnik, P. C.** Aminoacyl-tRNA synthetases catalyze AMP ⟶ ADP ⟶ ATP exchange reactions, indicating labile covalent enzyme-amino acid intermediates. *Proc. Natl. Acad. Sci. USA* 84:7891–7895; 1987.

304 **Szundi, I.; Stoeckenius, W.** Effect of lipid surface charges on the purple-to-blue transition of bacteriorhodopsin. *Proc. Natl. Acad. Sci. USA* 84:3681–3684; 1987.

305 **Wadsö, I.** Heats of hydrolysis of acetates and thiolacetates in aqueous solution. *Acta Chem. Scand.* 16:487–494; 1962.

306 **Wolfenden, R.** Waterlogged molecules. *Science* 222:1087–1093; 1983.

### *Cellular and Molecular Biology*

307 **Augustin, S.; Müller, M. W.; Schweyen, R. J.** Reverse self-splicing of Group II intron RNAs *in vitro. Nature* 343:383–386; 1990.

308 **Baldauf, S. L.; Palmer, J. D.** Evolutionary transfer of the chloroplast *tuf*A gene to the nucleus. *Nature* 344:262–265; 1990.

309 **Baudhuin, P.; Peeters-Joris, C.; Bartholeyns, J.** Hepatic nucleases: 2. Association of polyadenylase, alkaline ribonuclease and deoxyribonuclease with rat-liver mitochondria. *Eur. J. Biochem.* 57:213–220; 1975.

310 **Blobel, G.** Intracellular protein topogenesis. *Proc. Natl. Acad. Sci. USA* 77:1496–1500; 1980.

311 **Blum, B.; Bakalara, N.; Simpson, L.** A model for RNA editing in kinetoplastid mitochondria: "guide" RNA molecules transcribed from maxicircle DNA provide the edited information. *Cell* 60:189–198; 1990.

312 **Brody, E.; Abelson, J.** The "spliceosome": yeast pre-messenger RNA associates with a 40S complex in a splicing-dependent reaction. *Science* 228:963–967; 1985.

313 **Cairns, J.; Overbaugh, J.; Miller, S.** The origin of mutants. *Nature* 335:142–145; 1988.

314 **Cech, T. R.; Zaug, A. J.; Grabowski, P. J.** *In vitro* splicing of the ribosomal RNA precursor of *Tetrahymena:* involvement of a guanosine nucleotide in the excision of the intervening sequence. *Cell* 27:487–496; 1981.

315 **Cedergren, R.; Grosjean, H.; Larue, B.** Primordial reading of genetic information. *BioSystems* 19:259–266; 1986.

316 **Chesebro, B.; Race, R.; Wehrly, K.; Nishio, J.; Bloom, M.; Lechner, D.; Bergstrom, S.; Robbins, K.; Mayer, L.; Keith, J. M.; Garon, C.; Haase, A.** Identification of scrapie prion protein-specific mRNA in scrapie-infected and uninfected brain. *Nature* 315:331–333; 1985.

317 **Chu, F. K.; Maley, G. F.; Maley, F.; Belfort, M.** Intervening sequence in the thymidylate synthase gene of bacteriophage T4. *Proc. Natl. Acad. Sci. USA* 81:3049–3053; 1984.

318 **Covello, P. S.; Gray, M. W.** RNA editing in plant mitochondria. *Nature* 341:662–666; 1989.

319 **Crothers, D. M.; Seno, T.; Söll, D. G.** Is there a discriminator site in transfer RNA? *Proc. Natl. Acad. Sci. USA* 69:3063–3067; 1972.

320 **Doudna, J. A.; Szostak, J. W.** RNA-catalysed synthesis of complementary-strand RNA. *Nature* 339:519–522; 1989.

321 **Francklyn, C.; Schimmel, P.** Aminoacylation of RNA minihelices with alanine. *Nature* 337:478–481; 1989.

322 **Gō, M.** Correlation of DNA exonic regions with protein structural units in haemoglobin. *Nature* 291:90–92; 1981.

322a **Gould, S. J.; Keller, G.-A.; Schneider, M.; Howell, S. H.; Garrard, L. J.; Goodman, J. M.; Distel, B.; Tabak, H.; Subramani, S.** Peroxisomal protein import is conserved between yeast, plants, insects and mammals. *EMBO J.* 9:85–90; 1990.

323 **Grabowski, P. J.; Seiler, S. R.; Sharp, P. A.** A multicomponent complex is involved in the splicing of messenger RNA precursors. *Cell* 42:345–353; 1985.

324 **Gualberto, J. M.; Lamattina, L.; Bonnard, G.; Weil, J.-H.; Grienenberger, J.-M.** RNA editing in wheat mitochondria results in the conservation of protein sequences. *Nature* 341:660–662; 1989.

325 **Hall, J. L.; Ramanis, Z.; Luck, D. J. L.** Basal body/centriolar DNA: molecular genetic studies in *Chlamydomonas*. *Cell* 59:121–132; 1989.

326 **Hermes, J. D.; Blacklow, S. C.; Knowles, J. R.** Searching sequence space by definably random mutagenesis: improving the catalytic potency of an enzyme. *Proc. Natl. Acad. Sci. USA* 87:696–700; 1990.

327 **Hiesel, R.; Wissinger, B.; Schuster, W.; Brennicke, A.** RNA editing in plant mitochondria. *Science* 246:1632–1634; 1989.

328 **Hou, Y.-M.; Schimmel, P.** A simple structural feature is a major determinant of the identity of a transfer RNA. *Nature* 333:140–145; 1988.

329 **Hou, Y.-M.; Schimmel, P.** Evidence that a major determinant for the identity of a transfer RNA is conserved in evolution. *Biochemistry* 28:6800–6804; 1989.

330 **Huang, B.; Ramanis, Z.; Dutcher, S. K.; Luck, D. J. L.** Uniflagellar mutants of *Chlamydomonas:* evidence for the role of basal bodies in transmission of positional information. *Cell* 29:745–753; 1982.

331 **King, M. P.; Attardi, G.** Injection of mitochondria into human cells leads to a rapid replacement of the endogenous mitochondrial DNA. *Cell* 52:811–819; 1988.

332 **Lazowska, J.; Jacq, C.; Slonimski, P. P.** Sequence of introns and flanking exons in wild-type and *box3* mutants of cytochrome b reveals an interlaced splicing protein coded by an intron. *Cell* 22:333–348; 1980.

333 **Leinfelder, W.; Zehelein, E.; Mandrand-Berthelot, M.-A.; Böck, A.** Gene for a novel tRNA species that accepts L-serine and cotranslationally inserts selenocysteine. *Nature* 331:723–725; 1988.

334 **Maniatis, T.; Reed, R.** The role of small nuclear ribonucleoprotein particles in pre-mRNA splicing. *Nature* 325:673–678; 1987.

335 **McClain, W. H.; Chen, Y.-M.; Foss, K.; Schneider, J.** Association of transfer RNA acceptor identity with a helical irregularity. *Science* 242:1681–1684; 1988.

336 **McClain, W. H.; Foss, K.** Changing the identity of a tRNA by introducing a G-U wobble pair near the 3′ acceptor end. *Science* 240:793–796; 1988.

337 **Michel, F.; Lang, B. F.** Mitochondrial class II introns encode proteins related to the reverse transcriptases of retroviruses. *Nature* 316:641–643; 1985.

338 **Mörl, M.; Schmelzer, C.** Integration of group II intron bl1 into a foreign RNA by reversal of the self-splicing reaction in vitro. *Cell* 60:629–636; 1990.

339 **Muramatsu, T.; Nishikawa, K.; Nemoto, F.; Kuchino, Y.; Nishimura, S.; Miyazawa, T.; Yokoyama, S.** Codon and amino-acid specificities of a transfer RNA are both converted by a single post-transcriptional modification. *Nature* 336:179–181; 1988.

340 **Newport, J.** Nuclear reconstitution in vitro: stages of assembly around protein-free DNA. *Cell* 48:205–217; 1987.

341 **Oesch, B.; Westaway, D.; Walchli, M.; McKinley, M. P.; Kent, S. B. H.; Aebersold, R.; Barry, R. A.; Tempst, P.; Teplow, D. P.; Hood, L. E.; Prusiner, S. B.; Weissmann, C.** A cellular gene encodes scrapie PrP 27–30 protein. *Cell* 40:735–746; 1985.

342 **Orgel, L. E.** Selection in vitro. *Proc. R. Soc. London B* 205:435–442; 1979.

343 **Pain, D.; Kanwar, Y. S.; Blobel, G.** Identification of a receptor for protein import into chloroplasts and its localization to envelope contact zones. *Nature* 331:232–237; 1988.

344 **Ramanis, Z.; Luck, D. J. L.** Loci affecting flagellar assembly and function map to an unusual linkage group in *Chlamydomonas reinhardtii. Proc. Natl. Acad. Sci. USA* 83:423–426; 1986.

345 **Ratzkin, B.; Carbon, J.** Functional expression of cloned yeast DNA in *Escherichia coli. Proc. Natl. Acad. Sci. USA* 74:487–491; 1977.

346 **Robertson, H. D.; Branch, A. D.; Dahlberg, J. E.** Focusing on the nature of the scrapie agent. *Cell* 40:725–727; 1985.

347 **Rochaix, J. D.; Rahire, M.; Michel, F.** The chloroplast ribosomal intron of *Chlamydomonas reinhardtii* codes for a polypeptide related to mitochondrial maturases. *Nucleic Acids Res.* 13:975–984; 1985.

348 **Rogers, J.** Exon shuffling and intron insertion in serine protease genes. *Nature* 315:458–459; 1985.

349 **Rould, M. A.; Perona, J. J.; Söll, D.; Steitz, T. A.** Structure of *E. coli* glutaminyl-tRNA synthetase complexed with $tRNA^{Gln}$ and ATP at 2.8 Å resolution. *Science* 246:1135–1142; 1989.

350 **Schön, A.; Kannangara, C. G.; Gough, S.; Söll, D.** Protein biosynthesis in organelles requires misaminoacylation of tRNA. *Nature* 331:187–190; 1988.

351 **Schulman, L. H.; Pelka, H.** Anticodon switching changes the identity of methionine and valine transfer RNAs. *Science* 242:765–768; 1988.

352 **Schulman, L. H.; Pelka, H.** The anticodon contains a major element of the identity of arginine transfer RNAs. *Science* 246:1595–1597; 1989.

353 **Shippen-Lentz, D.; Blackburn, E. H.** Functional evidence for an RNA template in telomerase. *Science* 247:546–552; 1990.

354 **Shub, D. A.; Gott, J. M.; Xu, M.-Q.; Lang, B. F.; Michel, F.; Tomaschewski, J.; Pedersen-Lane, J.; Belfort, M.** Structural conservation among three homologous introns of bacteriophage T4 and the group I introns of eukaryotes. *Proc. Natl. Acad. Sci. USA* 85:1151–1155; 1988.

355 **Simon, S. F.; Blobel, G.; Zimmerberg, J.** Large aqueous channels in membrane vesicles derived from the rough endoplasmic reticulum of canine pancreas or the plasma membrane of *Escherichia coli. Proc. Natl. Acad. Sci. USA* 86:6176–6180; 1989.

356 **Sly, W. S.; Fischer, H. D.** The phosphomannosyl recognition system for intracellular and intercellular transport of lysosomal enzymes. *J. Cell. Biochem.* 18:67–85, 1982.

357 **Smit, J.; Nikaido, H.** Outer membrane of gram-negative bacteria: XVIII. Electron microscopic studies on porin insertion sites and growth of cell surface of *Salmonella typhimurium*. *J. Bacteriol.* 135:687–702; 1978.

358 **Struhl, K.; Cameron, J. R.; Davies, R. W.** Functional genetic expression of eukaryotic DNA in *Escherichia coli*. *Proc. Natl. Acad. Sci. USA* 73:1471–1475; 1976.

359 **Südhof, T. C.; Goldstein, J. L.; Brown, M. S.; Russell, D. W.** The LDL receptor gene: a mosaic of exons shared with different proteins. *Science* 228:815–822; 1985.

360 **Südhof, T. C.; Russell, D. W.; Goldstein, J. L.; Brown, M. S.; Sanchez-Pescador, R.; Bell, G. I.** Cassette of eight exons shared by genes for LDL receptor and EGF precursor. *Science* 228:893–895; 1985.

361 **Tonegawa, S.** Somatic generation of antibody diversity. *Nature* 302:575–581; 1983.

362 **Vestweber, D.; Brunner, J.; Baker, A.; Schatz, G.** A 42K outer-membrane protein is a component of the yeast mitochondrial protein import site. *Nature* 341:205–209; 1989.

363 **Visscher, J.; Bakker, C. G.; van der Woerd, R.; Schwartz, A. W.** Template-directed oligomerization catalyzed by a polynucleotide analog. *Science* 244:329–331; 1989.

364 **Voordouw, G.; Brenner, S.** Nucleotide sequence of the gene encoding the hydrogenase from *Desulfovibrio vulgaris* (Hildenborough). *Eur. J. Biochem.* 148:515–520; 1985.

365 **Weiner, A. M.; Maizels, N.** tRNA-like structures tag the 3′ ends of genomic RNA molecules for replication: implications for the origin of protein synthesis. *Proc. Natl. Acad. Sci. USA* 84:7383–7387; 1987.

366 **Weiner, H.; Stitt, M.; Heldt, H. W.** Subcellular compartmentation of pyrophosphate and alkaline pyrophosphatase in leaves. *Biochim. Biophys. Acta* 893:13–21; 1987.

367 **Wilcox, M.; Nirenberg, M.** Transfer RNA as a cofactor coupling amino acid synthesis with that of protein. *Proc. Natl. Acad. Sci. USA* 61:229–236; 1968.

368 **Yarus, M.** A specific amino acid binding site composed of RNA. *Science* 240: 1751–1758; 1988.

369 **Yu, G.-L.; Bradley, J. D.; Attardi, L. D.; Blackburn, E. H.** In vivo alteration of telomere sequences and senescence caused by mutated *Tetrahymena* telomerase RNAs. *Nature* 344:126–132; 1990.

370 **Zakut, R.; Shani, M.; Givol, D.; Neuman, S.; Yaffe, D.; Nudel, U.** Nucleotide sequence of the rat skeletal muscle actin gene. *Nature* 298:857–859; 1982.

### *Origin of Life*

371 **Allen, D. A.; Wickramasinghe, D. T.** Discovery of organic grains in comet Wilson. *Nature* 329:615–616; 1987.

372 **Borowska, Z. K.; Mauzerall, D. C.** Efficient near ultraviolet light induced formation of hydrogen by ferrous hydroxide. *Orig. Life* 17:251–259; 1987.

373 **Braterman, P. S.; Cairns-Smith, A. G.; Sloper, R. W.** Photo-oxidation of hydrated $Fe^{2+}$—significance for banded iron formations. *Nature* 303:163–164; 1983.

374 **Choughuley, A. S. U.; Lemmon, R. M.** Production of cysteic acid, taurine and cystamine under primitive earth conditions. *Nature* 210:628–629; 1966.

375 **Corliss, J. B.; Dymond, J.; Gordon, L. I.; Edmond, J. M.; von Herzen, R. P.; Ballard, R. D.; Green, K.; Williams, D.; Bainbridge, A.; Crane, K.; van Andel, T. H.** Submarine thermal springs on the Galapagos rift. *Science* 203:1073–1083; 1979.

376 **Eigen, M.** Experiments on biogenesis. In: Yoshida, Z.-I.; Ise, N., eds. *Biomimetic chemistry*. Tokyo: Kodansha Ltd.; 1983:51–78.

377 **Eigen, M.; Lindemann, B. F.; Tietze, M.; Winkler-Oswatitsch, R.; Dress, A.; van Haeseler, A.** How old is the genetic code? Statistical geometry of tRNA provides an answer. *Science* 244:673–679; 1989.

378 **Eigen, M.; Winkler-Oswatitsch, R.** Transfer-RNA, an early gene. *Naturwissenschaften* 68:282–292; 1981.

379 **Huebner, W. F.** First polymer in space identified in comet Halley. *Science* 237:628–630; 1987.

380 **Joyce, G. F.; Schwartz, A. W.; Miller, S. L.; Orgel, L. E.** The case for an ancestral genetic system involving simple analogues of the nucleotides. *Proc. Natl. Acad. Sci. USA* 84:4398–4402; 1987.

381 **Maher, K. A.; Stevenson, D. J.** Impact frustration of the origin of life. *Nature* 331:612–614; 1988.

382 **Miller, S. L.** A production of amino acids under possible primitive earth conditions. *Science* 117:528–529; 1953.

383 **Miller, S. L.; Bada, J. L.** Submarine hot springs and the origin of life. *Nature* 334:609–611; 1988.

384 **Mitchell, D. L.; Lin, R. P.; Anderson, K. A.; Carlson, C. W.; Curtis, D. W.; Korth, A.; Reme, H.; Sauvaud, J. A.; d'Uston, C.; Mendis, D. A.** Evidence for chain molecules enriched in carbon, hydrogen, and oxygen in comet Halley. *Science* 237:626–628; 1987.

385 **Oro, J.; Kimball, A. P.** Synthesis of purines under possible primitive earth conditions: I. Adenine from hydrogen cyanide. *Arch. Biochem. Biophys.* 94:217–227; 1961.

386 **Oro, J.; Kimball, A. P.** Synthesis of purines under possible primitive earth conditions: II. Purine intermediates from hydrogen cyanide. *Arch. Biochem. Biophys.* 96:293–313; 1962.

387 **Overbeck, V. R.; Fogleman, G.** Impacts and the origin of life. *Nature* 339:434; 1989.

388 **Schidlowski, M.** A 3,800-million-year isotopic record of life from carbon in sedimentary rocks. *Nature* 333:313–318; 1988.

389 **Shapiro, R.** The improbability of prebiotic nucleic acid synthesis. *Orig. Life* 14:565–570; 1984.

390 **Shapiro, R.** Prebiotic ribose synthesis: a critical analysis. *Orig. Life Evol. Biosp.* 18:71–85; 1988.

391 **Sleep, N. H.; Zahnle, K. J.; Kasting, J. F.; Morowitz, H. J.** Annihilation of ecosystems by large asteroid impacts on the early earth. *Nature* 342:139–142; 1989.

392 **Weber, A. L.** Formation of pyrophosphate, tripolyphosphate, and phosphorylimidazole with the thioester, N,S-diacetylcysteamine, as the condensing agent. *J. Mol. Evol.* 18:24–29; 1981.

393 **Weber, A. L.** Formation of pyrophosphate on hydroxyapatite with thioesters as condensing agents. *BioSystems* 15:183–189; 1982.

394 **Weber, A. L.** Prebiotic formation of "energy-rich" thioesters from glyceraldehyde and N-acetylcysteine. *Orig. Life* 15:17–27; 1984.

*Evolution*

395 **Baroin, A.; Perasso, R.; Qu, L.-H.; Brugerolle, G.; Bachellerie, J.-P.; Adoutte, A.** Partial phylogeny of the unicellular eukaryotes based on rapid sequencing of a portion of 28S ribosomal RNA. *Proc. Natl. Acad. Sci. USA* 85:3474–3478; 1988.

396 **Dus, K. M.** Camphor hydroxylase of *Pseudomonas putida:* vestiges of sequence homology in cytochrome P-$450_{CAM}$, putidaredoxin, and related proteins. *Proc. Natl. Acad. Sci. USA* 81:1664–1668; 1984.

397 **Eck, R. V.; Dayhoff, M. O.** Evolution of the structure of ferredoxin based on living relics of primitive amino acid sequences. *Science* 152:363–366; 1966.

398 **Gouy, M.; Li, W.-H.** Phylogenetic analysis based on rRNA sequences supports the archaebacterial rather than the eocyte tree. *Nature* 339:145–147; 1989.

399 **Küntzel, K.; Köchel, H. G.** Evolution of rRNA and origin of mitochondria. *Nature* 293:751–755; 1981.

400 **Lake, J. A.** Origin of the eukaryotic nucleus determined by rate-invariant analysis of rRNA sequences. *Nature* 331:184–186; 1988.

400a **Manhart, J. R.; Palmer, J. D.** The gain of two chloroplast tRNA introns marks the green algal ancestors of land plants. *Nature* 345:268–270; 1990.

401 **Mikelsaar, R.** A view of early cellular evolution. *J. Mol. Evol.* 25:168–183; 1987.

402 **Morden, C. W.; Golden, S. S.** *psbA* genes indicate common ancestry of prochlorophytes and chloroplasts. *Nature* 337:382–385; 1989. Corrected in *Nature* 339:400; 1989.

403 **Perasso, R.; Baroin, A.; Qu, L.-H.; Bachellerie, J.-P.; Adoutte, A.** Origin of the algae. *Nature* 339:142–144; 1989.

404 **Salemme, F. R.; Miller, M. D.; Jordan, S. R.** Structural conveyence during protein evolution. *Proc. Natl. Acad. Sci. USA* 74:2820–2824; 1977.

405 **Sargent, M.; Zahn, R.; Walters, B.; Gupta, R.; Kaine, B.** Nucleotide sequence of the 18S rDNA from the microalga *Nanochlorum eucaryotum. Nucleic Acids Res.* 16:4156; 1988.

406 **Schwartz, R. M.; Dayhoff, M. O.** Origins of prokaryotes, eukaryotes, mitochondria, and chloroplasts. *Science* 199:395–403; 1978.

407 **Sogin, M. L.; Gunderson, J. H.; Elwood, H. J.; Alonso, R. A.; Peattie, D. A.** Phylogenetic meaning of the kingdom concept: an unusual ribosomal RNA from *Giardia lamblia. Science* 243:75–77; 1989.

408 **Spiegelman, S.** An in vitro analysis of a replicating molecule. *Amer. Scient.* 55:221–264; 1967.

409 **Stewart, K. D.; Mattox, K. R.** The case for a polyphyletic origin of mitochondria: morphological and molecular comparisons. *J. Mol. Evol.* 21:54–57; 1984.

410 **Turner, S.; Burger-Wiersma, T.; Giovannoni, S. J.; Mur, L. R.; Pace, N. R.** The relationship of a prochlorophyte *Prochlorothrix hollandica* to green chloroplasts. *Nature* 337:380–382; 1989.

411 **Vossbrinck, C. R.; Maddox, J. V.; Friedman, S.; Debrunner-Vossbrinck, B. A.; Woese, C. R.** Ribosomal RNA sequence suggests microsporidia are extremely ancient eukaryotes. *Nature* 326:411–414; 1987.

412 **Yang, D.; Oyaizu, Y.; Oyaizu, H.; Olsen, G. J.; Woese, C. R.** Mitochondrial origins. *Proc. Natl. Acad. Sci. USA* 82:4443–4447; 1985.

### *Microbiology*

413 **Brugerolle, G.** Contribution a l'étude cytologique et phylétique des diplozoaires (Zoomastigophorea, Diplozoa, Dangeard 1910): VI. Caractères généraux des diplozoaires. *Protistologia* 11:111–118; 1975.

414 **Burger-Wiersma, T.; Veenhuis, M.; Korthals, H. J.; Van de Wiel, C. C. M.; Mur, L. R.** A new prokaryote containing chlorophylls *a* and *b*. *Nature* 320:262–264; 1986.

415 **Fauré-Frémiet, E.; Rouiller, C.** Etude au microscope électronique d'une bactérie sulfureuse, *Thiovulum majus* Hinze. *Exp. Cell Res.* 14:29–46; 1958.

416 **Zahn, R. K.** A green alga with minimal eukaryotic features: *Nanochlorum eucaryotum*. *Orig. Life* 13:289–303; 1984.

## MINIREVIEWS, COMMENTARIES, LETTERS

### *Molecular Biology*

417 **Cech, T. R.** Ribozyme self-replication? *Nature* 339:507–508; 1989.

418 **de Duve, C.** The second genetic code. *Nature* 333:117–118; 1988.

419 **Eisen, H.** RNA editing: who's on first? *Cell* 53:331–332; 1988.

420 **Guyer, R. L.; Koshland, D. E., Jr.** The molecule of the year. *Science* 246:1543–1546; 1989.

421 **Lamond, A. I.** RNA editing and the mysterious undercover genes of trypanosomatid mitochondria. *Trends Biochem. Sci.* 13:283–284; 1988.

422 **Maizels, N.; Weiner, A.** In search of a template. *Nature* 334:469–470; 1988.

422a **North, G.** Expanding the RNA repertoire. *Nature* 345:576–578; 1990.

423 **Schulman, L. H.; Abelson, J.** Recent excitement in understanding transfer RNA identity. *Science* 240:1591–1592; 1988.

424 **Struhl, K.** Helix-turn-helix, zinc-finger, and leucine-zipper motifs for eukaryotic transcriptional regulatory proteins. *Trends Biochem. Sci.* 14:137–140; 1989.

425 **Temin, H. M.** Retrons in bacteria. *Nature* 339:254–255; 1989.

426 **Yarus, M.** tRNA identity: a hair of the dogma that bit us. *Cell* 55:739–741; 1988.

*Origin of Life*

427 **Blake, C.** Exons—present from the beginning? *Nature* 306:535–537; 1983.

428 **de Duve, C.** Did God make RNA? *Nature* 336:209–210; 1988.

429 **Doolittle, W. F.** Genes in pieces: were they ever together? *Nature* 272:581–582; 1978.

430 **Gilbert, W.** Why genes in pieces? *Nature* 271:501; 1978.

431 **Gilbert, W.** Genes-in-pieces revisited. *Science* 228:823–824; 1985.

432 **Gilbert, W.** The RNA world. *Nature* 319:618; 1986.

432a **Gilbert, W.; Marchionni, M.; McKnight, G.** On the antiquity of introns. *Cell* 46:151–154; 1986.

433 **Green, M. R.** Mobile RNA catalysts. *Nature* 336:716–718; 1988.

434 **Mehta, N. G.** An alternative view of the origin of life. *Nature* 324:415–416; 1986.

435 **Rennie, J.** In the beginning, evidence grows that RNA was the first self-made molecule. *Sci. Am.* 261(3):28, 32; 1989.

435a **Waldrop, M. M.** Did life really start out in an RNA world? *Science* 246:1248–1249; 1989.

*Evolution*

436 **Cavalier-Smith, T.** Eukaryotes with no mitochondria. *Nature* 326:332–333; 1987.

437 **Cavalier-Smith, T.** Archaebacteria and archezoa. *Nature* 339:100–101; 1989.

438 **Lake, J. A.** An alternative to archaebacterial dogma. *Nature* 319:626; 1986.

439 **Lake, J. A.** In defence of bacterial phylogeny. *Nature* 321:657–658; 1986.

440 **Lewin, R.** Prochlorophyta as a proposed new division of algae. *Nature* 261:697–698; 1976.

441 **Lewin, R.** The unmasking of mitochondrial Eve. *Science* 238:24–26; 1987.

442 **Penny, D.** What, if anything, is Prochloron? *Nature* 337:304–305; 1989.

443 **Walsby, A. E.** Origins of chloroplasts. *Nature* 320:212; 1986.

444 **Woese, C. R.; Pace, N. R.; Olsen, G. J.** Are arguments against archaebacteria valid? *Nature* 320:401–402; 1986.

445 **Zillig, W.** Archaebacterial status quo is defended. *Nature* 320:220; 1986.

# INDEX OF AUTHORS

# INDEX OF SUBJECTS